PENGUIN BOO

BLESSED UNI

Paul Hawken is an environmentalist, entrepreneur, journalist, and best-selling author of six previous books. He is the architect and leading proponent of reform with respect to ecological practices. He currently operates a nonprofit organization and lives on Cascade Creek in northern California. Visit www.paulhawken.com and www.blessedunrest.com.

Praise for *Blessed Unrest*

"Paul Hawken has created a wondrous experience—a book that magically weaves together the world across time and place. From the wisdom of ancients to our modern predicament. From dead fish to the industrial giants that killed them. From endangered peoples defending nature against the onslaught to their new friends and allies in the capitals of capitalism. Within this vast tableau, the book tells the dramatic story of people rising to resist—a global coming together mobilized to change the world and save it."
—William Greider, author of *The Soul of Capitalism:
Opening Paths to a Moral Economy*

"This is first full account of the real news of our time, and it's exactly the opposite of the official account. The movers and shakers on our planet aren't the billionaires and the generals—they are the incredible numbers of people around the world filled with love for neighbor and for the earth who are resisting, remaking, restoring, renewing, revitalizing. This powerful and lovely book is their story—our story—and it's high time someone's told it. Nothing you read for years to come will fill you with more hope and more determination."
—Bill McKibben, author of *Deep Economy* and *The End of Nature*

"*Blessed Unrest* is a beautiful, soulful, crucial book. It is a manifesto of hope for the 21st century grounded squarely in the hearts of engaged people around the planet. Paul Hawken chronicles and testifies on behalf of this 'movement with no name' with his charismatic intelligence and insight. This book makes the invisible visible. I believe Hawken when he says we are part of the Earth's immune system each time we exercise our active compassion in the name of social justice and ecological health. I love this book. It is a field guide for all that is possible."
—Terry Tempest Williams, author of *The Open Space of Democracy*

"Many books describe the world in ways that break our hearts. *Blessed Unrest* invokes a heartbreak from which light pours. Paul Hawken is stupendously well informed. He is also critical without rancor, intuitive without woowoo, and poetic or hard-minded as the case requires. This is a work of enormous love and consequence. Every compassion-driven soul who reads it will be stunned by the scope and power of the movement we've inadvertently formed. When, inevitably, my daughters someday feel their hearts broken by the wounded world they have inherited, I will be handing them this book of books."　　　　　　　—David James Duncan, author of *The Brothers K* and *God Laughs & Plays*

"Paul Hawken's writings are always at the cutting edge of environmental thought, original, surprising and shot through with optimism. *Blessed Unrest* is an uplifting perspective, engendering wonder and hope. For all of us that are squirreling away in our individual small ways, it is inspiring to realize that millions of us can add up to an irresistible force. Read this book and shout 'Hallelujah!'"　　　　　　　—David Suzuki, author of *The Sacred Balance: Rediscovering Our Place in Nature*

"On one side the four horsemen of the apocalypse; on the other a vast and nameless uprising of peoples and organizations fighting for justice, places, communities, diversity, and health—the planetary immune system. Paul Hawken's *Blessed Unrest* is not just a good book, it is a necessary book, wise, eloquent, perceptive, sober, and timely but above all, hopeful. A landmark!"　　　　　　　—David Orr, author of *Design on the Edge* and *The Last Refuge*

"Paul Hawken has written an important and significant book—intelligent, compelling, moving, and hopeful. In the broad sweep of a history of diffuse and seemingly unconnected events and people, he has found emergent pattern. That pattern, amazing simultaneously in its intricacy and simplicity, gives clarity to the direction humankind is moving in its struggle for survival. Read and regain a sense of optimism for our grandchildren's grandchildren; and be motivated to ensure that they inherit a restored earth and an equitable society."　　　　　　　—Ray Anderson, CEO of Interface

"Paul Hawken is at the top of his storytelling art in *Blessed Unrest*. By revealing the twin heart of the environmental and social justice movements, he helps us know ourselves in a new way—as competent members of the natural world, intent on recovering from our stumble as a species. Each page yields surprise and 'of course!' recognition for what has been swelling beneath our collective ground for over 100 years. I read it in a single sitting, hungry for the next piece of the puzzle, the next 6-degrees-of-separation coincidence. Hawken makes these invisible truths obvious through impeccably researched tales told in the bell-clear prose of a statesman poet. In this chronicle of the groundswell with no name, we have found our Tocqueville, our Twain, and our Sinclair."
—Janine Benyus, author of *Biomimicry: Innovation Inspired by Nature.*

BLESSED UNREST

How the Largest Social
Movement in History
Is Restoring Grace, Justice,
and Beauty to the World

PAUL HAWKEN

PENGUIN BOOKS

PENGUIN BOOKS
Published by the Penguin Group
Penguin Group (USA) Inc., 375 Hudson Street, New York, New York 10014, U.S.A.
Penguin Group (Canada), 90 Eglinton Avenue East, Suite 700, Toronto,
Ontario, Canada M4P 2Y3 (a division of Pearson Penguin Canada Inc.)
Penguin Books Ltd, 80 Strand, London WC2R 0RL, England
Penguin Ireland, 25 St Stephen's Green, Dublin 2, Ireland (a division of Penguin Books Ltd)
Penguin Group (Australia), 250 Camberwell Road, Camberwell,
Victoria 3124, Australia (a division of Pearson Australia Group Pty Ltd)
Penguin Books India Pvt Ltd, 11 Community Centre, Panchsheel Park, New Delhi – 110 017, India
Penguin Group (NZ), 67 Apollo Drive, Rosedale, North Shore 0632,
New Zealand (a division of Pearson New Zealand Ltd)
Penguin Books (South Africa) (Pty) Ltd, 24 Sturdee Avenue,
Rosebank, Johannesburg 2196, South Africa

Penguin Books Ltd, Registered Offices:
80 Strand, London WC2R 0RL, England

First published in the United States of America by Viking Penguin,
a member of Penguin Group (USA) Inc. 2007
Published in Penguin Books 2008

20 19 18 17 16 15 14 13 12 11

THE LIBRARY OF CONGRESS HAS CATALOGED THE HARDCOVER EDITION AS FOLLOWS:
Hawken, Paul.
Blessed unrest : how the largest movement in the world came into being and why no one saw it
coming / Paul Hawken.
p. cm.
Includes bibliographical references and index.
ISBN 978-0-670-03852-7 (hc.)
ISBN 978-0-14-311365-2 (pbk.)
1. Environmentalism.   2. Environmental justice.   I. Title.
GE195.H388 2007
333.72—dc22   2006101145

Printed in the United States of America
Designed by Nancy Resnick

To Anuradha Mittal, whose grace, heart,
and intelligence inspired this work

CONTENTS

BLESSED UNREST

O ver the past fifteen years I have given nearly one thousand talks about the environment, and every time I have done so I have felt like a tightrope performer struggling to maintain perfect balance. To be sure, people are curious to know what is happening to their world, but no speaker wants to leave an auditorium depressed, however dark and frightening a tomorrow is predicted by the science that studies the rate of environmental loss. To be sanguine about the future, however, requires a plausible basis for constructive action: you cannot describe possibilities for that future unless the present problem is accurately defined. Bridging the chasm between the two was always a challenge, but audiences kindly ignored my intellectual vertigo and over time provided a rare perspective instead. After every speech a smaller crowd would gather to talk, ask questions, and exchange business cards. These people were typically working on the most salient issues of our day: climate change, poverty, deforestation, peace, water, hunger, conservation, human rights. They came from the nonprofit and non-governmental world, also known as civil society; they looked after rivers and bays, educated consumers about sustainable agriculture, retrofitted houses with solar panels, lobbied state legislatures about pollution, fought against corporate-weighted trade policies, worked to green inner cities, and taught children about the environment. Quite simply, they had dedicated themselves to trying to safeguard nature and ensure justice. Although this was the 1990s, and the media largely ignored them, in those small meetings I had a chance to listen to their concerns. They were students, grandmothers, teenagers, tribe members, businesspeople, architects, teachers, retired professors, and worried mothers and fathers. Because I was itinerant, and the organizations they represented were rooted in their communities, over the years I began to grasp the diversity of these groups and their cumulative number. My interlocutors had a lot to say. They were informed, imaginative, and vital, and offered ideas, information, and insight. To a great extent *Blessed Unrest* is their gift to me.

My new friends would thrust articles and books into my hands, tuck small gifts into my knapsack, or pass along proposals for green companies. A Native American taught me that the division between ecology and human rights was an artificial one, that the environmental and social justice movements addressed two sides of a single larger dilemma. The way we harm the earth affects all people, and how we treat one another is reflected in how we treat the earth. As my talks began to mirror my deeper understanding, the hands offering business cards grew more diverse. I would get from five to thirty such cards per speech, and after being on the road for a week or two would return home with a few hundred of them stuffed into various pockets. I would lay them out on the table in my kitchen, read the names, look at the logos, envisage the missions, and marvel at the scope and diversity of what groups were doing on behalf of others. Later, I would store them in drawers or paper bags as keepsakes of the journey. Over the course of years the number of cards mounted into the thousands, and whenever I glanced at them, I came back to one question: Did anyone truly appreciate how many groups and organizations were engaged in progressive causes? At first, this was a matter of curiosity on my part, but it slowly grew into a hunch that something larger was afoot, a significant social movement that was eluding the radar of mainstream culture.

So, curious, I began to count. I looked at government records for different countries and, using various methods to approximate the number of environmental and social justice groups from tax census data, I initially estimated a total of 30,000 environmental organizations around the globe; when I added social justice and indigenous peoples' rights organizations, the number exceeded 100,000. I then researched to see if there had ever been any equals to this movement in scale or scope, but I couldn't find anything, past or present. The more I probed, the more I unearthed, and the numbers continued to climb, as I discovered lists, indexes, and small databases specific to certain sectors or geographic areas. In trying to pick up a stone, I found the exposed tip of a much larger geological formation. I soon realized that my initial estimate of 100,000 organizations was off by at least a factor of ten, and I now believe there are over one—and maybe even two—million organizations working toward ecological sustainability and social justice.

By any conventional definition, this vast collection of committed individuals does not constitute a movement. Movements have leaders and ideologies. People *join* movements, study their tracts, and identify themselves with a group. They read the biography of the founder(s) or listen to them perorate on tape or in person. Movements, in short, have followers. This movement, however, doesn't fit the standard model. It is dispersed, inchoate, and fiercely independent. It has no manifesto or doctrine, no overriding authority to check with. It is taking shape in schoolrooms, farms, jungles, villages, companies, deserts, fisheries, slums—and yes, even fancy New York hotels. One of its distinctive features is that it is tentatively emerging as a global humanitarian movement arising from the bottom up. Historically social movements have arisen primarily in response to injustice, inequities, and corruption. Those woes still remain legion, joined by a new condition that has no precedent: the planet has a life-threatening disease, marked by massive ecological degradation and rapid climate change. As I counted the vast number of organizations it crossed my mind that perhaps I was witnessing the growth of something organic, if not biologic. Rather than a movement in the conventional sense, could it be an instinctive, collective response to threat? Is it atomized for reasons that are innate to its purpose? How does it function? How fast is it growing? How is it connected? Why is it largely ignored? Does it have a history? Can it successfully address the issues that governments are failing to: energy, jobs, conservation, poverty, and global warming? Will it become centralized, or will it continue to be dispersed and cede its power to ideologies and fundamentalism?

I sought a name for the movement, but none exists. I met people who wanted to structure or organize it—a difficult task, since it would easily be the most complex association of human beings ever assembled. Many outside the movement critique it as powerless, but that assessment does not stop its growth. When describing it to politicians, academics, and businesspeople, I found that many believe they are already familiar with this movement, how it works, what it consists of, and its approximate size. They base their conclusion on media reports about Amnesty International, the Sierra Club, Oxfam, or other venerable institutions. They may be directly acquainted with a few smaller organizations and may even sit on the board of a local group. For them and others the movement is small, known, and circumscribed, a new type of charity, with a sprinkling of ragtag activists who occasionally give it a bad name. People inside the movement can also underestimate it, basing their judgment on only the organizations they are linked to, even though their networks can only encompass a fraction of the whole. But after spending years researching this phenomenon, including creating with my colleagues a global database of its constituent organizations, I have come to these conclusions: this is the largest social movement in all of human history. No one knows its scope, and how it functions is more mysterious than what meets the eye.

What *does* meet the eye is compelling: coherent, organic, self-organized congregations involving tens of millions of people dedicated to change. When asked at colleges if I am pessimistic or optimistic about the future, my answer is always the same: If you look at the science that describes what is happening on earth today and aren't pessimistic, you don't have the correct data. If you meet the people in this unnamed movement and aren't optimistic, you haven't got a heart. What I see are ordinary and some not-so-ordinary individuals willing to confront despair, power, and incalculable odds in an attempt to restore some semblance of grace, justice, and beauty to this world. In the not-so-ordinary category, contrast ex-president Bill Clinton and sitting president George W. Bush. As I write this, Bush is on TV snarled in a skein of untruths as he tries to keep the lid on a nightmarish war fed by inept and misguided ambition; simultaneously the Clinton Global Initiative (which is a nongovernmental organization) met in New York and raised $7.3 billion in three days to combat global warming, injustice, intolerance, and poverty. Of the two initiatives, war and peace, which addresses root causes? Which has momentum? Which does not offend the world? Which is open to new ideas? The poet Adrienne Rich wrote, "My heart is moved by all I cannot save: / So much has been destroyed / I have cast my lot with those / who, age after age, perversely, / with no extraordinary power, / reconstitute the world."[1] There could be no better description of the audiences I met in my lectures.

This is the story without apologies of what is going *right* on this planet, narratives of imagination and conviction, not defeatist accounts about the limits. Wrong is an addictive, repetitive story; Right is where the movement is. There is a rabbinical teaching that holds that if the world is ending and the Messiah arrives, you first plant a tree and then see if the story is true. Islam has a similar teaching that tells adherents that if they have a palm cutting in their hand on Judgment Day, plant the cutting. Inspiration is not garnered from the recitation of what is flawed; it resides, rather, in humanity's willingness to restore, redress, reform, rebuild, recover, reimagine, and reconsider. "Consider" (*con sidere*) means "with the stars"; reconsider means to rejoin the movement of heaven and life. The emphasis here is on humanity's *intention*, because humans are frail and imperfect. People are not always literate or educated. Most families in the world are impoverished and may suffer from chronic illnesses. The poor cannot always get the right foods for proper nutrition, and must struggle to feed and educate their young. If citizens with such burdens can rise above their quotidian difficulties and act with the clear intent to confront exploitation and bring about restoration, then something powerful is afoot. And it is not just the poor, but people of all races and classes everywhere in the world. "One day you finally knew what you had to do, and began, though the voices around you kept shouting their bad advice,"[2] is Mary Oliver's description of moving away from the profane toward a deep sense of connectedness to the living world.

Although the six o'clock news is usually concerned with the death of strangers, millions of people work on behalf of strangers. This altruism has religious, even mythic origins and very practical eighteenth-century roots. Abolitionists were the first group to create a national and global movement to defend the rights of those they did not know. Until that time, no citizen group had ever filed a grievance except as it related to itself.[3] Conservative spokesmen ridiculed the abolitionists then, just as conservatives taunt liberals, progressives, do-gooders, and activists today by making those four terms pejoratives. Healing the wounds of the earth and its people does not require saintliness or a political party, only gumption and persistence. It is not a liberal or conservative activity; it is a sacred act. It is a massive enterprise undertaken by ordinary citizens everywhere, not by self-appointed governments or oligarchies.

Blessed Unrest is an exploration of this movement—its participants, its aims, and its ideals. I have been a part of it for decades, so I cannot claim to be the detached journalist skeptically prodding my subjects. I hope what follows is the expression of a deep listening. The subtitle of the book— *how the largest social movement in history is restoring grace, justice, and beauty to the world*—cannot be answered by one person. Like anyone, I have a perspective based on biases accumulated over time and a network of friends and peers who color my judgment. However, I wrote this book primarily to discover what I *don't* know. Part of what I learned concerns an older quiescent history that is reemerging, what poet Gary Snyder calls *the great underground,* a current of humanity that dates back to the Paleolithic. Its lineage can be traced back to healers, priestesses, philosophers, monks, rabbis, poets, and artists "who speak for the planet, for other species, for interdependence, a life that courses under and through and around empires."[4] At the same time, much of what I learned is new. Groups are *intertwingling*—there are no words to exactly describe the complexity of this web of relationships.[5] The Internet and other communication technologies have revolutionized what is possible for small groups to accomplish and are accordingly changing the loci of power. There have always been networks of powerful people, but until recently it has never been possible for the entire world to be connected.

The chapter "Blessed Unrest" is an overview that describes how this movement differs from previous social movements, particularly with respect to ideology. The organizations in the movement arise one by one, generally with no predetermined vision for the world, and craft their goals without reference to orthodoxy. For some historians and analysts, movements only exist when they have an ideological or religious core of beliefs. And movements certainly don't exist in a vacuum: a strong leader(s) is an earmark of a movement and often its intellectual pivot point, even if deceased. The movement I describe here has neither, and so represents a completely different form of social phenomenon.

The next three chapters are glimpses of some of the movement's roots. One cannot do justice to its history in a clutch of books, much less a few chapters. America has been the home of some of the most important progressive efforts in history—women's suffrage, abolition, civil rights, food safety—but you would not know that, given the narrowness of scope of today's education. My survey reflects the views of a North American because it is the only history I can adequately present. This bias is important to acknowledge, because global history is invariably skewed when seen through the eyes of Western culture, no matter how hard one tries to be objective. There are other histories, African and Native American, English and Japanese, Brazilian and Mediterranean, all equally valid, and all with their own particular inflections. In India, for example, environmentalism is a social justice movement, concerned with the rights of people to the land and its bounty. In 1991 Sunita Narain, the director of the Center for Science and the Environment in New Delhi, called global warming environmental colonialism, and was one of the first to question whether environmental management should be based on human rights rather than legal convention. In the United States the environmental movement found itself faced with a backlash when it was accused of placing the rights of the animals and plants on the land before those of people. Ron Dellums, an African-American congressman from Oakland, California, asked the Sierra Club, "I know you care about black bears but do you care about black people?"[6] In Germany the green movement became an organized political party, and its members now hold positions at the highest echelons of government. In the global South, environmentalism is a movement of the poor, with peasants leading campaigns that include land reform, trade rights, and corporate hegemony. The environmental movement began in England as a series of public health campaigns during the Industrial Revolution. In Italy, it concerns the dynamics between *la città* and *la campagna;* in South Africa, it is inextricably bound to social justice issues embedded in the country's history.[7] My purpose in recounting some of the threads of the past is not merely to extol great personages such as Darwin, Gandhi, Rachel Carson, or Thoreau, but to recognize the importance of connection and coincidence. Long ago, small and seemingly inconsequential actions took place that eventually changed the world—outcomes the original actors might never have imagined. One such occurrence was Emerson's encounter with the Jussieu family of scientists in Paris, a little remarked-upon event whose influence, as we will see, eventually wends itself into the civil rights movement 123 years later. In a time when people feel powerless, a history of altruism can be a balm because it reveals the power of helpful and humble acts, a reminder that constructive changes in human affairs arise from intention, not coercion.

"Indigene" and "We Interrupt This Empire" concern globalization. "Indigene" is concerned with indigenous cultures. Their traditional lands represent the greatest remaining sanctuaries of life on earth, and resource-hungry corporations are commercializing and destroying these biological arks. The cultures that have coevolved with these environments are resisting the encroachment, uniting with alliances of nonprofits to bring accountability and limits to unchecked development. "We Interrupt This Empire" focuses on organizations that are engaged in protecting citizens, workers, and environments from the juggernaut of free market fundamentalism.

The final two chapters look at the entire movement from two perspectives. "Immunity" uses the cellular metaphor of how an organism defends itself as a plausible way to describe the collective activity of the movement. The immune system is the most complex system in the body and provides a useful model for examining the properties of these groups. The terms *environment* and *social justice* encompass innovative organizations that are redolent with ideas and inventive techniques, and a few are explored here. I also consider the weakness of the movement, how its multiplicity and diversity may hobble it as the world descends into violence and disorder. "Restoration" describes the biological principles that inform all forms of life, including human beings, and uses these principles as a framework to bring a different vocabulary to the movement. In biologist Janine Benyus's quintessential summation, "life creates the conditions that are conducive to life." It is fair to ask whether that might not be a suitable organizing principle for all human activity, from economics to trade to how we build our cities. While it is risky to rely on life sciences to explain social phenomena, it is equally risky to assume that the standard language that has served to chronicle past social movements is sufficient to describe this one. The individuals featured in this book all try to do good, but this book is not only about doing good. It is about people who want to save the entire sacred, cellular basis of existence— the entire planet and all its inconceivable diversity. In total, the book is inadvertently optimistic, an odd thing in these bleak times. I didn't intend it; optimism discovered me.

BLESSED UNREST

There is vitality, a life force, an energy, a quickening that is translated through you into action, and because there is only one of you in all time, this expression is unique.... You have to keep open and aware directly to the urges that motivate you. Keep the channel open.... [There is] no satisfaction whatever at any time. There is only a queer, divine dissatisfaction, a blessed unrest that keeps us marching and makes us more alive than the others.

—Martha Graham to Agnes de Mille, *Dance to the Piper*[1]

How is one to live a moral and compassionate existence when one is fully aware of the blood, the horror inherent in life, when one finds darkness not only in one's culture but within oneself? If there is a stage at which an individual life becomes truly adult, it must be when one grasps the irony in its unfolding and accepts responsibility for a life lived in the midst of such paradox. One must live in the middle of contradiction, because if all contradiction were eliminated at once life would collapse. There are simply no answers to some of the great pressing questions. You continue to live them out, making your life a worthy expression of leaning into the light.   —Barry Lopez, *Arctic Dreams*

I am large, I contain multitudes.   —Walt Whitman, "Song of Myself"

C layton Thomas-Müller speaks to a community gathering of the Cree nation about waste sites on their native land in northern Alberta, toxic lakes so big you can see them from outer space. Shi Lihong, founder of Wild China, films documentaries with her husband on migrants displaced by construction of large dams. Rosalina Tuyuc Velásquez, a member of the Maya-Kaqchukel people, fights for full accountability for tens of thousands of victims of death squads in Guatemala. Rodrigo Baggio retrieves discarded computers from New York, London, and Toronto, and installs them in the favelas of Brazil, where he and his staff teach computer skills to poor children. Biologist Janine Benyus speaks to 1,200 executives at a business forum in Queensland about biologically inspired industrial development. Paul Sykes, a volunteer for the National Audubon Society, completes his fifty-second Christmas Bird Count in Little Creek, Virginia, joining fifty thousand others who tally 70 million birds on one day. Sumita Dasgupta leads students, engineers, journalists, farmers, and Adivasis (tribal people) on a ten-day trek through Gujarat exploring the rebirth of ancient rainwater harvesting and catchment systems that bring life back to drought-prone areas of India. Silas Kpanan'Ayoung Siakor exposes links between the genocidal policies of President Charles Taylor and illegal logging in Liberia, resulting in international sanctions and the introduction of certified, sustainable timber policies.

These eight individuals, who may never meet and come to know one another, are part of a coalescence comprising hundreds of thousands of organizations. It claims no special powers and arises in small discrete ways, like blades of grass after a rain. The movement grows and spreads in every city and country, and involves virtually every tribe, culture, language, and religion, from Mongolians to Uzbeks to Tamils. It is composed of families in India, students in Australia, farmers in France, the landless in Brazil, the Bananeras of Honduras, the "poors" of Durban, villagers in Irian Jaya, indigenous tribes of Bolivia, and housewives in Japan. Its leaders are farmers, zoologists, shoemakers, and poets. It provides support and meaning to billions of people in the world. The movement can't be divided because it is so atomized—a collection of small pieces, loosely joined.[2] It forms, dissipates, and then regathers quickly, without central leadership, command, or control. Rather than seeking dominance, this unnamed movement strives to disperse concentrations of power. It has been capable of bringing down governments, companies, and leaders through witnessing, informing, and massing. The quickening of the movement in recent years has come about through information technologies becoming increasingly accessible and affordable to people everywhere. Its clout resides in its ideas, not in force.

Picture the collective presence of all human beings as an organism. Pervading that organism are intelligent activities, humanity's immune response to resist and heal the effects of political corruption, economic disease, and ecological degradation, whether they are the result of free-market, religious, or political ideologies. In a world grown too complex for constrictive ideologies, even the very word *movement* to describe such a process may be limiting. Writer and activist Naomi Klein calls it "the movement of movements," but for lack of a better term I will stick with *movement* here because I believe all its components are beginning to converge.

The movement has three basic roots: environmental activism, social justice initiatives, and indigenous cultures' resistance to globalization, all of which have become intertwined. Collectively, it expresses the needs of the majority of people on earth to sustain the environment, wage peace, democratize decision making and policy, reinvent public governance piece by piece from the bottom up, and improve their lives—women, children, and the poor. Throughout history, armies, corporations, religious rulers, and political zealots have overpowered the majority world, which in our upside-down world we consider to be minorities.[3]

The definition of the term *social justice* has been debated for centuries. In this book, social justice means the implementation and realization of human rights as defined by the Universal Declaration of Human Rights ratified by the General Assembly of the United Nations in 1948, with the addition of the right to a productive, safe, and clean environment; the right to security from political tyranny; and the right to live and express one's own culture. The thirty Articles of the Declaration proper state that people have a right to freedom and liberty, that no one shall be enslaved or held in servitude or subjected to torture or cruel and degrading punishment. Everyone has the right to be recognized as a person by law, not subject to arbitrary detention or exile. People have the right to receive asylum if persecuted in their home country. Citizens in all countries have the right to education, livelihood, and fair working conditions, including the right to join a trade union and receive a living wage.

Human rights, by definition, apply to everyone who belongs to our species, wherever they are found. Concern for human rights is not a recent phenomenon, though its history has been grossly uneven, a chronicle written over time in the face of tyranny, barbarism, and setbacks.[4] Historical landmarks of human rights date from deep into the past, from Hammurabi's Code in ancient Babylon (in what is now Iraq) to the Enlightenment philosophies of the West. Between those times, Buddhist and Hindu teachings brought forth precepts of ecological responsibility; Confucius proposed a form of universal education; the Greeks practiced democracy and first raised the idea of natural laws pertaining to human rights; and early Islam established limitations on the authority of rulers, and advocated an independent judiciary to protect and respect human dignity.[5] Historically, however, religious and political movements that invoke the common good have tended to exclude certain groups: women, homosexuals, lower castes, enslaved people, or those born with disabilities, and many continue to do so to this day. Groups that populate the movement, in contrast, largely understand human rights from the viewpoint of the oppressed, not from that of the elite.

Along with gross violations of human rights are other endless indignities that billions endure: loss of water for agriculture, theft of local resources by government and corporations, incursions of mining companies that pollute, political corruption and hijacking of governance, lack of health care and education, big dams that have displaced millions of poor people, loss of land, trade policies that bankrupt small farmers, and more. What people want in their place is universal: security, the ability to support their families, educational opportunities, nutritious and affordable food, clean water, sanitation, and access to health care. According to more than 190 nations in the world, these are not entitlements; they are rights.

The movement for equity and environmental sustainability comes as global conditions are changing dramatically and becoming more demanding. We are the first generation to live on earth to witness a doubling of population in our lifetime. The babies born within the next thirty hours of your reading this sentence will replace the 250,000 people lost in the tragic tsunami of December 26, 2004. Nearly 3 billion more people will join the current population of 6.6 billion within fifty years, and the world has yet to figure out how to take care of those already here. By the middle of this century, resources available per person will drop at least by half. Since the eighteenth century, many labor-saving processes and methods have been instituted, and although human productivity has soared, hundreds of millions of potential workers and contributors to society feel overlooked and unneeded. Every week 1.4 million people pour into the world's slums to join a metastatic mass of squatters.[6] The stark, bland statistic that cites 3 billion people receiving less than $2 a day is heard so often that it no longer elicits a response, but it remains true, compelling, and appalling.

To those calling for more trade and global economic growth to salve such damning deficiencies, it should be pointed out that both have been promoted vigorously throughout the past twenty-five years, yet during that time inequities in the world have only worsened. "It's as though the people . . . have been rounded up and loaded onto two convoys of trucks (a huge big one and tiny little one) that have set off in resolutely opposite directions," writes Arundhati Roy. "The tiny convoy is on its way to a glittering destination somewhere near the top of the world. The other convoy just melts into the darkness."[7] With the industrial capitalist system busily dividing the world in two, its priorities do not encompass either justice or the environment. Most of the world's children are poor, and most of the poor are children.[8] More than 1 billion people in the world want jobs and cannot get them; for those who are employed, twice that number do not receive a living wage. Two billion more people will join the workforce in the next twenty years. In the United States, teenage employment is at its lowest level since 1948. Only two-thirds of U.S. workers between twenty and twenty-four were employed in 2005.[9] Only one species on earth does not have full employment and that is *Homo sapiens*. When informed of the world's chronic unemployment, one third-grader asked, "Is all the work done?"

If there is a pervasive criticism of global capitalism that is shared by all actors in the movement, it is this observation: goods seem to have become more important, and are treated better, than people. What would a world look like if that emphasis were reversed? This book explores a movement cultivating innovative, sometimes brilliant, social technologies that would accomplish just that reversal by returning people to the heart of the world and of life. It comprises design as much as action, imagination as much as organization. It also entails a courageous defense of human rights. It stresses innovation with a focus on everyday life: the demands and pleasures of learning, taking care of others, preparing food, raising children, taking journeys, and doing meaningful work. These timeless ways of being human are threatened by global forces that do not consider people's deepest longings.

When I'm discussing the movement with academics or friends in the media, the first question they pose is usually the same: If it is so large, why isn't this movement more visible? By that they mean, why isn't it more visible to news media, especially TV? Although global in its scope, the movement generally remains unseen until it gathers to take part in demonstrations, whether in London, Prague, or New York, or at annual meetings of the World Social Forum, after which it seems to disappear again, reinforcing the perception that it is a will-o'-the-wisp. The movement doesn't fit neatly into any category in modern society, and what can't be visualized can't be named. In business, what isn't measured isn't managed; in the media, what isn't visible isn't reported. Media coverage of the death of Pope John Paul and the election of Pope Benedict easily surpassed all coverage devoted to this movement over the past ten years, yet the number of people directly working and indirectly involved with this movement is greater than the number of people active in the Catholic Church. The papacy has history and specificity; the movement is about the future.

For most people, to understand something new requires a cognitive antecedent. When members of the Me'en tribe in Ethiopia were shown a coloring book that included an illustration of a local antelope, they didn't recognize the animal. They would smell the paper, twist it in their hands, feel its texture, listen to its sound, and even taste it gingerly, but they couldn't discern any animal from its picture alone. When anthropologists transferred the drawing to cloth, a material with which the tribe was familiar, a few of the tribespeople could make out something. A twenty-year-old woman gazed at the outline as a scientist traced the animal with her finger, and although she could see a tail, leg, ears, and a horn, when asked what the illustration represented, she had no idea.[10] Scientific experiments repeatedly show that groups of educated, urbanized people pay no attention to unfamiliar objects directly in front of them if they focus too strongly on familiar ones. What we already know frames what we see, and what we see frames what we understand. The Industrial Revolution went unnamed for more than a century, in part because its developments did not fit conventional categories, but also because no one could define what was taking place, even though it was evident everywhere.

Another reason the movement is hard to identify is that it is not an outgrowth of any particular ideology. This is the first time in history that a large social movement is not bound together by an "ism." What unifies it is ideas, not ideologies. There is a vast difference between the two; ideas question and liberate, while ideologies justify and dictate.[11] Generally, when an ideology forms, it is based on a set of beliefs originally set forth by one person. Those beliefs become an ism when adepts, followers, and factotums create an organization to contain and disseminate what is now a mixture of fact and faith. As they expand, ideological movements divide and redivide. In Christianity, for example, there were first Gnostics, Jewish Christians, and Pauline Christians. In the eleventh century, the Eastern Orthodox churches parted ways with the Catholic Church over papal primacy. When Martin Luther nailed his Ninety-five Theses to the chapel door at the University of Wittenberg in 1517, it marked the beginning of the Reformation movement and the birth of Protestant denominations, which soon included Mennonites, Lutherans, Baptists, Anabaptists, Zwinglianists, Puritans, Presbyterians, Anglicans, and Calvinists. Some 101 years after Luther's protest, the Thirty Years War began, a protracted struggle between Catholics and Protestants that eventually led to the deaths of five million Germans—one of every three people in the country, most of them peasants. Today there are more than one thousand Christian denominations in the United States. Other notable examples of sectarianism are Shi'ia and Sunni Muslims, who are often violently divided. Marxism numbers among its ranks Leninists, Trotskyites, Stalinists, Khrushchevites, and Maoists.

Big ideologies arose in the nineteenth century, and dominated our beliefs about what was true, false, and even possible into the twentieth century. Leaders have used ideologies as varied as communism, capitalism, populism, materialism, fundamentalism, imperialism, colonialism, and socialism to prop up their regimes, recruit their armies, and defend their policies. The big three—capitalism, socialism, and communism—fought for control of our minds, territories, and resources throughout the twentieth century, and now are being replaced by terrorism and economic and religious fundamentalism.[12] Because we are educated to believe that salvation is found in the doctrines of a single system, we are naively susceptible to dissimulation and cant. Ideologies prey on these weaknesses and pervert them into blind loyalties, preventing diversity rather than nurturing natural evolution and the flourishing of ideas. Ecologists and biologists know that systems achieve stability and health through diversity, not uniformity. Ideologues take the opposite view.

The end of the twentieth century saw the collapse of big ideologies, and into that vacuum now flow a number of different forms of populism, which variously invoke the Bible, Allah, Ram, nationalism, or free markets as their basis for legitimacy. Neoconservatives, radical Islamists, the Christian Right, and economic fundamentalists share the ability to supply people with surrogates for failed ideologies.[13] These groups act aggressively on our behalf, it is contended, because they know what is best for us. Radical Islamists view those opposing their theocratic vision as infidels who can legitimately be killed. The Christian Right regards non-Christians as needing salvation and redemption that can be delivered by God's laws, which only it fully understands. Neoconservatives believe that ordinary citizens cannot be entrusted with the reins of power, that a small group of superior individuals should rule over the majority of inferiors, using religion and the perpetual threat of war to create a Potemkin village of populism. Supporters of corporate-led globalization want to impose their market-based rules and precepts on the entire planet, regardless of place, history, or culture, in the belief that economic growth is an unalloyed good, and that it is best accomplished with the minimization or elimination of interference from government. These groups share a fundamental distaste for democracy and seek expediency, not plebiscites. Just as religion creates God in its own image, the pseudo-populists want to create a world that mirrors their simplified imaginings.

In *Unpopular Essays* Bertrand Russell wrote, "Man is a credulous animal and must believe in something. In the absence of good grounds for belief, he will be satisfied with bad ones." Each type of pseudo-populism comes into being to improve or save its respective adherents from the absence of a moral or social framework, which means that even if we don't understand them fully, we are expected to place our faith in them. Although these pseudo-populist organizations are relatively small, they manage to influence governments, invoke terror, and control large sums of capital. Relatively speaking, none of them has a large following, but they do have intense ones. In a global context, all stand on the most narrow of pedestals, and all have successfully seized key levers of power.

Global civilization is endangered by these isms. Climatic stability may be lost for centuries to come, poverty increases, fisheries collapse, megacities teem with influxes of rural refugees, water tables fall, and hunger and malnutrition grow, even in the richest country in the world. The twentieth century saw the greatest rate of destruction to the environment in all recorded history. It was also the cruelest, harshest, and bloodiest century in history. Eighty million were slaughtered from the beginning of the century through World War II; since then, more than 23 million people (mostly civilians) have been killed in more than 149 wars.[14] Endless research speculates about the exercise of war, but little is concerned with the maintenance of peace. We study different forms of greed, including neoclassical economics, but rarely the harmonization of human needs. For every dollar spent on U.N. peacekeeping, $2,000 is expended for warmaking by member nations. Four of the five members of the U.N. Security Council, which has veto power over all U.N. resolutions, are the top weapons dealers in the world: the United States, the United Kingdom, France, and Russia.[15]

In contrast to the ideological struggles currently dominating global events and personal identity, a broad nonideological movement has come into being that does not invoke the masses' fantasized will but rather engages citizens' localized needs. This movement's key contribution is the rejection of one big idea in order to offer in its place thousands of practical and useful ones. Instead of isms it offers processes, concerns, and compassion. The movement demonstrates a pliable, resonant, and generous side of humanity. It does not aim for the utopian, which itself is just another ism, but is eminently pragmatic.

And it is impossible to pin down. Generalities that seek to define it are largely inaccurate. Understandably, the movement defies conventional typologies. Its liberal leaders are often devout; its conservative leaders propose radical solutions. The movement crosses over hoary, razor-wired political boundaries. Should the idea of using renewable sources to achieve localized energy independence be categorized as radical, conservative, ecological, good long-term economics, or socially equitable? If the movement in all its diversity has a common dream, it is process—in a word, democracy, but not the democracy practiced and corrupted by corporations and modern nation-states. It is, rather, a reimagination of public governance emerging from place, culture, and people. What binds its constituents is a modus operandi that could be called the autonomy of diversity. Groups with varying outlooks and discrete goals cooperate on key issues without subordinating themselves to another group. While the key to its strength and success is this very diversity, it also leaves the movement singularly vulnerable.

However adaptive, diversity can also prevent connection, cooperation, and effectiveness. Inevitably there is jockeying for position and territory, and lack of collaboration, especially when organizations are forced to compete for scarce resources. There is narcissism, when small groups begin to stare into the waters of just causes and imagine themselves to be saviors. And because they are led and managed by human beings, there is gossip, churlishness, and backbiting. Within the movement *are* some who are sophomoric, callow, and atavistic. Small, splinter "liberation" groups have committed crimes such as arson, and proudly boast of it. Strongly held beliefs can breed fanaticism as easily as genuine breakthroughs. Unfortunately every misstep made in the name of social justice or the environment has received a disproportionate amount of publicity. In 2005 the *Los Angeles Times* devoted one hundred times more coverage to a vandalistic spree by three unaffiliated students who damaged or destroyed 125 SUVs than it did to the landmark *U.N. Millennium Ecosystem Assessment*.

While issues grow in importance, a balkanized movement does not match the scale of the problems. This is particularly obvious with respect to climate change. On one hand, the practical implementation of hands-on energy reduction needs to be implemented on a local scale. But the major policy changes and initiatives that must be undertaken at the national and international levels with respect to public transportation, oil company subsidies, and renewable energy are stymied by the corruption of politicians and special interests; as yet there has been no coming together of organizations in a united front that can counter the massive scale and power of the global corporations and lobbyists that protect the status quo.

Another potentially negative aspect of diversity is that the movement is mocked and misunderstood for the sheer number of causes it espouses. As in the parable of the blind men and the elephant, it is impossible to fully comprehend the totality of the movement, which cannot be perceived as a whole. Describing only the parts they see, the media use labels like environmentalists, small farmers, mothers, special-interest groups, agitators, protesters, minorities, idealistic youth, the "Prius set," peasants, indigenous people, greens, academics, activists, aging hippies, liberals, and children. When the media first began to cover the Women's Suffrage movement, Elizabeth Cady Stanton wrote, "All the journalists, from Maine to Texas, seemed to strive with each other to see which could make our movement the most ridiculous."[16] Nothing has changed since then, but even if the stereotypes were true they would not be complete. Missing are the visions and values that make this movement active and pervasive.

Politicians and the media assess strength by the capacity for bruising single-mindedness, not by breadth of interests or goals. The NRA *is* powerful, but that doesn't mean that people who address human trafficking in Burma, loss of sea turtles, desertification, and climate change must be regarded as ineffectual. Unless a movement champions a specific objective, it is often dismissed as a children's crusade without clout. When you link issues such as waste incinerators, sweatshops, endocrine disruptors, water pollution, and mountaintop removal, you may be judged to have too much on your plate. In fact, you could be thinking, as Aldo Leopold advised, "like a mountain," perceiving the rich complexities of a system and the interrelatedness of social and environmental problems within it. If one analyzes the challenges facing the world in systemic terms, the intelligence of implementing permaculture, microlending, green taxes, ecological footprinting, and fair trade becomes evident.

Conversely, if problems are viewed atomistically, the same strategies and solutions can be dismissed as idealistic or impractical. Creating genetically modified organisms to address hunger, building pebble bed nuclear reactors to address global warming, or waging opportunistic wars to establish democracies are all forms of downstream thinking that predicated today's dilemmas because none of them address the sources of the problem. In software, such fixes are called kludges, workarounds to repair bugs, expedient patches that may work for a time. Kludges do not address root causes but layer on Band-Aids to modify undesirable effects, ultimately forming a distressed whole.[17] Fixing the intractable problems besetting the world will require a convergence of social intelligence and natural science, two qualities traditional politics lack.

The as yet undelivered promise of this movement is a network of organizations that offer solutions to disentangle what appear to be insoluble dilemmas: poverty, global climate change, terrorism, ecological degradation, polarization of income, loss of culture, and many more. The world seems to be looking for the big solution, which is itself part of the problem, since the most effective solutions are both local *and* systemic. Although the groups in the movement are autonomous, the coming together of different organizations to address an array of issues can effectively become a systemic approach. Although the movement may appear inchoate or naively ambitious, its underlying structure and communication techniques can, at times, create a collective social response that can challenge any institution in the world.

But how powerful can a movement ultimately be if it sidesteps tribal/

state/nation divisions and replaces them with a network of associations that has no center? Some believe that to address meaningfully existing political and economic institutions, the movement must centralize to present a viable alternative in scale. Many others would argue that traditional, hierarchical forms of organization that require homogeneous agendas and goals are outmoded throughout the modern world, and that this movement is a harbinger of change. The answer may lie in the middle; it will depend on many factors.

Even though the origins and purposes of the various groups comprising the movement are diverse, if you survey their principles, mission statements, or values, you find they do not conflict. As a test, I have randomly chosen groups, found their values statement or the like, then printed them up and posted them on a wall, one after another. This exercise was useful in helping identify the unity underlying the diversity. The movement does not agree on every topic, nor will it ever, because that degree of consensus would begin to transform it into an ideology. New groups do not check in with a master organization to see if their mission statement is acceptable. What its members do share is a basic set of fundamental understandings about the earth, how it functions, and the necessity of fairness and equity for all people dependent on the planet's life-giving systems.

Although many progressives regard the ideas of Friedrich Hayek, Nobel Prize–winning economist and éminence grise of free-market economics, as anathema, he was one of the first to recognize the dispersed nature of knowledge and the effectiveness of localization and of combining individual understanding. Since one person's knowledge can only represent a fragment of the totality of what is known, wisdom can be achieved when people combine what they have learned. Hayek felt that viable social institutions had to evolve (we might now say "coevolve") to confront the problems at hand rather than reflect theories at mind. During World War II some made moral distinctions between Nazi Germany and Communist Russia, but Hayek saw them both as identical examples of totalitarian states. His concerns about totalitarianism are applicable to the large multinational institutions of today: transnational corporations, the WTO, the World Bank, and the International Monetary Fund (IMF), to name a few. They can be viewed as totalitarian because they encompass economic development and human needs under a uniform rubric. Although Hayek did not anticipate corporate-sponsored totalitarianism fed by religious zealotry or corporate control of media invoking the mantra of free markets, he did foresee a remedy for the basic expression of the totalitarian impulse: ensuring that information and the right to make decisions are co-located. To achieve this, one can either move the information to the decision makers, or move decision making rights to the information. The movement strives to do both.[18] The earth's problems are everyone's problems, and what modern technology and the movement can achieve together is to distribute problem solving tools.[19]

Any contemplation of the history of the world says different things to different people. Because the accumulation of past records, events, and remembrances is so vast, no one person can encompass global history, not in this lifetime or many. We are left instead with frameworks, lenses through which historians arrange the past to create a coherent narrative. Two such lenses will likely dominate our future attempts to understand our past: social justice and humanity's relationship to the environment.[20] Both address exploitation, and both frame the history of people's attempts to free themselves from abuse. We face today a dilemma about what standard will constitute the most salient evidence of progress. Will it be single measures of material accumulation, such as GDP, or will it be the health of the earth and its inhabitants? Social justice and attending to the planet proceed in parallel; the abuse of one entails the exploitation of the other. Slaves, serfs, and the poor are the forests, soils, and oceans of society; each constitutes surplus value that has been exploited repeatedly by those in power, whether governments or multinational corporations.

Our fate will depend on how we understand and treat what is left of the planet's surpluses—its lands, oceans, species diversity, and people. The quiet hub of the new movement—its heart and soul—is indigenous culture. The acknowledgment of aboriginal cultures is not a romantic gesture or wistful plea, nor does it value Neolithic cultures above modern ones, or native spiritual practices above other sacred traditions. Just as a wheel cannot turn without a stationary hub, the movement reaches back to the deep and still roots of our collective history for its axle. Indigenous people have a different sense of time because they remember a different history, and that memory brings an uncommon appreciation of their place in time. Simply stated, they possess patience. Things come and go; conquests, ideas, and leaders arise and fall away. For indigenous people, in the time that defines one's life, the relationship one has to the earth is the constant and true gauge that determines the integrity of one's culture, the meaning of one's existence, and the peacefulness of one's heart. In most indigenous cultures there are no separate social and environmental movements because the two were never disaggregated. Every single particle, thought, and being, even our dreaming, is the environment, and what we do to one another is reflected on earth just as surely as what we do to the earth is reflected in our diseases and discontents. C. S. Lewis wrote, "What we call Man's power over Nature turns out to be a power exercised by some men over other men with Nature as its instrument." It is because of this split between people and nature that the social justice and environmental arms of the movement have arisen as separate, each with its own history. Indigenous cultures provide the basis for understanding the two as one.

Unlike indigenous cultures, whose worlds are local, intimate, familiar, we live in the age of giants. In one day alone we pump 85 million barrels of petroleum out of the ground, and then burn it up. And on the same day we spew the waste of 27 billion pounds of coal into the atmosphere. One hundred million displaced people now wander the earth without a home. One company, Wal-Mart, employs 1.8 million people. ExxonMobil made nearly $40 billion in profits in 2006, enough money to permanently supply pure clean drinking water to the 1 billion people who lack it. We have consumed 90 percent of all the big fish in the oceans. Bill Gates's home covers one and a half acres and cost nearly $100 million.

Not surprisingly, people don't know that they count in such a mal-ordered, destabilized world, don't know that they are of value. A healthy global civilization cannot be constructed without building blocks of meaning, which are hewn of rights and respect. What constitutes meaning for human beings are events, memories, and small dignities—gifts that rarely emerge from institutions, and never from theory. As the smaller parts of the world are knitted into one globalized unit, the one thing we can no longer afford is bigness. This means dismantling the big bombs, dams, ideologies, contradictions, wars, and mistakes.[21]

In the midst of such giants a worldwide gathering of ordinary and extraordinary people are reconstituting the notion of what it means to be a human being. While they are organizing themselves into the largest movement in the history of the world, the movement only happens one person at a time. But how does one become an environmentalist or human rights campaigner? There are no missionaries. There are no postings offering lessons. Concerned individuals have to work it out for themselves and find colleagues who will mentor them. Movements are the expression of changed attitudes, and how each person comes to realize his responsibility to a greater whole is a unique experience. All social justice organizations can trace their origins back some 220 years ago, when three-fourths of the world was enslaved in one form or another. In 1787 a dozen people began meeting in a small print shop in London to abolish the lucrative slave trade. They were reviled and dismissed by businessmen and politicians. It was argued that their crackpot ideas would bring down the English economy, eliminate growth and jobs, cost too much money, and lower the standard of living. Critics also pointed out that abolition was being promoted by a small group of self-appointed troublemakers and extremists who had no expertise in trade or commerce.[22] But the audaciousness of this first expression of civil society was eventually rewarded, and six decades later slavery was legally forbidden almost everywhere.

Today the world faces a task that is exponentially more difficult than the abolition of slavery: the prevention of irreversible losses of planetary capacity to support life. The arguments against the abolition of slavery that were proffered in the Houses of Parliament at the end of the eighteenth century are almost exactly the same as the arguments put forward today about why our economy can't move away from fossil fuels to renewable energy, provide living-wage jobs for all, or defend the skies, forests, and waters.[23] If we are to survive, every citizen must be enlisted to accomplish this task, and that will not be possible unless we cease the worldwide war on the poor and mark a road to recovery that brings respect, dignity, and self-worth to all. For better and worse, we now occupy a human planet, one in which most evolutionary forces are guided or misguided by our hand. Weather is not only a process that we're subject to but also a complex dynamic for which we have unwittingly and suddenly become stewards. Human agency will alter the fate of all living beings because no part of the planet is unaffected by our activity. Although the egregious levels of corruption and violence visited upon us in daily headlines suggest otherwise, we can be grateful that humanity is a learning organism.

On February 15, 2003, between 6 million and 10 million people took to the streets in eight hundred cities around the world to protest the U.S. invasion of Iraq. It was the largest coordinated public demonstration in history, with estimates of 2 million demonstrators marching in Rome alone. Two days later in *The New York Times*, Patrick Tyler wrote that the demonstrations were "reminders that there still may be two superpowers on the planet: United States and world public opinion." It was a good line, and others seized upon it, from U.N. secretary general Kofi Annan to Jonathan Schell. But can the two entities be aptly compared? There is only one superpower on earth, as defined by indices of military and financial might; compared to it the movement is an unarmed pauper, but it is driven by an infrapower, a stirring from below that could be described as a physiological response of the body politic. Every reader of this book is probably a part of it, even if indirectly. To see what is invisibly in front of us requires a shift in our conception of change and power.

A familiar biological tease argues that a hen is an egg's way of making a new egg. Likewise, have we evolved plants to create agriculture, or have plants used agriculturists to evolve themselves? From a coevolutionary perspective, both propositions are true. What is the difference between a squirrel burying acorns across the forest and humans planting potatoes across the globe? Who is master, and who is servant? Is it the acorn's or potato's idea to be nutritious, or the creature that buries them? Evolution is not about design or will; it is the outcome of constant endeavors made by organisms that want to survive and better themselves.[24] The collective result is intoxicatingly beautiful, rife with oddities, and surpassingly brilliant, yet no agent is in control. Evolution arises from the bottom up—so, too, does hope. When fire destroys a forest, the species and plants that were lost will reassert themselves over time. Seeds that have lain dormant for decades and that germinate only when subjected to intense heat will come to life, burst into foliage, and bloom in the spring. These plants may have deep taproots that bring up minerals, or broad leaves that create a canopy to help preserve topsoil from sun and rain. The older the forest, the more resilient its capacity to regenerate. Humanity is older than the oldest forest. Its capacity to adapt and restore is vastly underestimated. Evolution is optimism in action.

Being compelled to make more of ourselves is the human lot. This book asks whether a significant portion of humanity has found a new series of adaptive traits and stories more alluring than the ideological fundamentalisms that have caused us so much suffering. Stories told too often begin to lose their force, as do societies, but humankind can also create a new narrative. As William Kittredge writes, "A society capable of naming itself lives within its stories, inhabiting and furnishing them. We ride stories like rafts, or lay them out on the table like maps. They always, eventually, fail and have to be reinvented. The world is too complex for our forms ever to encompass for long."[25] How many new stories and groups will it take before the world recognizes its evolutionary potential, not just its baseness? Because stories are greater than we are, their capacious narratives give us wiggle room to dream. It is why children's eyes light up and gaze far off when we read them tales of elves, kings, and Ents. Our families and communities connect us to the old and new stories, and guide us to "lean into the light."

This movement is a new form of community and a new form of story. At what point in the future will the existence of 2 million, 3 million, or even 5 million citizen-led organizations shift our awareness to the possibility that we will have fundamentally changed the way human beings govern and organize themselves on earth? What are the characteristics of leadership required when power arises instead of descends? What would a democracy look like that was not ruled by a dominant minority? What would a world feel like that created solutions to our problems from the ground up? What if we are entering a transitional phase of human development where what *works* is invisible because most heads are turned to the past? What if some very basic values are being reinstilled worldwide and are fostering complex social webs of meaning that represent the future of governance? These are but a few of the questions collectively posed by a movement that has yet to recognize it is a movement.

THE LONG GREEN

No Sierra landscape that I have seen holds anything truly dead or dull, or any trace of what in manufactories is called rubbish or waste; everything is perfectly clean and pure and full of divine lessons. This quick, inevitable interest attaching to everything seems marvelous until the hand of God becomes visible; then it seems reasonable that what interests Him may well interest us. When we try to pick out anything by itself, we find it hitched to everything else in the universe.

—John Muir, *My First Summer in the Sierra*

We should be cutting lies instead of trees.   —Jerry Martien, *Salvage This*

The first generation—our own—to worry about global threats like nuclear proliferation and climate change is effectively ahistorical; for most activists and campaigners, environmentalism has little or no history, no lineage. It is a response to immediate conditions, a reaction to waste, pillage, damage, and underneath it all lies the deepest grief about the shattering of the world.[1] Although problems like erosion, deforestation, and heavy metal poisoning plagued previous generations, the past provides little practical guidance for present woes. Nevertheless, as the environmental movement merges with a parallel social justice movement and joins with indigenous movements, history can provide a valuable genealogy and the rootedness of ancestry, just as we look to the past to understand the development of other species. And the philosophers and thinkers who inform these three movements may eventually become as seminal to our moral and intellectual development as Voltaire, Jefferson, and Locke.

Looking back, we now understand that the post-Columbian world marked the beginning of environmental globalization, starting with the importation of insects, vermin, plants, trees, tubers, food, bacteria, and diseases and culminating in the global ubiquity of toxic chemicals. The European rabbit, which barely survived in Spain during the last ice age, became a migrant to every continent except Antarctica. It modified, if not ravaged, the environments where it became established, especially in Australia, where the rabbit population could reach a billion after a wet spring. Five hundred years later, DDT, though banned by the EPA in 1972, is distributed over the western United States in plumes of polluted air drifting across the Pacific, courtesy of Chinese agriculture. Even when environmental problems seem to be only regional, their effects cannot be contained.

Polychlorinated biphenyls (PCBs) made by Monsanto in the 1970s migrate up the twenty-first-century food chain in the Great Lakes, into a walleye, onto a grill, and finally into a pregnant woman, potentially altering her child's limbic system and behavior for the next seventy years. Regional salinization of ancient Middle East breadbaskets because of overzealous irrigation changed human history, as will the overzealous flooding of greenhouse gases into the stratosphere.[2] The type of epiphany John Muir experienced in the Sierra Nevada about the idea of everything being hitched to everything else arises today whenever we overstep environmental limits. That realization, nascent though it may be in a historical sense, is now almost universally accepted, especially with the advent of climate change. But it is not universally acted upon, which is why there is a need for environmentalism.

In the West, the appearance of an environmental movement paralleled the emergence of biology as a science. A systemic understanding of nature was common to most indigenous cultures, but proved difficult for Europeans, in part because of theocratic dogma. As recently as two hundred years ago the earth and all life upon it was believed to be a tableau, the result of God's emptying a bountiful ark across the continents and into the oceans. The cast of characters was fixed, and scenes didn't change. Natural science consisted primarily of taking an accurate inventory of life forms, and was exemplified by the Swedish botanist Carl von Linné (1707–1778), later classified "Linnaeus" at his request. He stated his goals plainly: "The study of natural history, simple, beautiful, and instructive, consists in the collection, arrangement, and exhibition of the various productions of earth."[3]

In the eighteenth century, English scientists were inundated with "productions" being carried back by the fleet from distant lands—bags of roots and corms, vials of seeds, outsize hirsute spiders, beetles suspended in Cretaceous amber, opalescent nautili, sweeping fans of vermilion coral, leathery shrunken heads, giant armadillo skeletons, eviscerated harpy eagles, and rodent fossils as big as hippopotami. Even a live and disoriented rhinoceros was once unloaded onto a London dock. Human curiosities such as Yámana teenager Jemmy Button from Patagonia were kidnapped from cultures around the world and considered legitimate objects for Europeans to collect and study.[4] Although several classification schemes were proposed to sort through this botanical and zoological treasure trove, a systemic and integrated classification method for both plants and animals wasn't accepted until 1735, when Linnaeus proposed one in an eleven-page pamphlet entitled *Systema Naturae.* Species belonged to genera, or groups, which were organized into families, and upward through orders, classes, and finally the three grand kingdoms of nature. The initial three kingdoms were animal, vegetable, and mineral; the Monera, Protista, and Fungi kingdoms were later added, while mineral was dropped. Linnaeus first grouped human beings with primates, naming us *Homo diurnis,* man of the day, in contrast to

Homo nocturnes, trogloditae, which included chimpanzees. There were also racist listings for *Homo monstrosus,* a category that included the mythological Patagonian giants (Tehuelche people), Hottentots (Khoikhoi people), and Eskimo (Inuit). From Linnaeus' perspective, there could be no new species, now or ever. The possibility that all of humanity had a common genetic origin in the African savannah was inconceivable. Species could give birth only to the same, and each single species must have been present at the beginning of creation, a unity that could be attributed only to an omniscient creator. Although Linnaeus later conceded that new species had arisen since creation, he believed them to be variations or hybrids from our Edenic origin. Human beings entered the world de novo, they were minted by the mind of God.

Not everyone bought into creationist dogma. In France, Georges-Louis de Buffon postulated some type of evolutionary change as early as 1749. Jean-Baptiste de Lamarck, who single-handedly invented the field of invertebrate zoology, added to Buffon's theory and described a mechanism by which evolution might take place. Lamarck was vilified for his views, and died in obscurity and poverty. But by the middle of the nineteenth century, the inflexibility with which science understood living organisms had necessarily been undermined by new discoveries. Species were not fixed like stars in the sky, nor were continents cemented in place: the physicists explained that the universe moved according to its own laws, without the Almighty's guiding hand, and geologist Charles Lyell (1797–1875) further destabilized church doctrine by suggesting that geological changes could be measured over millions of years—blasphemy to the accepted creationist date of 4004 BCE for the origins of existence. Walking in the footsteps of his grandfather Erasmus Darwin and alongside his peer Alfred Wallace, Charles Darwin delivered the coup de grâce to the concept of an activist God with his heretical theories that species, like planets and rocks, were also subject to natural laws. At that time, many naturalists were clergymen, who had the leisure time to ponder texts and explore woodlands. They understandably overlaid the biblical account of the fifth day of Genesis onto the world they saw: winged creatures and everything that creepeth were immutable, permanently in the form the Creator fashioned them. The heterodoxy of *On the Origin of Species* (1859) could not have been less welcome to the Church and was famously challenged by Samuel Wilberforce, bishop of Oxford, known as "Soapy Sam" for his slippery cunning in debates. Inviting Darwinian proponent Thomas Huxley to a three-day British Association debate in 1860 at Oxford, Wilberforce, with ridicule in mind, pointedly demanded to know which side of Huxley's family had descended from a monkey. As the sarcasm seeped through the chamber, Huxley (who coined the word *agnostic* to describe his religious views) slowly rose and reflected that he would rather be descended from an ape than from a man of intelligence who would use his position and gifts to obscure the truth. Legend has it that devout women fainted while aspiring scientists cheered. For the faithful, it was deicide. For young scientists, it was liberation.

What lay in shards at Oxford was the concept of a Divine Authority: What had been a natural world given form by God had an independent life of its own. By introducing the scientific equivalent of a basic Buddhist tenet—everything changes, always, without exception—Charles Darwin and his colleagues transformed human understanding of life and, just as important, introduced the leitmotif of unending change into our relationship with living systems. This dissolution of rigid categories has not ceased, as biology continues to surprise even itself by new discoveries of the fluidity of life.

In the 1970s two scientists began research that further toppled the apparent orderliness of biology. Rather than retracing the steps of Darwin's voyage on HMS *Beagle*, Peter and Rosemary Grant established a base in the Galápagos, concentrating on one island in particular, Daphne Major, and studying Darwin's finches for twenty years. Their goal was to investigate what *On the Origin of Species* never addressed—namely, the origin of a species. The Grants were meticulous in their pursuit. They mist-netted and banded virtually every finch on the island, measuring beaks, drawing weight, and analyzing daily diet. Rainfall, food production, mating, and birthing were catalogued, recorded in journals, and finally entered into databases back at Princeton, beginning in 1973. Because of the tricolor bands they used, the Grants could identify every bird on the island (over one thousand) individually, citing its species, mate, siblings, parents, nest, and diet.

Darwin's greatest revelations had occurred after he returned to England and discovered that his Galápagos specimens were divergent and unique from their mainland cousins. One hundred ten years later the Grants had a similar experience: differentiation was obvious even in the circumscribed area of the Galápagos. The thirteen species had different beaks, each of which suited their particular diet. There were needle-nosed, curved needle-nosed, long chain-nosed, curved chain-nosed, long parrothead grippers, diagonals, and heavy-duty linesman's pliers.[5] Evolution had not been measured or observed in the field until the Grant study. When the couple analyzed the data back at Princeton, it was revelatory to them and the scientific world. Their conclusion: Darwin had no idea how right he was about evolution or how wrong he was about its occurring over "long lapses of ages." Evolution was occurring constantly; the rate of change in the Galápagos finch population was rapid, if not feverish. Assumptions about rates of speciation were seriously inaccurate. Species were evolving from generation to generation; traits were pliable and quickly adopted. As additional scientists study organisms more painstakingly and over longer periods, a structuralist view of biological organization is giving way to an image of an adaptive nature defined by process and flow, not by static structure and taxonomy.[6] The rhythm and speed of evolution may be less like a predictable march than like the responsive, fluid movement of dance. The world may be more sensitive to change, be it natural or anthropogenic, more *hitched,* to use Muir's metaphor, than anyone had thought possible.[7]

As we understand more about species, we understand more about habitat. As we become more informed about biophysical changes to the planet, we better understand how industrial processes influence the metabolism of the planet. Perhaps the most difficult concept to grasp about climate change is how even minute changes in CO_2 levels can magnify to have such potent effects. But it is not just carbon dioxide that does damage. The large influence of small changes to our environment appears repeatedly. When tadpoles are exposed to the pesticide Atrazine at 1/30,000th of "safe" levels, 20 percent of them become hermaphroditic and sterile adults. Infinitesimal chemical exposure during development can have a drastically different effect from that at maturity.[8] If natural El Niño cycles of rain and drought influence annual changes in speciation in Galápagos finches, consider the myriad long-term impacts of combusting 10 trillion pounds of mercury-bearing coal every year, or overspraying farms and suburbs of California with Malathion to eliminate the Mediterranean fruit fly. The magnitude of such macro-activity creates countless micro-interactions that can't be tracked or monitored.

To roughly calculate the geometrical quickening of our footprint on the planet, consider that the population is 1,000 times greater today than it was 7,000 years ago.[9] Additionally, people use 100 to 1,000 times more resources and energy than their ancestors did. In sum, the earth today withstands at least 100,000 times the impact it did in 5000 BCE. In other words, we have the same impact in five minutes that our ancestors had in a year. Expanding the equation means that we have the same impact in one year as our ancestors did in 100,000 years. But it is not merely a question of overdrawing our natural capital account; it is also a matter of destroying the currency.

Given the rate of damage that 6.6 billion people cause, the surprising thing about environmental groups is that there are only tens of thousands of them. Should we survive our planet-wide spasm of heedless growth and harm, we may well look back at the nineteenth century, not the eighteenth, as the age of enlightenment, for it was then that Western cultures first began to understand and appreciate the interrelationship between human beings and nature, to fully grasp the magnificence and complexity of the living world and our place within it. Environmental historian Donald Worster suggests that the nineteenth century may come to be called the Age of Ecology, for the science and philosophies of that era are the foundation for today's environmental movement.[10]

In 1836, to the west of Concord, Massachusetts, a fiercely precocious six-year-old Emily Dickinson was growing up in the village of Amherst; the HMS *Beagle* returned to London from a four-year trip with its young naturalist, Charles Darwin; on Long Island, a lanky, seventeen-year-old autodidact named Walt Whitman began teaching in a one-room schoolhouse; and in Cambridge, Massachusetts, Ralph Waldo Emerson's *Nature* was published, a small book that became the spiritual and intellectual headwater of one branch of American environmentalism.

In Ralph Waldo Emerson's time, the university was a place to wrestle with big ideas. The ghettoization of academics had not yet been instituted; all branches of knowledge were seen as extending from one tree. The campus was a "universe-city," not a factory creating specialists for the job market, and aspiring scholars willingly grappled with profound philosophical questions. Emerson read at a time when the number of books available was smaller but the content more monumental, and many writings converged to inform *Nature*. Emerson read the Bhagavad Gita in 1831, a transformative experience that turned his prior slights of "Hindoos" into rapt admiration. He placed it on the same level of importance as Christian scripture, and thereafter explored the thought of Zoroaster and Confucius. He laboriously studied nearly all fifty-five volumes of Goethe's work in German (a language in which he was not fluent), and they became an intellectual talisman that remained close by his side the rest of his life. At Cambridge he read Latin, Greek, and Homeric writings; studied rhetoric and grammar; practiced oration; labored through physics; plunged into philosophy; penned essays (although never winning the coveted Bowdoin Prize for best essay); attended lectures by William Ellery Channing; wrote gothic tales; and composed numerous poems. The years that followed his graduation—some difficult, some tedious, some revelatory—saw no lapse in Emerson's consuming appetite for literature, ideas, and eventually science, threads from all of which prepared him for *Nature*.[11]

Emerson wrote at a time when nature was described in florid and grandiose terms, and picturesque phrases were tossed to readers like nougats. ("'Twas morning—the sun rose under the brightest auspices, and the thin, vaporous clouds that flitted in the heavens, continued gradually to flee away before the gentle morning breeze, that seemed wont to greet their golden visages with the soft rustle of its dewy wings.")[12] Such prose was typical of the Romantic movement, where Shelley's poetic clouds "laughed"; Oliver Goldsmith's tigers were "noxious quadrupeds"; and sensitive plants were spoken about as having animal characteristics.[13] At the other extreme from romanticism was the school that believed nature was a mechanism that could be mastered by human invention. A general fascination with the new methods of industrialism was projected back onto nature, just as computational terminology such as "hardwiring" and "programming" is employed today to describe human functions. In this view nature was a storehouse of goods to be opened and exploited. Neither the romantic nor the utilitarian stance was ultimately to be Emerson's ken. He did not dwell on the loss of nature and its resources at the hands of agriculture and industry, but concentrated instead on the ways nature informed ideas and truth.[14] His book-length essay was an attempt to reconnect ideas and language directly to the living world. In a foreshadowing lecture delivered on November 5, 1833, Emerson concluded: "Nature is a language and every new fact one learns is a new word; but it is not a language taken to pieces and dead in a dictionary, but the language put together into a most significant and universal sense. I wish to learn this language, not that I may know a new grammar, but that I may read the great book which is written in that tongue."[15]

Withal, there was a solid foundation of Yankee pragmatism in Emerson's spirituality, a determined individualism evident in all his writing. His opening lines in *Nature* ask why we should not "enjoy an original relation to the universe." Mystical in his own relation to nature, his writing and talks constantly spoke of nature as a path to self. "We must trust the perfection of the creation so far as to believe that whatever curiosity the order of things has awakened in our minds, the order of things can satisfy. Every man's condition is a solution in hieroglyphic to those inquiries he would put." Emerson's nature is a journey, a spiritual palette. Although his essay sold only a few hundred copies in the first years after it was published, it was widely reviewed and discussed in literate circles. Appearing at a time when the assaults and injustices of industrialism were becoming more obvious, especially in cities, Emerson's belief in reason wedded to nature and self-reliance was widely embraced.[16]

As Mary Oliver has commented, however, all of Emerson's wildness was in his head;[17] he left the details of empirically observing and experiencing to Henry David Thoreau. Thoreau had already read *Nature* twice when its author addressed the Harvard senior class; Emerson's land on Walden Pond was where he retreated and built a cabin when he was twenty-eight years old. The two years and two months he spent there were neither idyllic nor leisurely, as he read and wrote constantly and lived in chilly confines with few comforts. But he was not deprived, as the weekly Sunday suppers with the Emersons augmented the more meager fare at his cabin. He took long walks across neighbors' lands, and despite his hermetic image, engaged everyone he met gregariously and with great curiosity.

While at Walden, Thoreau wrote his first book, *A Week on the Concord and Merrimack Rivers*, a tribute to his brother, who had died of lockjaw. It was published after he had returned to Concord, and did not sell well. The publisher returned seven hundred copies of the edition, for which he billed Thoreau, prompting the novice writer to quip that his library contained nine hundred volumes, seven hundred authored by him. Thoreau went back to work as a surveyor and pencil maker for eight more years, all the while rewriting and editing seven drafts of *Walden*. It is worth recalling here the ending of *Walden* for its anthemic quality. While this concluding passage has become a familiar homily read at weddings and commencements, few if any progressive thinkers, including Martin Luther King and Gandhi, have not been genuinely guided by it. It could be said that these were Thoreau's operating instructions to the world's activists:

> I learned this, at least, by my experiment; that if one advances confidently in the direction of his dreams, and endeavors

to live the life which he has imagined, he will meet with a success unexpected in common hours. He will put some things behind, will pass an invisible boundary; new, universal, and more liberal laws will begin to establish themselves around and within him; or the old laws be expanded, and interpreted in his favor in a more liberal sense, and he will live with the license of a higher order of beings. In proportion as he simplifies his life, the laws of the universe will appear less complex, and solitude will not be solitude, nor poverty poverty, nor weakness weakness. If you have built castles in the air, your work need not be lost; that is where they should be. Now put the foundations under them.

His friend Bronson Alcott wrote of the man, "Thoreau is himself a wood, and its inhabitants. There is more in him of sod and shade and sky lights, of the genuine mold and moistures of the green grey earth, than in any person I know. . . . This man is the independent of independents—is, indeed, the sole signer of the Declaration, and a Revolution . . . himself."[18] In 1851 Thoreau delivered the now-famous lecture to the Concord Lyceum in which he first declared that the preservation of the world is to be found in wildness. Two years later, a little-noted occurrence in a distant wild forest changed history, when one of the largest trees in the world was cut down and another destroyed. That incident eventually inspired the creation of the National Park system and the American land conservation movement.

In the spring of 1852 August T. Dowd, a contract hunter hired to supply fresh meat for laborers digging canals, discovered a grove of mammoth sequoias in Calaveras County, California, while chasing a wounded grizzly bear into the backcountry. The trees were so massive, Dowd thought he was dreaming, deluded by some quirk of light or perception. When he returned to camp with stories of his discovery, his campmates scoffed at what they thought were tall tales.[19] To lure disbelievers, Dowd concocted a story the following Sunday about killing the biggest grizzly he had ever seen. Nearly the whole camp followed him across ravines and flats of sugar pine and manzanita, through canyons and ponderosa, until they at last beheld what became known as the Calaveras Grove of Big Trees, a parklike setting with nearly mythic botanical towers rising from the valley floor.[20] The trees were immense, their crowns towering over three hundred feet, their bases ninety feet in diameter. The raised ridges of the bark could enfold a child; caverns that could hold ten people had been carved at their bases by ancient fires.

And as is the case with sequoias, the grove was cathedral quiet, because the trees are insect resistant and thus virtually birdless. These were living organisms unlike anything anyone had imagined. An article in the *Sonoma Herald* soon reported on Dowd's discovery, and the story was quickly picked up by newspapers in San Francisco, London, and Edinburgh. Although it was not the first sighting of giant sequoias—the Miwok Indians had lived among the big trees for many centuries—it was the first published account of them.

The astonishing report soon brought curious journalists, adventurers, and visitors to the site, as well as entrepreneurs, loggers, and most notably George Gale and his business partners, who saw show business in the natural wonder that was the tallest of the trees: a 300-foot-tall creature that could be exhibited as a sideshow for "a trivial fee."[21] Of course, Gale couldn't display the whole tree, but he was determined to get it onto the ground and remove part of it. The quest to fell the sequoia did not go easily. After boring holes through its trunk with long augers, the loggers who had been hired laboriously sawed through the spaces in between. Concerned that a 300-foot-tall sequoia might fall without notice at any time, the men worked cautiously. After being cut all the way through, however, the tree remained upright. Wedges were pounded in from all sides, and the crew made a battering ram from nearby lumber to knock it over, but the tree stayed perfectly still. More than three weeks of effort passed, and it finally took a gale to blow it down, which took place in the middle of the night. The noise of its felling woke people in mining camps fifteen miles away; mud and rock dislodged by the impact flew ten stories into the air, spattering the trunks of neighboring trees. The Big Tree was estimated to be 2,500 years old and remained green for several years because its trunk contained so much water.[22] The promoters removed some of the bark, cut a few cross sections, and left the bulk of the tree where it fell. The trunk was later made smooth and was used as a bowling alley, and the stump became an outdoor dance floor that could comfortably accommodate sixteen couples.[23]

After the Calaveras Grove, other big sequoias were found. In Yosemite's Mariposa Grove, the largest standing tree, known as the Mother of the Forest, rose 363 feet. Nearby was an even taller one that had fallen some decades earlier, dubbed the Father of the Forest, which measured 440 feet long and 120 feet in diameter. The Mother of the Forest was soon to join its mate. Scaffolds rose to encircle it at a height of 110 feet, and workers then removed the eighteen-inch-thick bark in eight-foot sections. The tree lived on for another five years while its cambium layers turned white, and it remained standing like a ghost for many more. The pieces of thick bark were crated and shipped downriver to San Francisco, and by clipper ship to New York, where they were reassembled and exhibited in 1854 at New York's Union Club.[24] Flyers promising a "vegetable monster" ninety feet in diameter were distributed around New York: "Great Tree, Recently Felled Upon the Sierra Nevada, California, Now Placed For Public Exhibition, Admission: 25 cents." Afterward, the skeletal bark of the tree was shipped to the North Transept of the Crystal Palace in Sydenham, England, where the interior was fitted with tables and chairs to form a perfect drawing room. During the New York exhibit, *Walden, or Life in the Woods* was published.

Despite the hyperbolic advertising, the exhibitions did not create the commercial sensation their promoters sought. Many viewers thought the specimens were fakes, while those who did believe they were authentic were horrified at the killing of such a natural wonder. Horace Greeley, editor of the influential *New York Tribune*, called it vandalism and villainous speculation. The editor of *Gleason's Pictorial* wrote, "To our mind, it seems a cruel idea, a perfect desecration, to cut down such a splendid tree . . . what in the world could have possessed any mortal to embark in such a speculation with this mountain of wood."[25] In 1857 an incensed James Russell Lowell called for a society for the protection of trees, citing a seemingly hereditary antipathy Americans held for trees and Indians.

The Mother of the Forest eventually became a spectral totem for new land policies. As America settled into its nationhood and reveled in the expansiveness of its ocean-to-ocean breadth, Easterners began to consider what role government should play in land preservation. Poor sellers initially, works by Emerson and Thoreau were now beginning to be widely read, and their influence was spreading. Horace Greeley, disgusted by the desecration of the Calaveras Grove but fascinated by tales of marvels being brought back from California, traveled there in 1859. He made a grueling pilgrimage to Yosemite on horseback, leaving Bear Valley at dawn and arriving in Yosemite Valley well after dark. His stirrups were too small, he hadn't ridden a horse save for a few hours in the previous thirty years, and when he dismounted he couldn't walk. Finally bedded down in a small cabin at one in the morning, he arose sore, stiff, blistered, and hungry. He spent only one day in the valley, but sent a dispatch to his fellow Easterners that he "knew of no wonder on earth which can claim superiority over Yosemite," and called on the government to protect "the most beautiful trees on earth."[26]

Just as Greeley published his accounts of California in *An Overland Journey* in 1860, photographer Carleton Watkins was beginning preparations for a Yosemite trip for one of his patrons. To capture the scope of the landscape he had heard described, he worked with a cabinetmaker to build a massive camera, using eighteen by twenty-two-inch wet collodion glass plates. In July 1861, Watkins journeyed to Stockton by stage, then to Copperopolis and Murphy's Camp, and finally by mule to Yosemite. With him was one stereo camera; his mammoth glass plate-view camera, fitted with a new wide-angle Grubb Aplanatic lens; a demon's brew of chemicals, including acetic acid, potassium nitrate, silver nitrate, nitric acid, potassium cyanide, bichloride of mercury, and potassium sulphide; gun cotton; dozens of fragile glass plates; and a tripod sturdy enough to hold his bulky camera. While there Watkins made thirty plates and one hundred stereographs of Yosemite Valley and the big trees. By December of the following year his albumen prints were displayed at the Goupil Gallery in New York, drawing gaping stares, rave reviews, and expressions of disbelief. Watkins's images contrasted sharply with those of a show that had preceded it, Matthew Brady's record of the carnage of the Civil War. Emerson received Watkins's stereographs from another Unitarian minister, Thomas Starr King, who had written lovingly of Yosemite for the *Boston Evening Transcript*. Albert Bierstadt, inspired by Watkins's images, journeyed to Yosemite shortly after and painted sweeping vistas of the valley in his New York studio upon his return. The wealthy papermaker Zenas Crane commissioned a painting of the giant sequoias for his opulent mansion.

In June 1864, while the Union Army was fighting in Petersburg, Virginia, 131 miles from Washington, Lincoln signed the Yosemite Land Grant, deeding 39,000 acres of Yosemite and the Mariposa Big Tree Grove to the State of California for public use and recreation. It was an unparalleled piece of legislation, for it rested on the novel proposition that wilderness should be eternally preserved. It came, surprisingly, at a time when forests were still regarded as a danger to pioneers and settlers, home to natives and animals that stalked the darkness. It was clearing that made forests safe, not preservation, as could be seen in the elimination of most eastern woodlands from the bald cypress in Florida to the white pine of Michigan. The guaranteed preservation of western wilderness seemed like an atonement for the ecological ravages that had swept the East Coast for a century.[27] What was a remarkable achievement then seems more remarkable today: the destruction of nature, in this case the Mother of the Forest, catalyzed legislation that had almost universal support from Congress, business, and naturalists. No one argued that setting aside Yosemite would hinder economic growth. California senator John Coness sponsored the Yosemite Grant, and not coincidentally he owned a set of Carleton Watkins's Yosemite prints. Eight years later, Yellowstone became the first national park, followed eighteen years afterward by Yosemite, General Grant, and Sequoia national parks. It is difficult to underestimate the influence of Watkins's photographs, for without them, Lincoln and the Congress might not have been able to set aside the original land for Yosemite.

In 1994 I journeyed to the Kitlope, North America's largest uncut temperate rain forest, upriver east of Kitimat in British Columbia. I traveled with Haisla elders Gerald Amos and Ken Hall, and Ecotrust founder Spencer Beebe, six months after Premier Mike Harcourt had returned traditional land rights to the tribe. A Portland, Oregon–based nonprofit, Ecotrust had members who spent months roaming the tribal lands with youths whom they had equipped with laptops and GPS instruments, and tribal elders who place by place restored the traditional names of the landscape onto Haisla topographical maps. Having spent a measure of my early years roaming the peaks of the Sierra Nevada, the dry eastern valleys that sprawl down to Mono Lake, and the White Mountain area above Owens Valley, I thought I knew wilderness. The Kitlope changed my perception of that term. I felt as if I had been thrust into a painting from the Hudson River school, a preternatural, romantic dreamscape trumped up for wealthy Easterners who would never get past Chicago. But here it was, the real thing, glaciers, rainbows, and all. The five species of Pacific salmon (chinook, chum, coho, pink, and sockeye) underfoot in the shallows were fodder for the grizzlies and black bears denning in the old-growth spruce and cedar forests. In the glacier-fed waters, river otters peered curiously, wolf packs roamed at night, mountain goats capered on alpine croppings, blubbery seals feasted on the easy pickings a hundred miles from the sea, and eagles nested in evenly spaced sequences along every spawning tributary. In a setting like that you could easily imagine what Watkins, Dowd, Greeley, and others must have experienced when they came west, or the theophany that swept over John Muir when he first strode deep into the Sierra Nevada.

In 1860, two years before his death, Thoreau delivered his most important lecture, from a scientific viewpoint, "The Succession of Forest Trees," which he presented to the Middlesex Agricultural Society. It introduced for the first time the idea that over time a community of plants and animals is gradually transformed in response to changes in the environment. Thoreau had read Darwin's *Journal of the Voyage of the Beagle* while at Walden, and had been taken by Darwin's description of species distribution and succession on the different islands of the Archipelago.[28] Horace Greeley, recently returned from his travels to the West, had been in correspondence with Thoreau and vigorously contested his assertion that trees arise naturally from seeds dispersed by the agencies of animals, water, and wind. Greeley and others believed that wild plants under certain conditions were spontaneously generated. Their progressivism, as it was labeled, was a variation of creationist dicta, positing a series of advanced creations on earth, culminating with man as the ultimate one, all linked "by an abstract unity to the mind of God."[29] At the time, this topic was an issue of vital concern to botanists and laymen alike, with roots in theological principle. For progressivists such as Louis Agassiz, the Swiss-born Harvard zoologist, spontaneous creation was fundamental to science, just as science was fundamental to religion.[30]

The debate between Greeley and Thoreau on the issue of spontaneous generation was similar to one occurring contemporaneously at Oxford and Cambridge between Darwinian evolutionists and their creationist antagonists. Thoreau's concept of ecological succession was not far from Darwin's natural selection. Although Agassiz framed it in scientific terms, at heart this was a religious debate, with Agassiz, the dogmatist, weighing in on the theological side. (Agassiz's particular creationist bias carried over into race issues, where he was a proponent of Caucasian superiority, a view that undermined the efforts of President Lincoln and delighted the pro-slavery lobby.) The controversy between Thoreau and the progressivists was not a mere historical footnote. Over time, as Thoreau's observations were confirmed by science, and spontaneous generation joined the flat earth as an untenable theory, it seemed that the creationist bias that had handicapped Western science for centuries was finally put to rest. Yet, even as this book is written, creationist theories are being supported by U.S. senators, holders of cabinet posts, and senior officials in the EPA and Department of Interior.

In the nineteenth century there were two branches of the environmental movement in North America. Emerson and Thoreau could be credited with inspiring the idea of man and nature as one, with all divisions between them arbitrary and dangerous. Opposed to that view was the conviction that man is necessarily superior, a view rooted in the work of George Perkins Marsh, a polymath, multilinguist, historian, author, lawyer, politician, and diplomat. As a congressman from Vermont, he delivered in 1847 a groundbreaking lecture to the Agricultural Society of Rutland County on deforestation that became the basis for his book, *Man and Nature*, which was published in 1864, after he had left the country.[31] (Marsh had been appointed ambassador to Italy, for which he set sail in April 1861, never to return.) Despite Marsh's absence, and the enduring popularity of Thoreau and Emerson, *Man and Nature* realigned the relationship between government, business, and society, and in so doing was arguably the most influential book ever published on the environment. In his biography of Marsh, David Lowenthal writes, "More than Marsh dreamed, *Man and Nature* ushered in a revolution in the way people conceived their relations with the earth. His insights made a growing public aware of how massively humans transform their milieus. Many before Marsh had pondered the extent of our impact on one or another facet of nature. But most took it for granted that such impacts were largely benign, that malign effects were trivial or ephemeral. None before him had seen how ubiquitous and intertwined were these effects, both wanted and unwanted. Marsh was the first to conjoin all human agencies in a somber global picture. The sweep of his data, the clarity of his synthesis, and the force of his conclusion made *Man and Nature* an almost instant classic."[32]

It is tempting to read *Man and Nature* as a reflection of a pastoral era. Marsh's own family home in Vermont, for example, had been nearly idyllic, giving legitimacy to the pronouncement that "people from Woodstock had less incentive than others to yearn for heaven."[33] In fact the book was published during an intellectually, morally, and economically chaotic period in America. The country was convulsed by a civil war. Capitalism was thriving on the upheaval, with industry hastening to appropriate and exploit a divided nation's resources. General Sherman's March to the Sea left a sixty-by-three-hundred-mile swath of bridges, factories, railroads, and government buildings in ruin. Corruption was rife, as the U.S. Treasury was emptied into the private sector to sustain the war. With the wartime levy on lamp fuel creating a surge in demand for recently discovered petroleum, John D. Rockefeller built his first refinery in Ohio and began his quest to monopolize petroleum using merciless tactics to drive competitors out of business. (One of those competitors was renewable energy, which he battled just as ruthlessly as he did rival companies.) The historian Lewis Mumford writes that in the war years "the colors of American civilization abruptly changed. By the time the war was over, browns had spread everywhere: mediocre drabs, dingy chocolate browns, sooty browns that merged into black."[34] In such grim conditions, the meaning of life became less important than the means of life. There "was a necessity for inventive adaptation which turned men from the inner life to the outer one. . . . For lack of an harmonious system of concepts and feelings, this necessary change did not lead to an intelligent adaptation of the environment; in the planning of cities and the layout of railroads, highroads, farms, in the exploitation of mineral resources and the utilization of the land, a good part of our soils and cities were ruined; indeed, the new industrial towns were ruins from the beginning."[35]

It would not have been an auspicious time for a book about the environment and sustainability in the manner of Emerson and Thoreau, and indeed, the book *Man and Nature* took a different approach. Its tone is stern and cautionary, its message profound and clear: nature thrived without humans, and while mankind's intervention almost always caused permanent and destructive alterations to the earth, intervene we must.[36] Marsh had been raised along the Ottauquechee River at the foot of Mount Tom on the fertile bottomland of his father's farm. He was instructed at an early age how to identify trees, understand watersheds, and gauge the damage caused by logging and overgrazing. He witnessed the forests of the northeastern United States being clear-cut and razed in front of his eyes, and later, as ambassador to Turkey and Italy, saw what deforestation and overgrazing had wrought over longer historical periods: loss of topsoil, erratic water flows, the drying up of springs, greater temperature swings, loss of fertility, less rainfall, and failed civilizations. Marsh was the first to describe this interdependence of the environment and society. He understood that no civilization could endure without a healthy environment to support it, and none had survived the compounding effects of deforestation, the desiccation of watersheds, and soil erosion. Yet his writing insists that if humans do not take charge of nature, they will be subdued by it—a radically different stance from Emerson's assertion that we *are* nature. Marsh's biographer labels his ideas anachronistic, and indeed, it is hard to imagine the learned and scholarly George Perkins Marsh writing today that it is man's responsibilty to dominate nature. But ideas of dominance are far from anachronistic, and are arguably more politically powerful today than in the nineteenth century, although they now rest within the right wing and those opposed to environmental conservation.

Gifford Pinchot, who founded the U.S. Forest Service under Theodore Roosevelt, had been so influenced by *Man and Nature* that he bestowed upon himself the title "father of conservation," an honorifc that had formerly been given to Marsh. Pinchot was a significant influence on and worked side by side with the president. Roosevelt established the Tongass and the Chugach forest reserves; set several small islands aside as the Hawaiian Islands Bird Reservation; and launched the idea of wildlife refuges, beginning with Pelican Island and later Mosquite Inlet in Florida, Lake Malheur in Oregon, and Culebra Island in Puerto Rico. Under the auspices of the Antiquities Act, he signed the Grand Canyon National Monument into being in 1908, and created seventeen other national monuments, including Mt. Olympus in Washington, Montezuma Castle in Arizona, Gila Cliff Dwellings in New Mexico, Devil's Tower in Wyoming (properly known as Bear's Lodge by Native Americans), and Muir Woods in California. Pinchot coined the term "wise use," as a means to encourage mixed use of the national forests. (Ironically, *wise use* has become a catchphrase used by right-wingers to fight against environmental organizations and government set-asides.)

From the beginning, conservation was an idea firmly rooted in upper-class, white society, with many of its leaders and spokespersons graduates of either Yale or Harvard. It enjoyed widespread support from both sides of the political aisle as well as business, a fortuitous combination largely maintained from the nineteenth century all the way through the Kennedy, Johnson, and Nixon administrations. Notable leaders in thought and deed include Mary Hunter Austin, Nathaniel Southgate Shaler, Ellen Swallow Richards, William Temple Hornaday, George Oliver Shields, George Bird Grinnell, Liberty Hyde Bailey, and so many more. The heuristics that informed early environmentalism were privilege, sport hunting, and class. Mrs. Lovell White, a society matron in San Francisco, created the Sempervirens Club (later known as the Save the Redwoods League) and prevented the logging of the Calaveras Grove of Big Trees by gathering 1.5 million signatures and submitting them to President Roosevelt.

Today, conservation organizations include the Nature Conservancy, World Wildlife Fund, National Wildlife Federation, the National Geographic Society, and the Wilderness Society. They are well-financed organizations (as of 2005, The Nature Conservancy had $1.7 billion in liquid assets and $4.4 billion in total assets) and are not threatening to the establishment because they *are* the establishment. They are effective and have plenty of clout.

A more confrontational and grassroots approach within the environmental movement was initiated by David Brower, who became executive director of the Sierra Club in 1952. This group had been founded in 1892 by Gifford Pinchot's friend John Muir (for whom the Muir Woods were named) and Robert Underwood Johnson. Muir had published his first nature essay in 1865. He argued vigorously for the establishment of preserves and set-asides, and his new organization was dedicated wholly to protecting one in particular, the Sierra Nevada, where he had hiked for years with his friend Johnson. Sixty years later, Brower revived a quiescent Sierra Club, whose several thousand members were mostly academics and professionals who could be seen on weekend trails nattily attired in lederhosen and Tyrolean hats. Brower expanded—exploded—the club's original mandate, and his seventeen-year legacy as head of the Sierra Club illustrates how the modern movement has drawn from the past. As a habitué of Yosemite and cognizant of Watkins's influence, Brower took advantage of the relationship between the portrayal of beauty and the passing of legislation. Beginning in 1960 he edited and published a series of nineteen large-format books with images by some of the greatest contemporary nature photographers that were like exhibits without a museum. A printer in Verona, Italy, crafted them on silky clay-coated stock. They retailed for $25, equivalent to $160 today. It was conservation deluxe, with prefaces written by Brower, photos selected by Brower, and all the promotion undertaken by Brower. The series was a spectacular success. The most popular of the volumes, entitled *In Wildness Is the Preservation of the World*, was published exactly one hundred years after Carleton Watkins's first exhibition in New York and sold nearly nearly one million copies.

The series affected public sentiment just as powerfully as Watkins's stereo images, and Sierra Club membership soared to more than seventy thousand. The books influenced federal legislation and helped tip the scales in favor of the Sierra Club in its campaign to block several proposals by the Bureau of Reclamation to dam portions of the Grand Canyon. The publications and the club's tireless efforts helped create national parks and seashores in the North Cascades, the California Redwoods, Great Basin, Alaska, Cape Cod, Fire Island, and Point Reyes, and they helped pass the Wilderness Act of 1964 and prevent dams in national monuments and the Yukon. They also made Brower the bane of politicians and bureaucrats. Brower accused Floyd Dominy, legendary head of the Bureau of Reclamation, of wanting to make the Grand Canyon a bathtub, a phrase that stuck. When Domini responded that boaters would be able to explore the canyon walls more easily, Brower shot back with a full-page advertisement in *The New York Times* crafted by Jerry Mander and Howard Gossage asking whether we should flood the Sistine Chapel in order to get a better look at the ceiling, a political act that prompted the IRS to remove Sierra Club's tax-deductible status.

In 1969 David Brower was asked to resign from the Sierra Club by its more conservative board, which included photographer Ansel Adams. Undaunted, he went on to found other organizations, including Friends of the Earth (with chapters in seventy countries and one million members), and the Earth Island Institute, an incubator that has produced dozens of other groups, including the Rainforest Action Network, Urban Habitat, Sea Turtle Project, and the International Rivers Network. Brower's barely concealed disdain for armchair environmentalism is obvious from one of his quotes: "Polite conservationists leave no mark save the scars upon the Earth that could have been prevented had they stood their ground."

The second half of the twentieth century marked the beginning of the modern environmental movement, whose leaders were more iconoclastic and radical, in no small part because the damage to living systems was becoming more pronounced. In histories of this period, many names inevitably appear: William O. Douglas, Hazel Wolf, Aldo Leopold, Edward Abbey, Garret Hardin, Donella Meadows, Jacques Cousteau, Jane Goodall, Jim Lovelock, David Suzuki, and Stewart Brand. These are honored individuals, but the chronicles tend to overlook the achievements of Native Americans: Tom Goldtooth, JoAnn Tall, Grace Thorpe, John Mohawk, Oren Lyons, and Lorelei Means. Nor are there African Americans represented, figures like Carl Anthony, who pioneered urban and multicultural environmentalism. For most of the twentieth century, the environmental movement was a white people's cause in the United States.

THE RIGHTS OF BUSINESS

Miss Rachel Carson's reference to the selfishness of insecticide manufacturers probably reflects her Communist sympathies, like a lot of our writers these days. We can live without birds and animals, but, as the current market slump shows, we cannot live without business. As for insects, isn't it just like a woman to be scared to death of a few little bugs! As long as we have the H-bomb everything will be O.K.
P.S. She's probably a peace-nut too.[1]

—Letter to *The New Yorker* protesting the publication of *Silent Spring*

The Bhopal tragedy is a symbol of the cruelty of corporations against humanity. The day that we succeed in holding Dow liable for the continuing disaster in Bhopal will be good news for people all over the world. From that day on chemical corporations will think twice before peddling poisons and putting profits before the lives and health of people. We are not expendable. We are not flowers offered at the altar of profit and power. We are dancing flames committed to conquering darkness and to challenging those who threaten the planet and the magic and mystery of life.

—Rashida Bee, Bhopal survivor and organizer[2]

O n June 16, 1962, *The New Yorker* magazine published the first of three installments of a forthcoming book, *Silent Spring*, by biologist Rachel Carson. The magazine's legendary editor, William Shawn, was ecstatic about the series, telling Carson it was a "brilliant achievement . . . full of beauty and loveliness and depth of feeling."[3] Carson was already widely read. Her previous book, *The Sea Around Us*, spent thirty-nine weeks atop the *New York Times* best-seller list and was translated into thirty languages. But unlike her previous works, all of which were widely praised, *Silent Spring* created an uproar that has never truly subsided. Carson's argument stood firmly in the tradition of demands for social and environmental justice that extended back to concerns about environmental health during the Industrial Revolution. It also marked, almost inadvertently, a turning point in the unspoken elitism and racism of the early environmental movement. Her exposé of industry-sponsored poisoning of the environment brought for the first time a broad cross section of the population into the environmental dialogue. The *environment* now included people's bodies, mothers' milk, African Americans, farmworkers, and the poor, some of whom were just as polluted as the Cuyahoga River, which famously caught fire in 1969. But as the environmental movement gradually became more diverse in its membership and broader in its scope, it incrementally lost the support of business and politicians, and was even seen as their enemy, and was abandoned to fend for itself.

Rachel Carson's subject was chlorinated pesticides, which she came to because of a controversy about aerial spraying of DDT over Long Island and New England by the USDA in an effort to eradicate fire ants, gypsy moths, caterpillars, and mosquitoes. Carson read letters from angry residents describing the death of songbirds, bees, and grasshoppers, and soon afterward agreed to write the magazine series.[4] At the time, DDT could be purchased in bulk for fifty cents a pound. It had been a savior during World War II, the first war in which fewer combatants died of disease than from combat wounds. This was almost entirely the result of DDT sprays and dustings that killed typhus-carrying fleas. There was no gainsaying this fact, and Carson's analysis of the benefits *and* costs of DDT and other newer pesticides therefore ran contrary to the pesticide industry's triumphal claims. Based on scientific evidence, she believed that some of the new chemical compounds introduced after the war were killing birds, fish, and animals, as well as causing cancer and other diseases in human beings.

Silent Spring began with a "Fable for Tomorrow," a fictional essay describing a storybook town's hellish descent into a pesticide-poisoned reality:

> There was once a town in the heart of America where all life seemed to live in harmony with its surroundings. The town lay in the midst of a checkerboard of prosperous farms, with fields of grain and hillsides of orchards. . . . Then a strange blight crept over the area and everything began to change. Some evil spell had settled on the community: mysterious maladies swept the flocks of chickens; the cattle and sheep sickened and died. Everywhere was a shadow of death. The farmers spoke of much illness among their families. In the town the doctors had become more and more puzzled by new kinds of sickness appearing among their patients. There had been several sudden and unexplained deaths, not only among adults but even among children, who would be stricken suddenly while at play and die within a few hours.
>
> There was a strange stillness. The birds, for example—where had they gone? Many people spoke of them, puzzled and disturbed. The feeding stations in the backyards were deserted. The few birds seen anywhere were moribund; they trembled violently and could not fly. It was a spring without voices. On the mornings that had once throbbed with the dawn chorus of robins, catbirds, doves, jays, wrens and scores of other bird voices there was now no sound; only silence lay over the fields and woods and marsh.

Carson could not have devised an opening passage more likely to inflame her critics. Science writing was supposed to be objective and rigorous, without emotion. The admixture of fable and science enraged some scientists as well, but the qualities that made the book anathema to them made it engaging to the general public. Certainly the "shadow of death" that caused children to die within hours was excessive, but the book was no jeremiad, and Carson's prediction as to the eventual outcome of the uncontrolled use of these chemicals could not have been more convincing: "Can anyone believe it is possible to lay down such a barrage of poison on the surface of the earth without making it unfit for all life? . . . [M]an is a part of nature, and his war against nature is inevitably a war against himself."[5]

Before *Silent Spring*, corporations were attacked by reformers and social critics primarily for their rapaciousness and inhumane working conditions. In Carson, they were faced with a soft-spoken critic who alleged that their products shouldn't be made at all. Her goal was to reduce, if not eliminate, a new class of pesticides used in agriculture, even though she supported the moderate use of safe pesticides and biological control agents. For the first time, modern industry had been broadsided and outflanked by an environmentalist. Shocked and infuriated, it reacted with condemnation, assaults, and mockery. Food giants such as General Mills and Gerber's, the pest control industry, agribusiness, chemical companies, and government agencies such as the USDA worked separately and together to destroy Carson's reputation and credibility. With this seminal confrontation, industry and the public relations industry cut their teeth, preparing them for the battles ahead. They have never relented in their fight. They have long since perfected techniques to marginalize scientific data that conflict with their financial interests. Their basic approach to counter such troubling evidence is to foreshorten time by emphasizing imminent problems over long-term concerns. For example, while Carson hypothesized that it would take a century for the full effects of pesticides to be seen, pesticide makers warned of potential crop losses that could occur as soon as the next planting season. When the Kyoto Protocol was being negotiated in 1997, albeit with only a slim chance of being ratified in the United States, the fossil fuel and automobile industries likewise sponsored advertisements showing people forced into dangerous small cars, or contending with having no fuel at all.

Rather than countering the thesis of *Silent Spring* with facts, which it could not do, industry was forced by the book's popular acclaim to undermine it on an emotional level. Initially, it was ill equipped for this battle, but it soon found an appropriate tone: anger, infused by the metaphors of war. The use of agrochemicals became a security issue, critical to preventing hunger and famine. Just as strong nations stockpiled munitions, bombs, and aircraft to counter possible enemy attack, the agricultural industry had its own cache of weapons, chemicals delivered by aircraft and heroic men who guarded the safety of the nation's food supply by attacking enemy insects. The strategy sounds ludicrous, yet the tactics used to stifle *Silent Spring* were a harbinger of how industry would attack its critics in years to come, whether the product in question was a Corvair or a pack of Marlboros. To industry, Carson was not merely an annoying interloper, she was a naïf making the nation vulnerable to attack. The president of Montrose Chemical Corporation, at the time the largest manufacturer of DDT, fired one of the first of many salvos, charging that Carson was not a scientist but a "fanatical defender of the cult of the balance of nature." Defeating Carson was a key objective in what had become nothing less than an industrial holy war.[6]

That controversy established a basic dynamic between environmentalists and industry. Both used fear to engage the public, and the threats each warned of all had a basis in fact. While environmentalists were genuinely apprehensive about a toxic future, industry was alarmed about its own future in the form of sales. A mismatch in terms of scope, perhaps, but it was a psychological draw on the emotional level. Robert White-Stevens, the somewhat frantic spokesperson for American Cyanamid, exemplified the quality of rhetoric employed by industry: "The real threat, then, to the survival of man is not chemical but biological, in the shape of hordes of insects that can denude our forests, sweep over our crop lands, ravage our food supply and leave in their wake a train of destitution and hunger, conveying to an undernourished population the major diseases, scourges of mankind."[7] The denunciations were biblical in scope, apocalyptic in tone. Joining the counterattack, Monsanto satirized Carson's work in a pamphlet entitled "Desolate Spring," wherein a small town, similar to the one Carson imagined, sees all its plants and lives destroyed by ravenous insects. "The bugs were everywhere. Unseen. Unheard. Unbelievably universal. On or under every square foot of land, every square yard, every acre, and county, and state and region in the entire sweep of the United States. In every home and barn and apartment house and chicken coop, and in their timbers and foundations and furnishings. Beneath the ground, beneath the waters, on and in limbs and twigs and stalks, under rocks, inside trees and animals and other insects—and yes, inside man."[8]

Another tactic was to raise the provocative question: Who let a woman into the room? Such misogyny was the subtext of many reviews, critiques, and industry broadsides. Former secretary of agriculture Ezra Benson wanted to know, "Why [was] a spinster with no children so concerned with genetics?"[9] The baton was picked up by Dr. William Bean, who dismissed Carson's thesis in the *Archives of Internal Medicine*: "*Silent Spring*, which I read word for word with some trauma, kept reminding me of trying to win an argument with a woman. It can't be done."[10] *Time* magazine called it "an emotional and inaccurate outburst."

But it was the men who proved to be emotional and hysterical in their responses to the book. Norman Borlaug, who won the Nobel Peace Prize for developing nonlodging varieties of wheat, lost his composure at a U.N. conference on food: "The current vicious, hysterical propaganda against [pesticides], being promoted today by fear-provoking, irresponsible environmentalists, had its genesis in the best-selling, half-science, half-fiction novel *Silent Spring*. . . . If the uses of pesticides in the USA were to be completely banned, crop losses would probably soar to 50 percent, and food prices would increase four-fold to five-fold."[11] Interlaced with constant assaults on Carson's credibility were mentions of her nature worship, and assertions that she wasn't really an accepted scientist but a misguided amateur, that her writing was for popular consumption, that her thesis didn't pass the rigor of peer review, and that she "overstepped" her place as a female writer of popular books.[12] Congressman Jamie Whitten from Mississippi, then chair of the House Appropriations Subcommittee on Agriculture, condescendingly suggested that the book "move over from the non–science fiction section of the library to the science-fiction section, while we review the facts—in order that we may continue to enjoy an abundant life."[13] A brochure was ginned up by the National Agricultural Chemicals Association entitled "How to Answer Rachel Carson," which assured its readers that DDT would disappear from the human body in ninety days. It denounced her book as "more poisonous than the pesticides she condemned."

After *Silent Spring* was published, and for years to follow, no peer-reviewed studies were conducted to justify the overwrought claims of the book's critics. No one challenged the fact that DDT killed insects effectively, but hard measurements of yield and cost benefits simply did not exist. Longitudinal studies done later did show large increases in crop yields starting in the early 1950s and extending into the 1970s. The problem with these data was that increases were entirely attributed to pesticides, ignoring any improvements in fertilizers, machinery, hybrid varieties of seeds, irrigation, and other factors. One study that compared crop losses in 1936 and 1957 showed that the amount of losses due to insects had not changed. But what did start in the 1950s were large-scale payment programs to farmers designed to reduce crop surpluses and provide price supports. With this perverse incentive in place, farmers removed arable land from cultivation to qualify for subsidies, and worked their remaining fields more intensely, eliminating crop rotation and diversity, thus encouraging the spread of insects. New synthetic pesticides allowed such cultural practices, which farmers would have once seen as injurious and foolish. And because constant spraying kills beneficial insects along with destructive ones, pesticides have the opposite effect of what their promoters intend: they increase the insect population, and that population's pesticide resistance. The result is a self-defeating cycle for farmers, who have no choice but to spray, and who must constantly find new types of pesticides to fight off insects resistant to current varieties.[14]

Throughout the battle with her critics, Carson was waging a second one; while the world was arguing the merits of pesticides and whether they caused cancer, Rachel Carson's own cancer was spreading. While finishing *Silent Spring*, she had been hit by successive waves of disease and acute discomfort. As she checked and rechecked her research on the links between pesticides and cancer, she was diagnosed with a malignant tumor in her breast. After undergoing a radical mastectomy and removal of lymph glands, she spent months recovering before she could return to work on the book. Publication was delayed again as she dealt with pneumonia, ulcers, and the weakening side effects of radiation. Then came bladder problems, a staphylococcus infection, and severe phlebitis in both legs, which crippled her. Soon after, her knees and ankles became swollen and inflamed, and she was given steroidal treatments. As deadlines passed, she labored on, sometimes from bed, sometimes in a wheelchair. Knowing that the book would garner intense publicity and scrutiny, she swore her closest friends to secrecy and hid her condition from the world; the word *cancer* was never mentioned to her agent or editor. Five months after publication, following a whirlwind of tours, speeches, interviews, and appearances, she began heavy radiation treatment again, with its attendant pains, nausea, utter fatigue, and depression. By then the cancer had spread to her bones. Her heart condition also worsened, prompting Carson to hope that her heart would kill her before the cancer did.

In her remaining months of life she was invited to give major addresses across the country, though she could accept only a few. Confined to a wheelchair, she politely told her hosts it was a touch of arthritis. Seated or standing, she never spoke angrily or aggressively; her voice remained calm, measured, and dignified. The speech she gave to the Kaiser Foundation at the Fairmont Hotel in San Francisco on October 18, 1963, was her last—and the first time she called herself an ecologist. The time had come, she said, for human beings to "admit their kinship with other forms of life. . . . We must never forget the wholeness of that relationship. We cannot think of the living organism alone; nor can we think of the physical environment as a separate entity."[15] The day after that speech, David Brower and his wife, Anne, took Carson and her wheelchair to Muir Woods to see her first redwood tree. From there they drove to Rodeo Lagoon in Fort Cronkite, where they watched a flock of two hundred brown pelicans (*Pelecanus occidentalis*) with their seven-foot wingspan wheeling through the afternoon light. When Carson returned home later that week, the pain had become constant and she could no longer care for herself. She could barely hold a pencil and was never to walk again. She died five months later, leaving one-third of her estate to the Sierra Club, some of which was used to fund Brower's handsome series of folio books on the environment.

One month after Carson's San Francisco speech, five million dead fish floated to the surface of the lower Mississippi River. Although smaller fish kills had occurred in previous years because of pesticide runoff from cane fields, this was a degree of magnitude greater than anything seen before. The State of Louisiana asked the U.S. Public Health Service to investigate. The agency found the cause to be the pesticide endrin, and the source was endrin's manufacturer, Velsicol Chemical Corporation, which had illegally dumped the substance into a wastewater treatment plant in Memphis. As it happened, Velsicol's law firm had sent a letter to Carson's publisher one month before *Silent Spring*'s publication, claiming that the book would disparage its products, particularly chlordane and heptachlor, with the clear implication that they would litigate. Velsicol also threatened *The New Yorker*, and later the Audubon Society, with protracted litigation. The company's attorney, Louis McLean, suggested that Carson's goal was to "create the false impression that all business is grasping and immoral, and to reduce the use of agricultural chemicals in this country and the countries of Western Europe, so that our supply of food will be reduced to east-curtain parity."[16] In the late 1980s the EPA finally banned endrin, heptachlor, and chlordane, but Velsicol continued to produce those pesticides for another decade for export, even shipping to countries in which they were banned.

On April 14, 1964, Rachel Carson's wish was granted, and she died of cardiac arrest. Whether the attacks against her hastened her death will never be known, but the effect of *Silent Spring* was ultimately measurable in every human being in America. DDT belongs to a family of chemicals called chlorinated hydrocarbons that are not soluble in water but do dissolve in lipids (fat). Because of this property, they are not easily eliminated from human and animal bodies but are stored in fatty tissue, thus building up in the food chain. In 1942, before DDT was commercially introduced in the United States, its metabolite, DDE, was not found in human tissue. By 1950, the average level of toxicity for people living in the United States was 5.3 parts per million. When *Silent Spring* was published, DDT production peaked at 180 million pounds per year, and the amount measured in human tissue peaked at 12.6 ppm. Sixteen years later, body burden had fallen to 4.8 ppm.[17] Today, it is under 1 ppm.

Silent Spring transformed a few hundred quiet conservation groups predominantly concerned about birds, national parks, and hiking into a much larger and more vocal *movement*. Historians generally treat the environmental movement as a postwar phenomenon, but Carson reignited an issue that has concerned people for hundreds of years in ways conservation never did: public health. Her genius was to link the loss of human health with the mind-set of biological dominance, with the idea that business and science had a mandate to conquer and exploit nature.[18]

As early as 1949 Aldo Leopold's collection of essays, *A Sand County Almanac*, had exposed readers to the science of ecology, but the public had not yet grasped the first principle of Ecology 101: namely, that everything is connected. It was through the lens of human health that the connection between agricultural practices, food chains, avian life, and human cancer was finally made clear and laid at the feet of society to assume responsibility. Industry's reaction to *Silent Spring* brought the issue of public health to the fore, and two different parties argued for the right to defend it (three, if you count the government's anemic response): a quiet, determined science writer, and defensive corporate executives. Each accused the other of profiting from false claims.

Silent Spring ended a century-long accommodation between industry and the environment, enlarging the conceptual framework of the environmental movement from conservation to include human rights and the rights of all living beings. *Silent Spring* made the environment immediate by revealing the pollution inside our bodies, not just in nature. And it made clear that toxicity played no favorites, though industry did, with the poor often paying the highest price. *Silent Spring* was one of the first critiques to question what has come to be known as corporate junk science—for example, the "science" used by the tobacco industry in its decades-long deception of the American public, a tactic dubbed *biostitution* by Robert Kennedy Jr. Without intending to, Carson also challenged the very notion of corporate hegemony and authority. The environmental movement discovered that to protect the environment, it had to confront power, corruption, and mendacity in the world of commerce, a struggle that extended back through history and across the world. From the beginning, an environmental movement had to be an environmental *justice* movement, and an environmental justice movement was de facto a *social* justice movement. Two seemingly unrelated elements of history had become reengaged in the public mind because of Carson's work. The question that continues to reverberate to this day is whether human rights trump the rights of business, or vice versa, a conflict that has been ongoing for more than three hundred years.

If we examine the history of industrialism we can see a long struggle between human and commercial rights, one that sometimes took place in the public eye, but was more often submerged or ignored. In the early nineteenth century teenage girls in Sheffield were employed as bench grinders to sharpen knives, scissors, and cutlery. Inhalation of metallic dust turned their complexions muddy, made their breathing labored and their coughs full of thick mucus. Before they became adults, they could no longer stand or sleep, and died soon thereafter. Toddlers from the ages of two to six were employed to make lace. Children were enslaved in airless rooms, pits, coalholes, and around dangerous pulleys and belts, to make chains, pins, snuffers, and nails. They worked seventy-two-hour weeks, from five in the morning until six at night. Young boys were used as piecers, cleaners, blowers, polishers, scavengers, spinners, jiggers, and runners. A ten-year-old boy who made "bad" nails would have his ears nailed to an iron counter for the day and receive no pay. Even the girls were hit and kicked. Accidents were common; fingers, toes, and arms were crushed or severed. Working conditions were so filthy that disease affected everyone. Boys and girls were exposed to lead, mercury, dyes, coal dust, chlorate of potassium, and sulfur. Early death from asthma, tumors, consumption, and pneumonia was the rule. It was often reported, though never formally documented, that William Pitt, when informed that his manufacturers could not pay higher taxes for his wars due to high wages, said, "Then take the children."[19] Whether or not Pitt actually said this, England and Scotland did take the children, as well as their parents, who worked under conditions comparable to their offspring's. All were exploited by what Wordsworth called a commercial "outrage done to nature" that cut workers off from fresh air, clean water, and life.

An often misunderstood confrontation between corporate and human rights occurred during the rise and fall of the English group known as Luddites. In the early 1800s mechanical looms and frames began replacing skilled craftsmen in the English Midlands, primarily in Nottinghamshire, Derbyshire, Leicestershire, and Lancashire. Although the mechanical looms produced inferior goods, they could be operated by apprentices and unskilled labor, which drove down wages and reduced the size of the workforce. There was a rippling and devastating effect on employment and income throughout the Midlands as the new technology spread. In the spring of 1811 a frustrated, angry group of weavers and knitters in Nottingham, led by a mythical General Ned Ludd (named for a kindly dullard in the village of Ansley who had lived thirty years earlier) and his Army of Redressers, broke into textile factories and destroyed shearing and stocking frames. In the same year the prince regent issued an order barring textile trade with Napoleonic France and its allies, and an imminent war with the United States further slashed production. To make matters worse, several years of bad harvests and weather doubled and finally tripled the price of wheat. Many unemployed weavers were reduced to begging, while those still employed had their wages cut. Starving families aimed their desperation at the big mills.

The intention of these Luddites, as history has labeled them, however inchoately and naively expressed, was to recognize workers' rights, including the right to freely meet and associate, in the face of technological disruption. Today the term *neo-Luddite* has a negative connotation, and is used as a term of opprobrium for one who fears technology and innovation, implying that the original Luddites were an ignorant mob intent on quashing progress. This is an unfair characterization; Luddites were artisans, highly skilled workers steeped in craft traditions who took great pride in the textiles they produced. The fundamental question posed by these weavers and croppers was, "What is progress without full employment?" which has yet to be answered satisfactorily. For these workers, whose lives quickly deteriorated because of mechanical looms, unemployment, and soaring food prices, there was no transition into other jobs or retraining, no grievance process, no safety net. What they wanted was the right to engage in collective bargaining, and their grievances seem modest by contemporary standards: First, socks should not be made from inferior cloth produced by the wide stocking frames but should be seamless one-piece knits. Second, "colts," apprentices who had not fulfilled their seven-year legal training requirements, could not be employed to do the work of skilled weavers. Their demands were never taken seriously. When the right to speak collectively was further denied by an act of Parliament, they turned to property destruction to force employers to the bargaining table. The worst moment came with the torching of the Wray and Duncroff mill near Bolton. Twelve people were arrested, and four were put to death, including Abraham Charlston, reportedly twelve years old, who sobbed for his mother on the scaffold. The tragedy was underscored when a later investigation revealed that the Luddites had actually *refused* to take part in the plan for arson. Their numbers had been infiltrated by industry-paid spies who then proceeded to hire a mob from a nearby town to torch the factory in the name of the Luddites.[20]

When the movement spread to France, it gave us the word *sabotage*, derived from the practice of tossing wooden shoes (*sabots*) into moving machinery. British historian Eric Hobsbawm called such acts "collective bargaining by riot," a fair description of the tactics employed in the face of mill owners' intransigence and Luddite confusion.

It is helpful to consider that, although machine wrecking had a long history in mill towns, dating back more than a century, there was no historical context for what was still the beginning of the productivity revolution of the Industrial Age. The Luddite movement failed in no small part because mill owners felt threatened, and they were justified in their fears. Notes from General Ludd would arrive at night promising death to their guards, wives, and families if they did not destroy the mechanical "demons." By the end of 1811 a thousand frames had been broken. In 1812, Parliament passed the Frame Breaking Bill, making destruction of manufacturing equipment punishable by death. But mill owners were not accountable under the law for the death of workers if caused by overwork, toxins, neglect, unsafe working conditions, accidents, or disease.

Carson had no particular agenda against business per se, but she saw the results of the relentless drive for corporate primacy in one specific area—namely, agricultural pesticides. When President George H. W. Bush refused to sign the Convention on Biodiversity at the Earth Summit in 1992, explaining that it was his job to protect "business rights," he repeated the oft-heard complaint of corporations that liberals and do-gooders unjustifiably criticize commerce and stifle economic development. This logic has become cast in stone and is repeated endlessly, but when one looks back dispassionately over the past centuries, it is impossible to find a period when business didn't have a disproportionate share of rights in the world.

Business justifies these rights because of its indisputable argument that it creates value, a position that nevertheless neatly evades the other side of the issue: How much value does it destroy in the process of carrying out its activities? Whether value is taken from the environment in the form of resources or despoliation, or from people in terms of wages, conditions, or worker health, it is largely unaccounted for in the calculation of value. Rachel Carson's reluctant conclusion was that once-respected businesses were creating products that destroyed value. They were exceeding their license to operate, and creating a public health hazard that threatened the web of life. Business rights are illegitimate if they remove rights from others, if they are not reciprocal and mutual with the rights of citizens, and if they extirpate other forms of life. From an economic viewpoint, what citizens have been trying to do for two hundred years is to force business to pay full freight, to internalize their costs to society instead of externalizing them onto a river, a town, a single patient, or a whole generation.

The International Campaign for Justice in Bhopal is illustrative of the complexity, breadth, and doggedness of the movement in addressing this imbalance. For the media, Bhopal is ancient news. For hundreds of thousands of people who lost family members or have become chronically disabled, it is daily reality. When events slip beyond the horizon of media coverage, they disappear from public discourse: abuse of power thrives in silence, shrinks in the light. The primary goal of movement groups is to prevent the fading away into darkness of the issue by continually placing it in the public eye until justice has been served. Union Carbide built the Bhopal plant in 1979 to manufacture methyl isocyanate, an unstable and extremely hazardous compound used to make Sevin, a common pesticide used on cotton, corn, and vegetables. The facility was intended to further India's move toward agricultural self-reliance, and was brought about through an agreement by the Madhya Pradesh government and the Union Carbide and Carbon Corporation. Bhopal officials and the Indian subsidiary of Union Carbide opposed placing the plant in an urbanized area but were overruled by the U.S. owners because a less populated area was more expensive. Corners were cut and safety standards sidestepped to enable the company to export low-cost pesticides throughout Asia. Tanks containing methyl isocyanate were too large, overfilled, and unrefrigerated. Flare towers were inoperative and backups nonexistent, and softness in the pesticide market caused further cutbacks in safety staff. The company had no contingent safety plans; it did not apply the same standards of safety engineering to the design and construction of the plant as it used in the United States. Three months before the explosion at the Bhopal facility, internal audits conducted by staff from a sister plant in Virginia warned of the danger of a runaway reaction, a cautionary report that never got to India or the people of Bhopal. Twenty-one years after the fatal leak, the company has not made public the nature of the chemical releases or their toxicity, thereby rendering medical treatment frustrating and difficult.

It is estimated that at least 100,000 people have sustained persistent injuries, debilitating illnesses, and disabilities from gas inhalation, including birth defects and diseases of the lungs and eyes. Although Union Carbide CEO Warren Anderson announced that he took moral responsibility for the accident shortly after it occurred, in the end, the company backed away from his statement and chose to litigate, moving juridical venue to India, where damage awards are considerably smaller. Union Carbide ultimately negotiated a settlement of $470 million—a figure that worked out to $800 per plaintiff—while spending an additional $100 million on public relations, advertising, and legal fees. The victims of the Bhopal disaster were not consulted about the settlement imposed by the Indian Supreme Court, and no child under eighteen was allowed to file a claim. Sixteen years after the settlement, $330 million remains frozen in trust accounts. Because most of the claims and costs were covered by insurance, the company did not have to pay out the full amount, and took a relatively small charge against earnings of 43 cents a share in a year in which it booked profits of $4.88 per share. In 2002 Dow Chemical purchased the assets of Union Carbide.[21,22]

Bhopal illustrates all too tellingly the asymmetry of corporate and civil rights. Although Dow and Union Carbide are under criminal indictment in India, they refuse to respond to the charges. There is no system of justice to hold them accountable for the impact their mismanagement had on their victims' human rights, for their Indian subsidiary company was controlled by the parent company in shares and deed. In the plainest of language, Union Carbide had the right to extend itself into one of the more beautiful cities of India, but no requisite accountability. While the WTO and constituent trade ministers work diligently to undo restrictions on corporate opportunity, there is no equivalent international organization that addresses economic and corporate responsibility. If NGOs hold up caution signs to globalization, it is because thousands of other human rights violations have only added to the grief of Bhopal, wounds that give people pause throughout the world.

The arc of this discussion brings us inexorably to Exxon, and requires a historical flashback to the man who has become the poster child for corporate ruthlessness, John D. Rockefeller. Here again, a woman was the whistleblower for corporate malfeasance, and in this case it involved the corrupt, illegal practices of the world's richest man. Ida Tarbell's *The History of Standard Oil* (1904) contained a remarkable series of nineteen articles published over a two-year period in *McClure's* magazine. In it she recounted damning facts, stories, and anecdotes about the company, many received directly from one of its partners and longtime directors, Henry H. Rogers. A centimillionaire (tantamount to a multibillionaire today), who invested heavily in oil, minerals, and railroads, Rogers was naively forthcoming in his discussions with Tarbell, believing that his cooperation would ensure she got her facts right. She did, and the reading public was outraged, while to his dying day John D. Rockefeller seethed at the mention of Tarbell's name. Written in the calm and lucid prose later employed by Rachel Carson, *The History of Standard Oil* laid bare the corrupt and unethical practices that led to the company's dominance. Within seven years of the book's publication, the Standard Oil Trust had been dismembered into thirty-four separate companies under the Sherman Antitrust Act. The four largest became Exxon, Mobil, Chevron, and Amoco. Exxon merged with Mobil in 1998, Chevron purchased Texaco in 2000, and BP acquired Amoco in 1998. ExxonMobil and Chevron were among George W. Bush's biggest contributors when he was governor of Texas and when he ran for president, and had much to say about the government's policies on climate change through private meetings with Vice President Dick Cheney. ChevronTexaco named an oil tanker after Secretary of State Condoleezza Rice and has been accused of environmental and human rights abuses in Ecuador and Nigeria.

The pesticide companies that dogged Rachel Carson paid a young public relations flack named E. Bruce Harrison to lead the campaign against her under the banner of the National Agricultural Chemical Association. Harrison later started his own firm, which helped devise the oil and automobile industry campaign called the Global Climate Coalition—"a voice for business in the global climate debate"—whose main purpose was to kill the Kyoto Protocol in the United States. ExxonMobil was one of its biggest supporters. Harrison also is credited with creating the concept of greenwashing. He realized early on that a number of large New York- and Washington-based environmental organizations depended on direct mail campaigns to maintain their finances. These organizations were prime targets to cut deals with miscreant corporations, which would enable both the company and the nonprofit to use minor corporate concessions to enhance their respective images. One example was the deal between Environmental Defense and McDonald's that called for the fast-food chain to introduce recycled paper into its waste stream of French fry containers, tray liners, and napkins. Harrison became a wealthy man, his company was sold to the Ruder Finn agency in New York in 1996, and the GCC was disbanded in 2002 after widespread industry defections.

ExxonMobil once issued directives forbidding the use of the word *sustainability* in all internal or external communications, and has vigorously funded groups that fight or delay policy on CO_2 reduction. It appears that another of its goals is to corporatize how science is perceived and understood by the public, creating doubt and fear whenever possible, but always couched in the language of reason. To do so, Exxon funds so-called think tanks that work diligently to create skepticism, if not cynicism, about efforts to mitigate climate change.[23] One of the recipients of its largess was the Action Institute for the Study of Religious Liberty, which received $155,000, then labeled emission controls "a misguided attempt to solve a problem that may not even exist."[24] The American Council of Capital Formation received $250,000 and rewarded its sponsor with this proclamation: "Science questions must be addressed before the United States and its allies embark on a path as nonproductive as that of the Kyoto Protocol." The American Council on Science and Health ($90,000 donation) assured us that "Policymakers can safely take several decades" to respond to global warming. The American Enterprise Institute ($960,000) wrote, "Between hacky-sack games, enviro-moralists kick around the imminent apocalypse of global warming, brought on—they're sure—by the pollution of human industry and the mindless plunder of our shared heritage. . . . The most recent studies now cast major doubt on global warming itself—the basis for all the gloom and doom predictions."[25] The American Legislative Exchange Council ($712,000) reported that "global warming could actually save lives."[26]

Citizens for a Sound Economy ($302,150) claimed: "The science behind global warming is inconclusive, and to teach otherwise is fear mongering."[27] And the Reason Public Policy Institute ($230,000) "reasoned" that "the sun, not a gas, is primarily to 'blame' for global warming."[28] Perhaps the most perverse subversion of climate science comes from another ExxonMobil grantee, the Mercatus Institute, which honored writer Michael Crichton, whose 2005 book *State of Fear* equates global-warming experts with Nazis, as a peer of Upton Sinclair and Rachel Carson.[29] Between 1998 and 2004, ExxonMobil awarded $1.74 million to the Competitive Enterprise Institute (CEI), the granddaddy of skepticism-for-hire. CEI and thirty other organizations have been dubbed the Cold Earth Society, and collectively have received $8 million from ExxonMobil. CEI sued President Bill Clinton to prevent expansion of research on global climate change by the U.S. Global Change Research Program; it also counseled President Bush not to attend the U.N. Summit on Sustainable Development in 2002, the tenth anniversary of the original Earth Summit, at which his father, attending the first global meeting on the fate of the environment, found it more important to protect business rights. The CEI Web site adopts the tactics created thirty-six years earlier by the agricultural chemical industry: make doubt and uncertainty about science the conventional wisdom, make people afraid of environmental initiatives, make people afraid of environmentalists, make people afraid of anything and everything except the official word offered by corporate-sponsored think tanks:

> Although global warming has been described as the greatest threat facing mankind, the policies designed to address global warming actually pose a greater threat. The Kyoto Protocol and similar domestic schemes to ration carbon-based energy use would do little to slow carbon dioxide emissions, but would have enormous costs. These costs would eventually fall most heavily on the poorest nations in the world. Luckily, predictions of the extent of future warming are based on implausible scientific and economic assumptions, and the negative impacts of predicted warming have been vastly exaggerated. In the unlikely event that global warming turns out to be a problem, the correct approach is not energy rationing, but rather long-term technological transformation and building resiliency in societies

by increasing wealth. CEI has been a leader in the fight against the global warming scare.

The argument in favor of fossil fuels pretends to be about science and policy, but it more accurately involves a question of rights: the debate over business rights versus the rights of citizens is claiming public attention once again. When will a Rosa Parks of this issue stand up and take a seat at the front of the bus? That person would not step aside while Congress was being corrupted by corporate contributions. That person would question the rights of business and would ask that the concept of the separation of church and state be enlarged to include the separation of corporation and state. That person would call for third-party objective science as the basis of governmental policy, and would protest the politicization of the EPA. That person would demand complete transparency in how corporations donate their money, and would petition for legislation forbidding corporations to impersonate citizen groups, and so much more. Perhaps that person is Ray Anderson, founder of the textile and carpet company Interface, who became cochairman of the President's Council on Sustainable Development under Bill Clinton. More than any other CEO in the country, Ray Anderson has taken to heart the necessity of completely transforming an industrial company so that it not only is sustainable, but takes steps to restore what damage has been done in the past. Whoever he or she is, that person would have the courage to say that we have marched too long in lockstep with economic policies and assumptions that are harmful to the earth and the majority of its people, and that it's time we spoke truthfully about the consequences of our actions, about the enormous polarization of wealth, about how we treat others, about how economic globalization has become a race to the bottom enforced by rules that very few have agreed to. We'll celebrate that person in histories yet to be written—an honor I suspect will not be bestowed upon a single rich and powerful businessperson of today, who one hundred years from now is more likely to be remembered as a name on a building or foundation.

Just as ecology is the study of the relationship between living beings and their environment, human ecology examines the relationship between human systems and their environment. Concerns about worker health, living wages, equity, education, and basic human rights are inseparable from concerns about water, climate, soil, and biodiversity. The cri de coeur of environmentalists in Carson's time was the same as that of the Lancashire weavers, the same as in the time of Emerson, the same as in the time of 2005 Nobel Peace Prize winner Wangari Maathai of Kenya. It can be summed up in a single word: life. Life is *the* most fundamental human right, and all of the movements within the movement are dedicated to creating the conditions for life, conditions that include livelihood, food, security, peace, a stable environment, and freedom from external tyranny. Whenever and wherever that right is violated, human beings rise up. Today, they are rising up in record numbers, and in a collective body that is as often as not more sophisticated than the corporate and governmental institutions they address.

EMERSON'S SAVANTS

There is an answer from every corner of the globe . . . the enslaved, the sick, the disappointed, the poor, the unfortunate, the dying, the surviving cry out, it is here.

—Ralph Waldo Emerson, *Journals*[1]

We shall match your capacity to inflict suffering by our capacity to endure suffering.

—Martin Luther King Jr., *The Trumpet of Conscience*

Those who work selfishly for results are miserable.

—Sri Krishna, Bhagavad Gita, The Song of God

As I write, the world is surfeited with news stories of the poorest people in New Orleans, dazed and traumatized, wading through the flooded ruins of their lives. Save for a few mentions in the media, there was little discussion of global warming as a factor in Hurricane Katrina. Katrina was the twenty-fourth recorded Category Five hurricane (hurricanes that sustain winds greater than 155 mph), and only the fourth to have made landfall in the United States; it was the seventh to have maintained Category Five status longer than thirty hours. It was followed by another Category Five hurricane, Rita, the fourth most powerful hurricane on record, and soon after by Wilma, the strongest Atlantic hurricane ever recorded.

Katrina and the storms that followed cannot be directly linked to climate change because separately they are only incidents. But when such extraordinary weather events occur around the globe in a single year, they are more than a collection of incidents; they begin to comprise a pattern, because the buildup of equatorial heat and the oscillating flow of the global jet stream are all part of one system. In the same year as Katrina, Los Angeles received the most rain in a century, capped by a two-foot snowfall in the mountains; winds peaking at 124 miles per hour forced the closure of nuclear power plants in Scandinavia; a record-setting drought in Portugal created conditions for the worst wildfires in the country's history; thirty-seven inches of rain fell on Mumbai in twenty-four hours, killing a thousand people and dislocating 10 million more; and the Missouri River flowed at its lowest level since records were kept.[2] In Mumbai as in New Orleans, the poor suffered greatly.

Although poverty and climate change seem unrelated, they have common roots for the simple reason that we *are* nature, literally, in every molecule and neuron. We contain clay, minerals, and water; are powered by sunshine through plant life; and are intricately bound to all other species, from fungi to marsupials to bacteria. In our lungs are oxygen molecules breathed by every type of creature ever to have lived on earth, along with the very hydrogen and oxygen atoms that Jesus, Confucius, and Rachel Carson breathed. In the West, the revelation of this unity captured the interest of some of the brightest minds of the nineteenth century.

Although greatly resisted by orthodox religion, the explosion of scientific discovery that revealed the interconnectedness of life set off a parallel philosophical search for its meaning. Nature was the central metaphor of the nineteenth century; "question authority" was its motto. A new radical inquiry into nature included an attempt to define the nature of the mind. If there is a web of life, if human beings are a part of a vast living system, what are the moral implications? How could anyone claim absolute knowledge of the truth? What moral right could a government or group of people claim to justify using violence to impose their "truth" on another? What if any were the distinctions between nature and human nature? These queries called into question assumptions about war, slavery, and authoritarian religion and government, while giving rise to new concepts of justice and individual rights, all of which now started to be viewed through a new lens of interdependence and mutuality. These were the fundamental principles of ecology, a science as yet unnamed, and through its lens the beginnings of a social justice movement also could be discovered. And once again, a critical contribution can be traced back to Ralph Waldo Emerson.

On Christmas Day 1832, a despondent Emerson set sail for Europe. The death of his young wife the previous year had made him pensive and moody. His interest in the formal church had waned and then withered. He gave a series of lectures on the Gospels three months after Ellen's death, but they were to be his last biblical pronouncements. By his own admission, he was debilitated by grief.[3] Sailing into the teeth of a nor'easter, he left America, the ministry, his church, and theology, and landed in Malta in February 1833 at the start of a journey that would bring him in contact with Samuel Taylor Coleridge, Thomas Carlyle, and William Wordsworth.

In July Emerson visited the Jardin des Plantes and the Cabinet of Natural History in Paris. The Jardin was not a botanical garden for weekend strolling but a major European research center for the study of medicine, food, and silviculture. The gardens were the hub for the work of Antoine Laurent de Jussieu, a scientist who, with his uncle Bernard Jussieu, created the first natural system of botanical classification, a scientific milestone that became the template and precursor of Darwin's theory of evolution. Jussieu's system was based on observation and knowledge of the whole plant. It required an understanding of relationship and morphology, as well as the willingness and ability to see connections between species in and over time. The mind of the young Emerson was animated by the exquisite relationship between living things displayed at the Jardin, and his visit to the adjoining Cabinet of Natural History was similarly moving. Birds and other fauna were arrayed in sweeping palettes of color and form, with the obvious implication of evolution uniting them.

It was Emerson's first encounter with the web of life. He found himself standing on the threshold of the as-yet-unnamed discipline of ecology, not as a scientist but as a philosopher. The precision of Jussieu's classification system revealed, if only implicitly, how life was connected. Mental curtains fell, the division between human and other life forms vanished, and the interdependence of all life was realized.[4] *Epiphanic* may be too strong a word for this period in Emerson's life, yet his journal entries indicate moments of near-rhapsodic insight. He imagined religion, science, and nature as one field of thought, with no lacunae or elisions. Like William James, Emerson was less interested in religion per se than in the religious experience.[5] The human mind came to represent for Emerson the possibility and magnificence of nature, and nature in turn encompassed all human activity.[6] He anticipated modern neuroscience in seeing that the human mind was not some tabula rasa on which we inscribe our culture, but in the words of Stephanie Pace Marshall, a "magnificent, pattern-seeking, complex living system whose structures are not fixed."[7] His first lecture upon returning to America was not to the Second Church in Boston, the ministry of his father, but to the Boston Society of Natural History.

Almost as a testament to his discovery of mutual interdependence, Emerson planted seeds that would develop into what were, and continue to be, two disparate concepts that animate our daily existence: how we treat nature and how we treat one another—the foundations of environmental and social justice. In school we are taught the Yankee Emerson, the stalwart clergyman who is the author of essays on character, prudence, and self-reliance. It's as if Calvinist America had cherry-picked his writings so as to exclude the Unitarian dreamer. In fact, his thinking is steeped in a universality that eludes formal categories, which has been responsible for his appeal's magnifying with time. The morality he proposed came from perceiving and receiving nature, not from established codes or judgments: *I have confidence in laws of morals as of botany. I have planted maize in my field*

every June for seventeen years and I never knew it to come up strychnine. My
parsley, beet, turnip, carrot, buck-thorn, chestnut, acorn, are as sure. I believe
that justice produces justice, and injustice injustice.[8]

Emerson unknowingly happened upon the perfect carrier for his ideas in
the person of the young Harvard student Henry David Thoreau. Emerson,
the nonconformist, found his match in young Thoreau, who had been
accepted to the college in 1833, the same year William Whewell coined
the word "scientist." In his senior year, the twenty-year-old Thoreau read
Nature, a mesmerizing, unorthodox text, not once but twice. Books that
influence us often contain ideas we already recognize due to intellectual
preparation or predilection; rather than casting thunderbolts from the blue,
they may unlock sensibilities from within. For Thoreau, *Nature* was one of
those books. He once quipped to Emerson that Harvard taught all the
branches of knowledge but "none of the roots."[9] In *Nature*, he had found
his way below the surface to the source of the philosophical humus.

Robert Richardson, a biographer of both Emerson and Thoreau, writes:
"*Nature* is also a manifesto for transcendentalism, the American version of
German philosophical idealism which had as a pair of cornerstones the belief that intuition was a valid mode of knowing and was necessary as a coun-
terbalance to experience. Most interesting of all for Thoreau is Emerson's
insistence in *Nature* on a line of thought as old as Stoicism: that the individual, in searching for a reliable ethical standpoint, for an answer to the
question of how one should live one's life, had to turn not to God, not to the
polis or state, and not to society, but to nature for a usable answer. Stoicism
taught, and Emerson was teaching, that the laws of nature were the same as
the laws of human nature and that man could base a good life, a just life, on
nature." When Emerson and Thoreau met after the latter's graduation, he
offered what may have been his most laudable suggestion to the young
thinker: "Keep a journal," which Thoreau did for seven thousand pages to
the end of his life.

Nine years later, one of the twenty-nine-year-old's journal entries described the night he spent in Middlesex County Jail after being arrested by
neighbor Sam Staples for not paying his 6-bit poll tax, a significant sum given
that his whole cabin at Walden Pond cost $28.12. On July 23, 24, or 25,
1846 (the exact date is unknown), Thoreau was apprehended as he walked
into town to pick up some shoes that were being resoled.[10] Although he had
previously refused to pay poll taxes because they prevented African Americans

and the poor from voting, what he sought to protest this year was the war with Mexico.

The Mexican-American War, which gave birth to the infamous phrase "manifest destiny," unleashed expansionist fever throughout the country. After Mexico gained independence from Spain in 1821, the new government failed to sort out the rights and responsibilities of its own territories, including Texas. As Anglos moved into and settled the area, tensions grew and eventually boiled over, culminating in a series of skirmishes for Texas's own independence. Texas eventually won its freedom and joined the Union in 1845 under Democratic president James Polk, for whom this "great desert" was a means to a bigger objective: California. To win it, he had to gain control over a large parcel of Mexican territory that included the southernmost portions of Texas between the Rio Nueces and the Rio Grande, New Mexico, Arizona, Utah, Nevada, and parts of Colorado.

While the Polk administration had already made up its mind to go to war for the land, the question remained how to win over Congress and the American people to the cause. The one politically viable way to achieve Polk's expansionist dreams was to provoke Mexico to launch an attack on American soil, an event that would provide moral justification and incite patriotic fervor. That was accomplished by sending an expeditionary lure into a region historically governed and populated by Mexicans. As Congress was voting on the annexation of Texas, Polk stationed General Zachary Taylor on the banks of the Rio Grande, ready to intercept and engage Mexican soldiers at any opportunity.

Spoiling for a fight, Taylor's forces arrived in late March 1846, set up camp, commenced building a fort, and pointed their artillery across the river at the town of Matamoros. Four weeks later, one of the general's patrols was attacked when it stumbled upon an encampment of two thousand Mexican soldiers, and sixteen men were killed. The lure set by Polk and Taylor worked perfectly, because it could be reported that Mexicans fired first on outmanned American troops. A distant forebear of mine, Ethan Allen Hitchcock, while serving as a lieutenant colonel under General Taylor, kept a diary that provides an insider's view of the destined path to war. Colonel Hitchcock, upon receiving orders to muster, wrote, "Violence leads to violence, and if this movement of ours does not lead to others and to bloodshed, I am much mistaken."[11] When news came of the inevitable fatal skirmish, Polk famously declared, "American blood has been shed on

American soil." (Abraham Lincoln later asked which spot of America, in fact, had the blood been shed upon.) The resolution to declare war was debated in the House for thirty minutes, and in the Senate for a full day, May 13. Thoreau went to jail that July, a radical response to the war that would change world history.

Jail was not, as it happened, a great sacrifice for Thoreau. He was bailed out the following morning, probably by his sister, and was given hot chocolate to take the sting out of his overnight stay. What is symbolically important about his brief incarceration is that very few people went to jail on principle in the mid-nineteenth century. Eighteen months later, on January 26, 1848, Thoreau delivered a lecture, based in part on that experience, entitled "The Rights and Duties of the Individual in Relation to the Government," which was published the next year under the title "Resistance to Civil Government." It was a direct repudiation of "The Duty of Submission to Civil Government," William Paley's fourth chapter in Book Six of *The Principles of Moral and Political Philosophy*. (Paley also bears the ignominy of being the intellectual godfather of intelligent design.) Thoreau's text fully amplifies the meaning of his resistance, for here he flatly states that when government jails people unjustly, the just man should be incarcerated. "I did not for a moment feel confined . . . ," was his retort to the authorities who had seized him.

The legend of a vexed Emerson's asking Thoreau what he was doing in jail, and Thoreau's riposte—"Waldo, what are you doing out there?"—is apocryphal.[12] Nevertheless, it points to a deepening rift in their relationship. Emerson did not approve of what he saw as pointless resistance. Perhaps it was a reaction to his own inadequate response to the war, but it also reflected his reluctance to disturb the social web. Thoreau, for his part, fiercely believed in human interdependency, a belief that called out for a willingness to respond to moral imperatives, however distant. He did not, as do most people, distinguish between the act of paying a tax and the acts the tax pays for. (The war was rife with horrors, including the atrocities of Texas Rangers who raped and pillaged village after village.) The connectedness Emerson experienced in the natural world, Thoreau saw in the human world. It was a case of the student, Thoreau, taking the teacher at his word—act as if everything is connected. Thoreau's "Resistance" essay was as passionate a manifesto as Emerson's *Nature*, arguing that there are no buffers between the individual's conscience and one's submission to taxation and the injustice it finances in the world.[13] "One's tax, or one's consumer dollar does not simply disappear into the coffers of a faraway government or a spectral corporation; it funds acts that may be immoral, and the funder must therefore acknowledge his role as an accomplice," writes Evan Carton, amplifying the idea that Thoreau's "usefulness as revolutionary remains intact."[14]

It is convenient to assign Thoreau the single, definitive role of pacifist, with the understanding that his actions and writings were forerunners of and models for nonviolent resistance. In "Resistance to Civil Government," Thoreau is indeed a pacifist, counseling passive means to resist injustice, but his other work and the positions he took were not always congruent with that stance. The concept of nonresistance was much in discussion around Concord in the 1840s. William Lloyd Garrison, abolitionist publisher of *The Liberator*, easily the most radical newspaper of the nineteenth century, wrote an essay in 1837 stating that force should not be met by force, an argument that greatly influenced Leo Tolstoy's rejection of violence, fifty-six years later, in his book *The Kingdom of God Is Within You*. Debating at the Concord Lyceum the question of whether forcible resistance is ever justified,[15] however, Thoreau chose the affirmative: certain circumstances do warrant the use of force.

Only a few years later John Brown's insurrection would meet Thoreau's criteria for justifiable force. When Congress opened up Kansas and Nebraska to slavery in 1854, Brown and three of his sons, fearing that slavery would spread to the rest of the West, joined other abolitionists encamped in the territories to prevent it. Pro-slavery zealots saw abolitionists as weak, nonviolent adversaries, but did not discount the fact that they posed a grave threat. On May 21, 1856, hundreds of white supremacists from Missouri descended on the anti-slavery town of Lawrence, Kansas, looting and burning homes, printing presses, and businesses. The following day, the senior senator from South Carolina, Preston Brooks, knocked Massachusetts senator Charles Sumner unconscious with his gold-knobbed cane on the Senate floor after Sumner delivered an anti-slavery speech. For Brown, these were the final insults, and with his sons and three others, he went the next day to pro-slavery territory, where he ordered and watched the execution of three leaders. That act forever changed the Southern perception of abolitionists as powerless, and in the roiling debate that followed the violence, Thoreau was an avid supporter of Brown's cause.

Still, "Resistance to Civil Government" remains as relevant today as when written, and stands as Thoreau's most influential legacy. In it he denounces the expansionist Mexican War as the work of a few individuals "using the standing government as their tool . . ." Thoreau decries democracies based on majority rule, because they allow the many to be manipulated by the few, negating justice and conscience in favor of expediency. The essay is also a staunch defense of self: "Must the citizen ever for a moment, or in the least degree, resign his conscience to the legislator? Why has every man a conscience then? I think that we should be men first, and subjects afterwards. . . . The only obligation which I have a right to assume is to do at any time what I think right. It is truly enough said that a corporation has no conscience."

The publisher of Thoreau's collected works posthumously retitled this essay as *Civil Disobedience* in 1866, even though the word *civil* is used in the essay only four times, and *disobedience* and *obedience* never. Some contend that Thoreau himself renamed the piece before his death in 1862, but no one knows the true source of the phrase or who bears responsibility for the now-famous title.[16] Regardless of who bears that responsibility, the essay's concept of civil disobedience may have changed the life of a Hindu lawyer in Durban, South Africa, named Mohandas Karamchand Gandhi.

Gandhi's fight against apartheid and racism sought to make nonviolent resistance an active, not passive, process. Angered and humiliated by the passage of the Asiatic Registration Act in the Transvaal colony of South Africa, legislation that required Indians to be fingerprinted and documented and to carry their registration with them at all times, Gandhi was determined to resist it. The law handed him an unintended tool: nonregistrants could be fined and jailed. On September 11, 1906, more than one thousand Indians met in a protest at the Empire Theatre in Johannesburg, where they passed a resolution defying the government: by unanimous vote, Muslim and Hindu alike agreed to be arrested and jailed rather than obey the law. In his journals Gandhi questioned the repercussions if he was to be arrested: "But I must confess that even I myself had not then understood all the implications of the resolution I had helped to frame; nor had I gauged all the possible conclusions to which they might lead."[17] He later described that night as the advent of the satyagraha movement, but at the time neither the word *satyagraha* nor a specific set of tactics existed, only the intent to fight injustice actively but peacefully.

Satyagraha is a three-step instrument of change, not a mere act of protest. The first step is to object to an unjust law or policy, and petition for its removal. Absent an acceptable response, the second step is to break the law. The third step is to undergo the consequences, be they arrest, violence, abuse, or privation. *Satyagraha* means "holding firmly to the truth" in

Hindi, and by following the process, remaining focused and calm within one's belief. The seeds of satyagraha were sown in Gandhi long before he came to South Africa. In his home state of Gujarat, cultural tradition emphasized the role of suffering as a means to draw attention to an issue or dispute. An aggrieved party would sometimes bring shame to an offending party by fasting in front of his office or home,[18] a tactic to which Gandhi would turn at the most troubling moments in his later years. Although Gandhi was convinced that an oppressor could more likely be induced to see what was right than be forced to fight back, he wasn't a pacifist in a conventional sense. Satyagraha was a tactical stratagem, an instrument of disruption that could undermine and endanger the interests and objectives of the state.

Scholars argue whether Gandhi first read *Civil Disobedience* in 1906 or 1907, but whatever the case, it strengthened his resolve and influenced a fledgling movement.[19] Throughout his life, Gandhi quoted the piece in letters, interviews, and articles as a buttress for the philosophy of satyagraha. While both men made arrest a badge of honor in their struggle against injustice, Gandhi introduced it as an instrument of a mass movement rather than as a matter of individual conscience. Thoreau's principled dismissal of the ignominy and confinements of jail was also echoed faithfully by Gandhi from that point on. His first arrest and jailing occurred in January 1908; when apprehended a second time nine months later for instigating a mass burning of registration certificates, he took Ruskin and Thoreau to jail with him so that he "could find arguments in favor of our fight."[20]

It could be said that in his influence on Gandhi, Thoreau was returning a favor, for the Bhagavad Gita and the Upanishads had demonstrably influenced his own thinking. Although Gandhi cited other influences on his thought—Tolstoy, the Sermon on the Mount, Socrates, and Ruskin—Thoreau was for him the one teacher besides Jesus who was both pragmatist and activist, a doer and not just a thinker. Thoreau was the philosopher who, in Gandhi's words, "taught nothing he was not prepared to practice in himself."[21]

On December 1, 1955, nearly fifty years after the first satyagraha protests, Rosa Parks boarded an evening bus in Montgomery, Alabama, taking a seat in the fifth row, the first in the "colored" section of the bus. When the white section filled she was asked by the white driver to give her seat up, which she declined to do, despite being met with threats. With that refusal she set off a cascading series of events that changed America.

Parks's arrest came after years of degrading treatment of black passengers. In 1953, Jo Ann Robinson, a teacher at Alabama State College and head of the Women's Political Council, had met with city commissioners and demanded an end to the humiliating policies of discrimination; six months after that meeting, the Council informed the city that it was considering mounting a boycott of the bus system.

Bus drivers would sometimes throw black riders' change and transfers on the floor; on cold, rainy evenings the bus would speed past them as they waited at stops. Black men who paid their fare would be asked to step back out onto the street and board through the rear, but when they did, the bus would leave without them. One young black veteran of the Korean War, in uniform, was shot dead by a policeman in a dispute over a ten-cent fare.[22]

Although Parks's decision was courageous and unplanned, it was not impulsive. She was not too physically tired to move to the back of the bus, as has often been reported; what she *was* tired of was being demeaned and humiliated. She had grown up in a racist world. Klansmen could often be heard in the woods behind her grandparents' Alabama farm. She was shunted around, as were all blacks, to inferior rooms and separate doors, elevators, and water fountains with "colored" scrawled on them. When the bus arrived that night in 1955, she was taken aback to see that the driver was James Blake, who had mortified her twelve years earlier by dragging her off his bus because she wouldn't enter through the rear door.[23] Although she now hesitated about getting on, she bravely stepped aboard—a moment for which she had been trained and prepared. She had worked for years for Clifford and Virginia Durr, white aristocratic Southerners whose families went back generations. Clifford Durr had served in the Roosevelt and Truman administrations on the Federal Communications Commission. When asked to have his staff sign loyalty oaths, he walked away from his job. His wife had spent the 1940s touring with African American activist Dr. Mary Bethune to abolish the poll tax (the same one that Thoreau wouldn't pay). The year after the 1954 Supreme Court ruling making public school segregation illegal, Parks was given a scholarship by Virginia Durr to attend a summer course at the Highlander Folk School in Tennessee on the implementation of *Brown v. Board of Education*.[24] The school was started by Myles Horton in the 1930s to help economically disadvantaged whites but by the early 1950s had begun to focus on civil rights. Horton, an admirer of Gandhi, used his school to train its pupils in how to achieve integration and civil rights. Parks's experience there was the first time in her forty-two years that she did not feel hostility from a group of white people, and she reveled in the morning smells of coffee and bacon, knowing the food had been cooked and served by whites.[25]

On December 5, the day of Parks's court appearance, Jo Ann Robinson decided to call the bus boycott. That afternoon, black ministers formed a group called the Montgomery Improvement Association to support the boycott and elected as its president an eloquent young Baptist minister from the Dexter Street Church named Martin Luther King, Jr. Although he accepted, he had initially been reluctant, and had spent the weekend pondering the implications of the boycott, both professionally and for the community. Today, it is easy to look back and imagine that initiating a boycott was a straightforward matter. But at that time, the segregated South was a different place from what it is today. Behind the mannerly speech and outward politeness was a heightened tension that was conveyed in the body language, in the eyes, and in any number of dismissive gestures. And beneath it ran an even deeper current, one of latent and explosive violence, even mayhem. Months before, in Mississippi, fourteen-year-old Emmett Till had unintentionally whistled at a white woman shopkeeper (he had a speech defect from polio) and was lynched three nights later by a party led by the woman's husband. He was mutilated, castrated, and shot, his skull crushed beyond recognition. The lynch mob was arrested, tried, and set free. This incident, though highly publicized, was not anomalous; there had been on average one lynching per week in the ninety-year period since Reconstruction.[26] In every community in the South, poor whites took it upon themselves to be enforcers of the apartheid system while the middle class averted its eyes. Anytime a march, demonstration, or gathering occurred, blacks would be stared down and called out by people who knew them. Serving as leader of any type of resistance was effectively an invitation to physical violence.

That night, at a community meeting convened to decide whether to proceed with the boycott, King, with only thirty minutes to prepare, delivered his first civil rights speech, a homily of simple declarations starting with a theme that he would return to later in his career, that black people were "sick and tired of being sick and tired":

> There comes a time. [long pause] There comes a time when
> people get tired—tired of being segregated and humiliated, tired
> of being kicked about by the brutal feet of oppression. We had

no alternative but to protest. For many years we have shown amazing patience. We have sometimes given our white brothers the feeling that we like the way we are being treated. But we come here tonight to be saved from that patience that makes us patient with anything less than freedom and justice. [longer pause] One of the great glories of democracy is the right to protest for right. The white Citizen's Council and the Ku Klux Klan are protesting for the perpetuation of injustice in the community. We are protesting for the birth of justice in the community. Their methods lead to violence and lawlessness. But in our protest there will be no cross burnings. No white person will be taken from his home by a hooded Negro and brutally murdered. There will be no threats and intimidation. Our method will be that of persuasion, not coercion. We will only say to the people: "Let your conscience be your guide." Our actions must be guided by the deepest principles of our Christian faith. Love must be our regulating ideal. Once again we must hear the words of Jesus echoing across the centuries: "Love your enemies; bless them that curse you, and pray for them that despitefully use you." [The audience is on its feet shouting affirmatively.] If we fail to do this our protest will end up as a meaningless drama on the stage of history, and its memory will be shrouded with the ugly garments of shame. In spite of mistreatment that we have confronted, we must not become bitter and end up hating our white brothers. As Booker T. Washington said: "Let no man pull you down so low as to make you hate him." [The audience is cheering and shouting.] If you will protest courageously and yet with dignity and Christian love, when the history books are written in future generations the historians will have to pause and say, "There lived a great people—a black people—who injected new meaning and dignity into the veins of civilization." That is our challenge and our overwhelming responsibility.[27]

Although King's sermon on the opening night of the boycott counseled peacefulness and nonviolence, his knowledge of nonviolent resistance was then limited. The language of his speech was still conventionally militant, as he spoke of the "weapons of protest," "tools of coercion," a God that would "break the backbone" of power and "slap you out of the orbits" of your relationships if He was not obeyed.

When King's house was bombed on January 30, 1956, two months into the boycott and the eighth anniversary of Gandhi's assassination, hundreds of black people, some carrying handguns, knives, and rifles, came close to a race riot against the white police and racist mayor who had arrived at the scene.[28] King placated the crowd that night, but soon thereafter armed men were guarding his house. The only boycott-related death was that of a white woman, Juliette Morgan, a librarian and Montgomery native who had written a letter to the *Montgomery Advertiser* on December 12, 1955, that perceptively compared the boycott to Gandhi's Salt March. The boycotters "seem to have taken a lesson from Gandhi—and our own Thoreau, who influenced Gandhi. Their own task is greater than Gandhi's, however, for they have greater prejudice to overcome." Morgan was taunted and harassed by townspeople, ostracized by former friends, and finally found an escape from her misery by taking her own life.[29]

From that point until his assassination over a dozen years later, King was anxious and feared for his safety. With premonitions of his own death he applied for a gun permit but was turned down by the local sheriff. He then presented his request to the governor of the state, "Big Jim" Folsom, who also declined it.[30] Nevertheless, he eventually did obtain a gun, and armed bodyguards surrounded him and other leaders day and night, turning his home into an arsenal. Not until late February of 1956 did King begin his education in nonviolence, initially under the tutelage of Bayard Rustin, who schooled him in the Gandhian revolution that had overthrown British rule in India. Rustin was adamant that a nonviolent movement could not tolerate the presence of gun-toting guards, and that the firearms inside the King household had to be removed. King was displeased and confused; he then understood nonviolence as meaning only not initiating violence. He felt that he and other blacks had the right to defend themselves. As he told an interviewer at that time, "When a chicken's head is cut off, it struggles most when it's about to die. A whale puts up its biggest fight after it has been harpooned. It's the same thing with the southern white man. Maybe it's good to shed a little blood. What needs to be done is for a couple of those white men to lose some blood. Then the federal government will step in."[31]

Rustin counseled King and his wife, Coretta, and before he left town he introduced them to Glenn Smiley, a Methodist minister who had studied nonviolence since 1942. When Smiley asked King if he knew of or had read Gandhi, King demurred, saying that his admiration for the man was not matched by any particular knowledge of him or his work. Immediately Smiley presented him with Gandhi's *Autobiography*, Thoreau's *Civil Disobedience*, and Richard Gregg's *The Power of Nonviolence*.[32] When asked after the boycott ended which books most influenced him, King cited all three works. By the time King had been introduced to them, nonviolence had gone through two decades of experimentation and field research in America, led by pacifists in the War Resisters League and the Fellowship of Reconciliation.[33] King unintentionally fashioned a new nonviolent movement undergirded by compassion but armed with his particular Baptist eloquence.[34] He wielded that sword for twelve years and two months, until he was murdered outside his room at a Memphis motel.

The strands that connect Emerson to Thoreau, Thoreau to Gandhi and King, and Gandhi to Parks and King, all express Emerson's faith that thought "was sent and meant for participation in the world . . . ."[35] History tends by its very nature to obscure the mundane acts that are the harbingers of change. What if Thoreau had not attended Emerson's lecture or read his book? What if Thoreau had paid his poll tax? What if the title of his essay had not been changed to *Civil Disobedience*? What if Virginia Durr had not helped Rosa Parks to get to Highlander? What if an editor at the *Indian Times* had not given Thoreau's essay to Gandhi? What if Rosa Parks had been too intimidated by driver James Blake that December night? What if Ralph Abernathy, the civil rights veteran who had actively promoted King's candidacy to lead the Montgomery boycott, had wanted that role for himself, a position he probably could have gained?[36]

We are made aware of the proverbial forks in the road of life from an early age. Whether at commencement or from the pulpit, we are told there is a convenient path, and a less traveled road of integrity. From a Buddhist perspective, the adage is infinitely true. We face such forks a million times a day, even in the space of a breath. Life is permeated with possibility at every instant. What distinguishes one life from another is intention, the one thing that we can control. Rosa Parks's intentions were deep and unswerving, as were King's, Thoreau's, and Gandhi's; so, too, were Jo Ann Robinson's and Virginia Durr's. While the events of the world were out of their control, their resolve was not.

Maybe the best way to understand the future implications of the movement's daily actions is to remember Emerson's moral botany: corn seeds produce corn; justice creates justice; and kindness fosters generosity. How do we sow our seeds when large, well-intentioned institutions and intolerant ideologies that purport to be our salvation cause so much damage? One sure way is through smallness, grace, and locality. Individuals start where they stand and, in Antonio Machado's poetic dictum, make the road by walking. Thoreau insisted in *Civil Disobedience* that if only one man withdrew his support from an unjust government, it would begin a cycle that would reverberate and grow. For him there were no inconsequential acts, only consequential inaction: "For it matters not how small the beginning may seem to be: what is once well done is done forever."[37]

INDIGENE

It might help if we non-Aboriginal Australians imagined ourselves dispossessed of the land we lived on for 50,000 years, and then imagined ourselves told that it had never been ours. Imagine if ours was the oldest culture in the world and we were told that it was worthless. Imagine if we had resisted this settlement, suffered and died in the defense of our land, and then were told in history books that we had given it up without a fight. Imagine if non-Aboriginal Australians had served their country in peace and war and were then ignored in history books. Imagine if our feats on the sporting field had inspired admiration and patriotism and yet did nothing to diminish prejudice. Imagine if our spiritual life was denied and ridiculed. Imagine if we had suffered the injustice and then were blamed for it.

—Paul Keating, prime minister of Australia, Redfern Park, December 10, 1993, at the launch of Australia's celebration of the International Year of the World's Indigenous People

Chevron has neglected us. They have neglected us for a long time. For example, any time spills occur, they don't do proper cleanup or pay compensation. Our roofs are destroyed by acid rain from their chemicals. No good drinking water in our rivers. Our fish are killed on a daily basis by their chemicals, even the fishes we catch in our rivers, they smell of crude oil. Chevron know the right thing to do, they intimidate us with soldiers, police, navy and tell us that cases of spill are caused by us. They have been threatening us that if we make noise, they will stop production and leave our community and we will suffer, as if we have benefited from them. Before the '70s, when we were here without Chevron, life was natural and sweet, we were happy. When we go to the rivers for fishing or forest for hunting, we used to catch all sorts of fishes and bush animals. Today, the experience is sad. I am suggesting that they should leave our community completely and never come back again. See, in our community we have girls, small girls from Lagos, Warri, Benin City, Enugu, Imo, Osun, and other parts of Nigeria here every day and night running after the white men and staff of Chevron, they are doing prostitution, and spreading all sorts of diseases. The story is too long and too sad. Tell Chevron that we are no longer slaves; even slaves realize their condition and fight for their freedom.

—Mrs. Felicia Itsero, 67, Ijaw mother and grandmother from the Gbaramatu clan, Niger River Delta[1]

We have this extraordinary conceit in the West that while we've been hard at work in the creation of technological wizardry and innovation, somehow the other cultures of the world have been intellectually idle. Nothing could be further from the truth. Nor is this difference due to some sort of inherent Western superiority. We now know to be true biologically what we've always dreamed to be true philosophically, and that is that we are all brothers and sisters. We are all, by definition, cut from the same genetic cloth. That means every single human society and culture, by definition, shares the same raw mental acuity, the same intellectual capacity. And whether that raw genius is placed in the service of technological wizardry or unraveling the complex threads of memory inherent in a myth is simply a matter of choice and cultural orientation.

—Wade Davis, "The Ethnosphere and the Academy"[2]

The past grows longer, and the future grows shorter.

—White Feather, Hopi elder of the Bear Clan

On the Feast Day of the Eleven Thousand Virgins in 1520, three caravels captained by Fernão de Magalhães rounded a large headland at the fifty-second parallel in what is now southeastern Chile and discovered the rumored southwestern passage to the Pacific Ocean, an *estrecho* that would later bear his name. The fleet sailed into an enchanted but unforgiving land: steel blue glaciers calved massive icefalls into the fjords; Andean condors wheeled across the deep canyons blanketed by impenetrable forests; gusty williwaws swept down the ice-covered cordillera; milky blue rivers drained into the frigid channel; southern right whales and orcas breached in front of the boats; penguins crowded the beaches; and yawning two-ton elephant seals lolled on the sand, occasionally shuddering off their coat of flies. On the southern bluffs, amid the dense beech tree forest, naked copper-colored Indians watched the arrival of Magalhães's ships. It was October, still wintry in the southern hemisphere, but the rapt audience, slathered in seal blubber and an occasional animal skin, was indifferent to the gales that chilled Magalhães's sailors to the bone. Tucked within the windblown trees, these people had dwelled at the "end of the earth" since the last ice age. True to their nomadic nature they maintained no permanent dwellings but preferred wigwams, traveling in tree-bark canoes. To ensure their survival, they carried a basket of embers everywhere, a constant lifeline to fire. When at sea, they placed coals on a pile of sand in the center of the canoe, a welcome source of heat for the women who would dive into the 48-degree water for shellfish. It was their smoky fires Magalhães saw across the archipelago that gave rise to the Spaniards' name for the region, Tierra de los Fuegos, which later became singular: Tierra del Fuego, the Land of Fire.

Collectively, the Fuegian tribes—the Selk'nam, Ona, Haush, Alacalufs, and Yámana—became the subject of wild speculation (cannibalism) and fantastic tales (Patagonian giants), and the victims of European violence, for the next three hundred years. In 1578 an awestruck Sir Francis Drake watched sleet melt on the skin of naked women while his crew froze in the icy storm. In the seventeenth century Dutch and Spanish parties came ashore and massacred natives like so many pests. In 1828 the British captain Robert Fitzroy arrived on a survey mission for the Royal Navy. In reprisal for the theft of an Admiralty whaleboat, he took four hostages and demanded the return of his property.[3] Unable to find the boat after an extensive search, Fitzroy hatched a plan to bring his captives to England to transform them into a "civilized," churchgoing people. The four would be educated and christened, and then returned to Patagonia to establish an Anglican mission. Two of the men and the one girl were Alacaluf people; the adolescent boy was Yámana. It was customary to give natives whimsical Christian names that none too subtly diminished their status: El'leparu was named York Minster; the teenager, Orundellico, became Jemmy Button; the nine-year-old girl Yokcushlu became Fuegia Basket; and the second Alacaluf was called Boat Memory, and deemed the smartest of all, although his real name was never known because he died of smallpox soon after his arrival. Each was named after an object, as indeed, in almost every respect, the Fuegians were treated by their captors as objects, specimens of the Linnaean "productions of earth" to be examined by experts at home. In London, they became minor celebrities, as rumors spread of a dissolute race of naked cannibals who lived in canoes and ate raw fish, if not one another. After nine months of religious schooling, they were summoned to appear at the court of King William IV and Queen Adelaide, where Fuegia Basket was presented with a lace bonnet, a ring, and a small dowry. To fulfill his goal of bringing Christianity to Tierra del Fuego, Fitzroy set sail on December 27, 1831, on the 240-ton bark HMS *Beagle* with seventy-four crew members, an Anglican catechist who would establish a mission with his three new converts, and a recent graduate of Cambridge, the young naturalist Charles Darwin.

Nearly a year later, when the *Beagle* had finally anchored at Good Success Bay at the tip of Tierra del Fuego, Darwin had his first encounter with native people and was aghast.[4] Men daubed with white and red paint ran along the shore yelling and gesticulating, their locks flying as they leaped over rocks and trees. Signal fires were ignited far down the headlands, but whether they were intended as welcome or warning, he could not tell. To Darwin, it seemed to be the ragged edge of the world, a minatory country full of savagery.[5] Up until that time he had only seen the Fuegians clad in waistcoats, hoop skirts, or trousers, not seal blubber, and was barely able to accept that the people he met were members of the human race, a conclusion that played no small part in his later theories of evolution: "I could not be-lieve how wide was the difference between a savage and civilized man: it is greater than between a domesticated and wild animal, inasmuch as in man there is a greater power of improvement."[6]

Captain Fitzroy's drawings of the Fuegians had been widely reproduced in England after the first voyage and reinforced the unconscious bigotry of the time. Fitzroy believed in phrenology—the study of the shape and size of the head as a means to reveal character and personality traits—and con-sciously or unconsciously projected onto his subjects fearful visages that conveyed the lowest form of human development. His portraits featured protruding foreheads, overhanging brows, oversize flared nostrils, bulbous lips, slack jaws, black straw hair, and sullen eyes set narrowly or far apart, with dark, predatory pupils staring intently ahead.[7] By way of telltale con-trast, the photographs taken by the 1883 French Mission Scientifique du Cap Horn of the Yámana, the Kawéskar, the Selk'nam, and the Aónikenk reveal a people who are remarkably similar in physiognomy to the Menba of Tibet or the yurt-dwelling herders of the Ordos region of Mongolia.[8,9] The men have strong jaws, high cheekbones, and raven hair. While many of the women were photographed at missions, clad in Western clothes, looking worried and crestfallen at their fate, those images that capture women and children outdoors in native dress or undress show a healthy and stunning people, as beautiful as women or children anywhere.

Although Darwin could be jubilant about the quirky prance of the native guanaco (a close relative of the llama and vicuña), his interest in the local human inhabitants was more guarded and puzzled. His powers of ob-servation failed him with the Fuegians, as he was flummoxed by many ele-ments of their culture, including their use of language. On one hand he was surprised by their ability to perfectly mimic English speakers, repeating en-tire sentences, right down to their accompanying facial expressions, coughs, tics, and accents. But because he could not easily distinguish words in their own native language, he concluded that they were merely repeating a few simple phrases and thus had a very small vocabulary: "Their language does not deserve to be called articulate: Capt Cook says it is like a man clearing his throat; to which may be added another very hoarse man trying to shout & a third encouraging a horse with that peculiar noise which is made in one side of the mouth. Imagine these sounds & a few gutturals mingled with them, & there will be as near an approximation to their language as any European may expect to obtain . . . I believe if the world was searched, no lower grade of man could be found."[10] The word *barbarian* derives from the Greek *barbarus,* meaning someone who babbles in another tongue; people who didn't speak Greek were considered "barbaric." What Darwin and no one knew at that time was that these "barbaric" savages dwelled in the spoken word and spent nights in their pit houses in animated and nuanced conversations.[11]

It fell to Thomas Bridges, an orphan adopted by a missionary family, to reveal the true complexity of one of the Fuegian languages, Yámana.[12] Working primarily with George Okkoko and his wife, he spent twenty-one years compiling a dictionary containing 32,430 words and inflections, a number that was comparable to Japanese vocabulary, before accounting for Chinese and English influences. Because Bridges died in 1898 before the dictionary was completed, we are left to imagine the sum total of Yámana vocabulary. The dictionary can be read as a document, as local science, as anthropology, or as metaphor. It is as sparse in nouns as Tierra del Fuego is in resources, but richer than English in verbs. As you turn the pages of this remarkable document, you realize that there seems to be a precise word to describe every moment of their life. To appreciate that level of intelligence required to understand and use 32,430 words, consider that Samuel Johnson's *Dictionary of the English Language,* published in 1755, contained 42,773 words. Shakespeare is credited with having used 29,066 different text words in his complete works, but in terms of truly distinct words, and disregarding overlapping usages, there are fewer than 20,000.[13] A well-educated American may have a vocabulary of 20,000 words but will use no more than 1,500 to 2,000 over the course of a week.[14] Half of the conversational vocabulary of an American teenager consists of fewer than forty words.

Yámana is a language of finesse and subtlety. It has sixty-one words for kin, compared to twenty-five in English. *Guratuku* means to marry someone selfishly or with impure intent; *taisasia* is to be covered up on the ground like eggs in a nest; *porapola* was a freshwater seaweed but also referred to striped bears; *mamihlapinatapai* indicates that two people are looking at each other, hoping the other will do something they want but that neither wants to do; *ondagumakona* means to pick mussels off clusters one by one from a boat, and cook and eat them at the same time; the word for depression was a crab molting its shell; *yámana* itself means highest form of life, living, to be alive.[15]

Darwin's judgment that "in this extreme part of the world man exists in a lower state of improvement than in any other part" was true from the perspective of Victorian England. Of the hundreds of tribes and cultures in the Americas, the Yámana were one of the more primitive. They were Neolithic, had no written language, and could be petty thieves with little regard for private property or personal boundaries. Yet what do we make of a tribe that used metaphor to describe mental disease, skeletons, and the passage of time? Was the Yámana language an unusual isolate, or was it commensurate with other native tongues? What of other languages lost since the conquest? Because few people had the tenacity and patience of Thomas Bridges, we know little if anything about extinct languages, or the thoughts that permeated the lives of the 10 million Indians who died as Spanish slaves.

Europeans likewise questioned the intelligence of the Yámana, ridiculing them for such things as not wearing clothing. But the local people knew better than to wear fabrics in the climate of Tierra del Fuego. Europeans who got caught out in wet clothes often died of pneumonia; when Yámana were forced to wear wool sweaters by the missionaries, they perished by the hundreds. It was not the Yámana and their sister tribes who reduced the 500 million guanaco that populated the preconquest Americas to an endangered species and decimated the seal and sea lion rookeries for pelts.

Today the Yámana themselves have virtually vanished. There are only two remaining native speakers, and they do not talk to each other. The last speakers *are* deceased in the case of Abipon, Chané, Kunza, Canichana, Itene, Jorá, Shinabo, Acroá, Lule Toconoté, Araram, Kamakan, Karipuna, Oti, Nukuini, Paranawát, Otuke, Kakauhua, Anserma, Cagua, Chibcha, and hundreds of other languages in the Americas. None of them succumbed through its own actions. Just as Fitzroy felt justified in displaying and exploiting his Yámana abductees in the name of civilizing them, a five-hundred-year-long history of extirpation, genocide, and colonization has resulted in a never-ending process of cultural cleansing for the supposed benefit of the victim. That process is as fatal today as it was in preceding centuries.

This is not to argue that languages or cultures can or should be preserved as specimens, for they inevitably atrophy and disappear if they are unable to respond dynamically to changes in politics, economics, or environments. When there is a massive die-off of species or languages, however, it is no longer change, but extermination.[16] Areas of the world that are the most biologically diverse are also the most diverse in language, yet the rate of language decline is greater even than that of species loss. Since the conquest, half of the world's languages have disappeared. Of the remaining living languages currently in use, more than three thousand are dying.[17] There are today 362 critically endangered bird and animal species and 438 critically endangered languages with fewer than fifty speakers.[18] Extant in the Americas and throughout the world are approximately 6,800 languages attached to cultures that, just like every family and society, want to survive.[19] It may sound strange to impute intent to a language, but only if language is considered an inert object. The term "living language" may be more accurate than imagined; languages may literally be alive, their cellular words constantly combining into grammatical life forms.[20] A language dies when it is not spoken to a new generation of children. At the rate of decline we are now experiencing, half of our living cultural heritage will disappear in a single generation. As cultures disappear at the rate of thirty a year, we find ourselves placing our species' cultural eggs in fewer and fewer baskets.

Indigenous cultures keenly sense that they are judged by the terms of a Darwinian bias: namely, that there are higher and lower peoples, and that because one's culture is not European or modern, one's history is not as distinguished or valuable. Anthropologist Wade Davis, who has studied and lived with indigenous cultures throughout the world and particularly in the Americas, argues convincingly that a reductionist view of language, one that views it solely in terms of vocabulary and grammar, misses the point. Language is nothing less than the living expression of a culture, part of what he calls an ethnosphere, "the sum total of all the thoughts, dreams, ideals, myths, intuitions, and inspirations brought into being by the imagination since the dawn of consciousness."[21] In the Yámana language, verbs, nouns, and phrases contain mores, rules, and traditions, a kind of applied science that has guided people's relationship with the land and its resources. Cultural expression represents centuries of learning by trial and error. Language also represents "breathtakingly intricate beauty," aesthetic and intellectual wealth contained within the invisible folds of sound.[22] Despite languages' preciousness, one perishes on average every two weeks.[23]

Nevertheless, there is strong bias against a polyglot world. In various myths, language profusion is a curse, a punishment from God, implying that in a perfect world we would speak in one voice. The biblical Babylonians built a Tower of Babel to reach the heavens, and the penalty was the Confusion of Tongues. Pre-Aztec Cholulans in what is now Mexico began construction of the largest pyramid in the world in 200 BCE, but legend has God toppling it because it rose too close to the stars. The Toltecs believed mankind built a tower to the next world after the deluge to protect themselves from future floods, and were scattered for their insolence, never again to understand one another. The canon of lingual unity is carried forward in a phenomenon like Esperanto, which arose from the belief that there would be greater understanding and peace if everyone spoke and understood a lingua franca, but a visit to present-day America or the killing fields of Cambodia can easily dispel that theory.[24]

Our official view of indigenous cultures has tended to be patronizing, if not simply dismissive. British historian Hugh Trevor-Roper wrote in 1965 that the function of native cultures "is to show to the present an image of the past from which by history it has escaped."[25] In 1987 a widely used textbook, *American History: A Survey*, written by three academic historians, stated that the invasion of the Americas "is the story of the creation of a civilization where none existed."[26] Christopher Hitchens, writing in *The Nation* on the Columbian quincentenary, took a dated but opposite tack, writing that the "Arawaks who were done in by Columbus's sailors, the Inca, the Comanche and the rest were not the original but only the most recent inhabitants. . . . 1492 was a very good year."[27] Neurobiologist Kenan Malik's essay "Let Them Die" states that any nostalgia for what cannot survive in the modern world is reactionary, multicultural liberalism. He argues that a language spoken by a few has no function, that it is more of a conceit than a culture, closer to a secret code than a means of communication, and that people should switch to dominant languages to achieve a better way of life in the "modern mainstream to which the rest of us belong."[28] As insensitive as these statements are, they illuminate the prevailing attitude toward indigenous relations to this day. Eurocentrism fosters, to some extent even dictates, assimilation of these "marginal" cultures. Missionaries and others believe that if indigenous people could read, pray, and have jobs, they would be happier, a vision comparable to mowing mountain meadows to make them into lawns.

A Western bias about belonging to a superior culture is valid only if we use selective yardsticks. Rather than assuming that people want to surrender to Western values (and overlooking the impossibility of everyone's jumping onto the bandwagon of rampant materialism, given planetary constraints), we would be wiser to consider the loss of language as yet another indicator of the worldwide collapse of ecosystems, as well as a product of the cultural hegemony that represses or punishes those who continue to use local language.[29] Ways of life are not abandoned so much as they are made impossible or are obliterated.

Indigenous people *do* want the conveniences of modern life—electricity, antibiotics, the Internet—and while speaking other languages provides that access, they do not wish it at the expense of losing their birthright. The Ojibwe drive pickup trucks, but they are still Ojibwe. People's land, songs, rituals, celebrations, and language are intricately interwoven, inseparable from their cultural identity. Indigenous people rarely give up their native language without external pressure or force;[30] Malik should ask Tibetans if they truly want to learn Mandarin. Wade Davis sees languages the way a biologist sees species diversity: "Distinct cultures represent unique visions of life itself, morally inspired and inherently right. And those different voices become part of the overall repertoire of humanity for coping with challenges confronting us in the future. As we drift toward a blandly amorphous, generic world, as cultures disappear and life becomes more uniform, we as a people and a species, and Earth itself, will be deeply impoverished."[31] If languages were fungible objects, their loss would have little meaning. But if languages are living things, inextricably intertwined with biological diversity, the loss of "verbal botanies" is irreparable, and a monoglot globe is as unthinkable as a world with only one species of tree, flower, and bird.[32]

The Fuegians could not appreciate the genius of Magalhães's feat in reaching their archipelago; the great explorer and his men could not understand the genius of the Fuegians in being there at all. Both sides can be excused for their bias, but can we make the same allowances for Trevor-Roper, Hitchens, the three historians, and many others, who still posit a "discovery" of the Americas by Columbus? Definitive on this subject is Charles Mann's recent book *1491*, which upends the previous conceptions of pre-Columbian history. Virtually every myth of America's past, whether Arcadian or primitive, is dislodged in light of recent archeological discoveries. Columbus died thinking he had found the western route to India; that is, he didn't really know where he was on any of his four voyages. The word *Indian* is our ongoing testament to his ignorance, and Native Americans counter that you cannot "discover" an inhabited land. The Ptolemaic maps Columbus used presumed there was only one landmass on earth, and that the Atlantic connected to Asia because on the third day of Genesis, God separated the water and the land, with the water being one-seventh, leaving no room for another ocean. What Columbus stumbled upon was an unknown continent that was in fact more populous than Europe, with bigger cities, more advanced medicine, superior agriculture, and healthier people—none of which he realized.

Even for those interested in the truth, it has taken centuries to fully appreciate the depth and breadth of New World cultures, because they imploded rapidly after the conquest. It is estimated that of the 90 million to 112 million people living on the American continents at the time of the conquest,[33] 98 percent of the native populations, from Patagonia to the uppermost reaches of the Yukon, died of disease, violence, and heartbreak within two hundred years.[34] The conquered Mexica and Inka were depopulated and traumatized by smallpox, typhus, measles, influenza, and diphtheria before they ever saw a soldier, victims of a pandemic greater than the Black Death of the fourteenth century. Hernán Cortés and Francisco Pizarro had most of their work done for them by disease vectors. Adding agony to tragedy, Columbus and his countrymen engaged in what we now call crimes against humanity: genocide, rape, murder, torture, and pillage, not only because they despised the cultures they encountered, but because native people were irrelevant to European culture, except as slaves and as a means to the end of finding gold.

One grasps for plausible descriptions of the conquerors' state of mind. On a single day, in front of the priest Bartolomé de las Casas, the Spanish raped and beheaded three thousand people.[35] "They cut off the legs of children who ran from them. They poured people full of boiling soap. They made bets as to who, with one sweep of his sword, could cut a person in half. They loosed dogs that 'devoured an Indian like a hog, at first sight, in less than a moment.' They used nursing infants for dog food."[36] It was a grievous meeting of cultures that tests the definition of what it means to be cultured, leaving both American continents a charnel house of misery, a five-hundred-year-old wound that has not healed.

In 1492, the Inka Empire extended 2,200 miles down the coast of South America and was the largest on earth, greater than any country in Europe, greater than the Ottoman Empire, the Ming Dynasty, or the lands of Ivan the Great in Russia.[37] The capital of the Mexica (Aztec) empire, Tenochtitlán, was larger than the biggest city in Europe; its markets, causeways, broad avenues, botanical gardens, temples, and immaculate sanitation system astounded the Spaniards. In North America, the Haudenosaunee, comprising today six nations—the Seneca, Cayuga, Onondaga, Oneida, Mohawk, and Tuscarora—had a representative parliament, complete with citizen referendums, that dated back to 1142. This governing structure continues to this day, internally administered by female clan heads.[38]

Before these came Mayan, Hopewell, and Olmec cultures. Meso-Americans created five thousand varieties of corn, beginning 7,500 to 12,000 years ago.[39] There are several theories as to the origin of corn, but scientists remain unsure how the wild cultivar was domesticated to become a staple. The Inkas had developed three thousand different types of potatoes suitable for every conceivable environment by the time of the Spanish invasion, which represented a model of agricultural sustainability, as the staple crop could defend itself against any number of blights and variations in soil and weather.[40] Today, corn is the largest grain crop by weight grown in the world, and three root vegetables developed in the Americas—potatoes, sweet potatoes, and cassava—are collectively the largest source of global calories. When you take into account an agriculture that also included coffee, cacao, tomatoes, avocado, peppers, cayenne, chilies, peanuts, cashews, tobacco, sunflower, safflower, vanilla, pineapple, papaya, blueberries, strawberries, passion fruit, pecans, butternut squash, pumpkin, zucchini, maple syrup, cranberry, tapioca (from cassava), and a whole assortment of beans, it is not difficult to concede that Amerindian farmers were the leading plant breeders in history. Europeans, who had gone chronically hungry for centuries, came to an edible landscape farmed by people who by and large were well fed.[41]

Agriculture *is* culture, and the Americas have been cultured—one might say gardened—for a long time. The romanticized notion of a pristine environment, the idea that white men discovered a virgin continent, was a fanciful one. Beginning in the Pleistocene, humans have altered the land to the benefit of themselves and other species. If you walk into a primary Amazonian forest with an ethnobotanist, you will find a landscape that has been transformed over thousands of years by the intervention of the native population. Forests were converted into silvicultural gardens (what is called agroforestry today), a dynamic and enduring relationship that supplied a year-round crop of medicines, fibers, fruits, and animals. If we could walk the tall-grass Buffalo Commons before the mass slaughter of ruminants and ungulates, we would find ourselves head high in grasses in one of the most fertile savannas in the world, sustained and kept productive by fire ecology. Evidence indicates that native people, from Alaska to Santa Barbara, as if responding to a continent-wide signal given four thousand years ago, pushed the pinniped population from onshore to offshore rookeries. This increased the population of seals and sea lions by eliminating predation from wolves and bears and, by making hunting in baidarkas and canoes more difficult, lowered mortality further. There *were* civilizational failures, such as the Anasazi, which were primarily drought and weather related, but on the whole, the pattern of interaction with the environment increased the productivity and biodiversity of natural systems without causing concomitant losses.

Technologically, Western culture dances to preeminence with an iPod plugged into both ears. When it comes to innovation, literature, and creativity, it is dazzling. The ability to go deep into the ocean or as far as the moon is spectacular, but as Robert Oppenheimer reminded us, being blessed with technological insight does not confer self-insight. If we measure Western culture by how it has treated people of different ethnicity and race, it is anathema. If we judge it by the treatment of its own people, including children, the elderly, and the poor, it is an embarrassment. And if we try to calibrate American superiority by its treatment of the environment, the United States is one of the least intelligent civilizations in the history of the planet. English, when scientific and technical terms are taken into account, has more than three million words. "Brilliant" is hardly sufficient to describe such an achievement. But how do you describe an American administration that will spend $1 trillion on winning a war for Iraqi oil while refusing to allocate any funds to reduce dependency on oil? For $1 trillion, the United States could have catalyzed the replacement of its entire automobile fleet with plug-in hybrid-electrics getting 500 mpg (cars running on batteries 90 percent of the time), powered by renewable energy and biodiesel. To invert Trevor-Roper's logic, indigenous cultures may show us an image of a future by which we can escape our present. If a culture does not become like us, it may not be a failure but a gift to what is now an uncertain future.

The difference between human evolution and that of wild animals is that culture is a critical and determining part of our environment. If overwhelming the ecological integrity of watersheds, forests, or seas is biological suicide, it is not a big step to imagine that overwhelming indigenous cultures is equally shortsighted. Ecosystems are created by the interaction of living organisms; each species within the system depends on the mutualism of those relationships to ensure its long-term survival. As we eliminate one culture every fortnight, we therefore court two types of extinction—cultural and biological. The blessing of the existence of many languages is that they interact, change, and grow. What would English be without mulligatawny, gestalt, and déjà vu, or tens of thousands of other imported words? Species colonize and migrate; so, too, do memes and words. Languages spawn argot, vernacular, and dialects, producing rich variants in a never-ending process. There is no stable ecosystem or language, but only systems growing and changing in a dynamic process of evolution. "Everything forgets. But not a language," wrote literary critic George Steiner.[42] Almost all indigenous people recorded life as a linguistic continuum, because their survival depended on such an intimate relationship to the land and one another. Their bodies were not something that merely carried their brain around, but an entire sensate organ wedded to their habitat and tribe. Their contribution to the world is not their lifestyle but experiential knowledge diligently gleaned from generations of interaction with the natural world.

Living within the biological constraints of the earth may be the most civilized activity a person can pursue, because it enables our successors to do the same. You cannot live within the carrying capacity of a region if, like Columbus, you don't know where you are. Most of the developed world lacks this knowledge. We have little understanding of where our water and food come from, the impacts of our cars and homes, the activities undertaken by others around the globe to support our lifestyle, and the effects we have on the environment and its people. John Maynard Keynes cautioned that we live our lives under the illusion of freedom but are likely to be "slaves to some defunct economist." Even that description understates the problem. The world may be caged by a defect of the entire economic profession—namely, the idea that we can assess value in banknotes, or that we can understand our relationship to the material world using an abstract metric rather than a biological one. The extraordinary advances made by Western societies will, in the end, be subservient to the land and what it can provide and teach. There are no economies of scale; there is only nature's economy. We cannot turn back the clock, or return to any prior state on the planet, but we will never know ourselves until we know where we are on this land. There is no reason that we cannot build an exquisitely designed economy that matches biology in its diversity, and integrates complexity rather than extinguishing it. In accomplishing this, there is much to be gained from those who have not forgotten the land.

The Mi'kmaq people know the world through sound. One of their tribal legends tells of three brothers, each of whom was granted a wish by a great magician named Glooskap. The first wanted to be tall so all the women could see him. The second wanted to be secure in the beauty of the forest and never work again. The third wanted to never fall ill and to live a long life. Glooskap called in Cuhkw the Earthquake, who shook the land, seized the three, and planted them in the forest. The first became the tallest pine tree and could be seen for miles around; the second became the pine tree with the deepest roots and withstood the wind or any storm; and the third became the oldest pine tree and lived for centuries. To this day, the Mi'kmaq name their pine trees by the sound of the wind soughing through the branches one hour before sunset in the fall. Elders can remember the prior names of native stands of pine, and detect how trees are changing due to environmental damage from acid rain and air pollution by comparing their names with the current sounds they make.[43]

In 1981 marine biologist R. E. Johannes published *Words of the Lagoon*, a seminal book detailing his work with Ngiraklang, second chief in the village of Ngeremlengui in Palau. When Johannes first went to Palau to learn about reef fisheries from indigenous people, his mates in the Australian scientific community joked that he had gone "tropo," a term reserved for someone who has spent too much time in the tropical sun. Ngiraklang, who was born in 1894, and educated with only a half year of schooling in carpentry, turned out to be a phenomenal scientist and marine biologist. For fisherfolk the cycles of the moon chart a map of the ocean. They will tell the height, strength, and timing of tides, and thus accessibility to lagoons and shallow inlets within reefs. They will also reveal the location of a school of fish, what kind of behavior they exhibit, and how susceptible they are to being netted or otherwise caught. Palauan fishermen know the best times to fish, and they always align with the moon. The night before the new moon on Namoluk Atoll is called Otolol, meaning "to swarm." Fish spawning during lunar cycles form large groups and are exceptionally easy to catch. Ngiraklang could not only identify more than three hundred different species of reef fish, but knew the lunar spawning cycles of more species of fish than were described in all of the marine biology literature in the world.[44] Villagers commented that Johannes was the first scientist who had inquired after their knowledge instead of telling them of his own.

Indigenous science is an observational science, recorded in myths, stories, teachings, and, in particular, language. The tenacity of indigenous cultures and their dedication to place is stunning. My favorite Wade Davis story concerns an Inuit elder. When Canada tried to force the Inuit into encampments on Baffin Island in the 1950s, one Inuit grandfather would have nothing to do with it. "The family took away all of his weapons and all his tools, hoping that would force him into the settlement. Did it? No. In the middle of an Arctic night with a blizzard blowing, the old man slipped out of the igloo into the darkness and simply pulled down his caribou hide and sealskin trousers and defecated into his hand. As the feces began to freeze he shaped it in the form of a blade. As the shit took shape he put a spray of saliva along one leading edge to create a sharp edge. When the implement was finally created from the cold, he used it to kill a dog. He skinned the dog and used the skin of the dog to improvise a harness and used the rib cage of the dead dog to improvise the sled, harnessed an adjacent dog, and then with shit-knife in belt disappeared over the ice flow."[45] The elder returned alive and well in the spring. Forty years later, the Canadian government relented and returned the Inuit lands, creating Nunavut, a territory the size of western Europe.

Extinction of species and cultures is driven by globalization, the pursuit of progress through resource extraction and economic expansion—the dream of the North, if you will, the dream of Columbus. Native people have remarked that, of the many promises made by white men, the only one they kept was the vow to take their land.[46] But invaders haven't succeeded in taking it all, and today approximately five thousand indigenous cultures are seeking to protect their homelands, which constitute one-fifth of the land surfaces on earth.[47] In many cases these are the least corrupted forests, mountains, and grasslands remaining on earth, holdouts to the ongoing fire sale of its resources.[48] The forces arrayed against them are political, economic, and military, abetted by arguments about progress and modernity. Indigenous nations, and indeed the whole of the movement without a name, are political actors of a completely different nature. They do not have fighter planes, armies, currency systems, or U.N. status. They hold fast to something that is of inestimable value: ancestral lands preserved over millennia by stories and culture. These lands are biological arks, a kindness to a spent world because most indigenous people, despite temptations and coercion and in the face of centuries of racism and genocide, refuse to be prodigal with the gifts that have been bestowed on them. For extractive industries, these biological arks represent sources of natural gas and oil, nuclear waste disposal sites, and deposits of coal, timber, water, and minerals. The value of these resources will only rise as demand increases, bringing geologists and lawyers to the last untouched areas of the earth. Most indigenous people have withstood the pressure and enticements to sell out and join the mainstream because they know the land is not theirs, and they have seen what happens when corporations "own" nature. A worldwide cultural siege is being undertaken by a global economy hooked on growth, and resistance to it represents the heart of the movement this book addresses.

For the developed world, there is a choice to be made: to promote economic policies that despoil indigenous lands, or to support cultures and the remaining biological sanctuaries. Native people do not have the luxury of such choices, because their survival is at stake. Those who support free-market fundamentalism have been able to make critics of globalization appear to be anti-progress, even to the extent of portraying themselves as allies of the poor. Indigenous land rights issues bring globalization into sharper focus and break it down into discrete, more comprehensible components. But in the case of indigenous sovereignty, globalization is a mismatch—complex trade rules, corporate interests, and international agreements are arrayed against traditional values in sparsely populated lands.

Native cultures continually have to counter nationalistic claims on their sovereignty. The Gwich'in people in Alaska confront ongoing threats to drill in the Arctic National Wildlife Refuge. In northern Alberta the Cree, Athabasca, Chipewyan, and Dene people face proposals to build the world's largest nuclear reactor to power expanded oil extraction from the Athabasca Oil Sands, the largest reserve in the world.[49] In Ecuador the World Bank and several international oil companies are helping finance a pipeline that cuts through the heart of the Mindo Nabillo Cloudforest Reserve, which in turn will precipitate oil drilling in the territories of the Secoya, Siona, and Cofan people. Already, because of exploration by Conoco and drilling by Chevron, the developed areas are marked by contaminated soil, toxic waste pits and rivers, air pollution, illegal logging, disease, crime, and prostitution. The combined population of the Gwich'in, Athabasca, Chipewyan, Secoya, Siona, Cofan, and Cree people numbers approximately 200,000, yet their traditional lands extend over millions of acres. In all these cases, indigenous populations are dealing with one or more national or subnational governments, but the larger force bearing upon their lands is globalization. The major investors in the three examples cited above are overseas or transnational corporations from Europe, India, China, and the United States, because all desperately need energy to support an unsustainable way of life.

A small minority of the world's population, measured in millions, has begun to stand in front of corporate offices, shareholder meetings, bulldozers, drill rigs, excavators, and waste pits, to say, in effect, *Stop, no further.* Ironically, the very force that works against indigenous people—globalization—can also work for them. Corporate kleptocracies can no longer conduct business in isolation in remote parts of the world; with sensing devices, satellites, video cameras, and the Internet, it has become more difficult to operate in secret. Indigenous organizations have been able to partner with conservation, fair trade, and human rights organizations in the North to create robust partnerships that have greatly amplified their influence and drawn media attention to their cause, leveraging the legitimacy of their resistance onto the world stage. Because of these alliances and new communication technologies, a remarkable rebirth is taking place among native peoples, and is arising largely unnoticed. They are organizing, networking, collaborating, and learning how to navigate the political highways of the global age.

The Dayak peoples of Borneo, which include Ibans, Kenyahs, Kelabits, Penans, and two hundred other riverine or hill-dwelling ethnic groups, face siltation, erosion, and destruction of their homeland from extensive clear-cutting of primary forests and water and crop pollution from oil companies. The Ijaw and Ogoni people of Nigeria have seen the rich Niger River delta devastated by oil pipeline ruptures, air and water pollution, toxic wastes in their rivers and fisheries, and fires from the accidents and flaring of gas. The Kogi of Colombia face extermination due to aerial spraying by U.S. planes of herbicide cocktails (Agent Green) designed to prevent the cultivation of coca, which the Kogi do not grow as a cash crop. The Wapishana of Guyana are challenging patents on their native foodstuffs by multinational corporations. The Garifuna of Honduras are protesting the construction of resorts and developments on expropriated land. The San people in Botswana have been banished from their ancestral lands of twenty thousand years, the Kalahari Gemsbok National Park and the Central Kalahari Game Reserve, in favor of tourism and diamond mining concessions, and now face extinction in resettlement camps. The Lenca of Honduras are fighting an IMF-promoted law that allows for removal of indigenous people who live adjacent to mines and mineral deposits. The Arhuaco of Colombia are subjected to guerrilla warfare between paramilitaries, drug traffickers, and the Colombian army within their lands. The Adnyamathanha Aborigines in the Flinders Range of Australia confront radiation spills from uranium mining.[50] In Chile and Argentina, Barrick Gold proposes an open-pit mine at Pascua Lama that will destroy portions of three different glaciers, literally blowing them up to uncover what will become a large open-pit gold mine at the headwaters of three rivers. The Maasai are being displaced by large-scale, export-oriented agriculture in Tanzania. The Bagyeli, Sara, Mass, Mundani, and Hakka people in Chad face oil wells, pipelines, and destruction of their way of life from petroleum exploration. Enormous oil fields were recently discovered under the tribal land of the Amazigh (Berber) people in Morocco, territories appropriated during French colonization. The Amazigh will receive nothing from Lonestar Oil Company, which is developing the concession under a contract from the national government. As an Amazigh shepherd remarked, "One day, they arrived on my land with diggers, and planted their flags. And said the land was now theirs."[51] The Anuak face upheaval and violence as the Ethiopian government makes way for oil exploration in their territories. The Embera-Katió of Colombia have been displaced by the Urrá Dam, which flooded their ancestral home. The Santa Isabel Dam in Brazil, built to supply energy for aluminum companies, will flood and destroy the traditional way of life of the riverine Surui-Aiwekar people, whose livelihood depends on fish and babassu nut gathering. The Sammi in Norway are threatened by military and hydroelectric projects. The list goes on and on.

The list of companies and agencies that legally or illegally impose their will on indigenous cultures is formidable.

OIL AND GAS
  Chevron
  Total S.A.
  ExxonMobil
  Petronas
  Talisman Energy
  Lundin Oil
  OMV
  PetroChina
  Royal Dutch Shell
  Nigerian National Petroleum
  AGIP
  Pinewood Resources
  Gambela Petroleum
  Lonestar Energy
  Kerr-McGee
  YPF Oil

BHP
Petrobras
Occidental Petroleum
Repsol
Harken Energy
Burlington Resources
Alberta Energy
Pemex
PETRAD
Cairn
Unocal
Hunt Oil
Yukos Oil
British Petroleum
Consorcio Norandino
Tractebel
Southern Energy
El Paso Energy
Halliburton
Schlumberger

TIMBER
Thanry
Bollore
Coron
Alpi
WTK
SESAM
Danzer
Stabach
La Forestière
World Bank
Shimmer
Shelman
Oriental Timber
Seabord
Moconá Forestry
Newman Lumber

MINING
  American Mineral Fields
  Eagles Wings Resources
  Tiomin Resources
  Rio Tinto
  Bechtel
  BHP
  Western Keltic Mines
  Inmet Mining
  Placer Dome
  Netetsu-Nippon
  Alcan
  INCO
  Newmont Mining
  Echo Bay Mines
  Benguet Mining
  Asarco
  Peabody Coal

GOLD
  Goldfields Ghana
  Barrick Gold
  Noranda
  Anmereosa Exploration Ghana
  Teberebie Goldfields
  Homestake Mining
  Omai Gold Mines
  Golden Star
  Mars Geosciences
  Nucon Resource
  Freeport McMoRan

DAMS AND WATERWAYS
  World Bank
  IFC
  OPIC
  Fortis
  Belize Electric

Alcoa
BHP Billiton
ENDESA
Cargill
ADM
Midland
Skanska
Vivendi
Norsk Hydro
Manitoba Hydro

PAPER
Pan African Paper Mills
Raiply Timber
Aracruz Celulose

COCA ERADICATION
U.S. Drug Enforcement Administration
Monsanto
DuPont
DynCorp

Contrast this list of corporations and agencies to the network of organizations promoting environmental and social justice for indigenous people in just one region, the Amazon:

Acción Ecológica
Acción por la Vida
Action and Communication Network for International
    Development
Amazon Alliance for Indigenous and Traditional Peoples of the
    Amazon Basin
Bank Information Center
Bolivian Forum on Environment and Development
Center of Documentation and Information Bolivia
Center for International Environmental Law (CIEL)
Center for International Policy—Demilitarization Program and
    Columbia Project

Center for Public Integrity
Centro de Estudios Jurídicos e Investigación Social
Centro de Medios Independientes de Ecuador
Coalizão Rios Vivos (a coalition of more than three hundred
   NGOs and indigenous communities)
Colectivo de Estudios Aplicados al Desarrollo Social
Comisión Ecuménica de Derechos Humanos
Confederación de Pueblos Indígenas de Bolivia (CIDOB)
Confederation of Indigenous Nationalities of Ecuador
   (CONAIE)
Coordinadora de Organizaciones Indígenas de Cuenca Amazonica
   (COICA)
Coordinadora Étnica de Santa Cruz (CESC)
Coordination of the Indigenous Organizations of the Brazilian
   Amazon (COAIB)
EarthRights International
EarthWays Foundation
El Frente de Defensa de la Amazonía
European Working Group on Amazonia
Federación Interprovincial de Nacionalidad Achuar del Ecuador
   (FINAE)
Forest Peoples Programme
Forests and European Union Resource Network (FERN)
Grupo de Trabalho Amazonica
Independent Federation of Shuar People of Ecuador (FIPSE)
Oil Watch
Organización de Nacionalidad Shiwiar de Pastaza de La Amazonia
   Ecuatoriana (ONSHIPAE)
Organización de la Nacionalidad Zapara del Ecuador (ONZAE)
Organization of Chiquatanos Indigenous Peoples
Pachamama Alliance
Pastoral Land Commission
Productividad Biosfera Medio Ambiente Bolivia
Rainforest Action Network
The Rainforest Information Centre
Resource Center of the Americas
Rettet den Regenwald (Germany)
Survival International

All of the NGOs listed above are assisted by and networked with San Francisco–based Amazon Watch. As noted in the next chapter, Amazon Watch is a type of organization that exists around the world—a *watch* group which monitors, communicates, litigates, challenges, and publicizes a specific place, species, company, or issue. When I first met Amazon Watch founder Atossa Soltani at a conference in Canada in the early 1990s, she was leaving her job at a nonprofit to start the new watchdog group, with a staff of one. Now, with Leila Salazar, Soltani works with local people in the Amazon and a network in the United States and Europe, making Amazon Watch a formidable force for justice on ancestral indigenous lands. Their first action was against Texaco in Ecuador, a campaign that continues today against Chevron, which merged with Texaco in 2004. Texaco came to Ecuador in 1964, having obtained a million-acre concession in the Amazon basin, an area known as el Oriente, near the Amazon's headwaters. There it began oil exploration in the native lands of the Cofan and other indigenous people. At first, straight roads were bulldozed out into the rain forest, and explosive charges were detonated every two hundred meters to determine the location and size of the oil deposits. In 1967 Texaco discovered a rich field and began bringing in equipment and workers by boat, helicopter, and transport planes. That discovery led Texaco to drill numerous exploratory wells, which resulted in the discharge of approximately one million gallons of tarry mud and formation water for each. When an operating well came into production, its slurry of oil, mud, and formation water was sent to a processing facility to separate the oil. Each processing facility generated 4.3 million gallons a day of toxic water, which was placed into unlined waste pits. Leaks, overflows from rain, and other discharges have spread the toxic fluids throughout the entire area of production. It is estimated that Texaco ultimately pumped 18.5 billion gallons of formation waters, as well as crude oil, into the region's waterways and into more than six hundred open and unlined waste pits.

Formation waters, which lie atop oil deposits, contain hydrocarbons, heavy metals, and salts. Oil well discharges contain polycyclic aromatic hydrocarbons (PAHs), a group of more than one hundred chemicals that includes benzene, benzopyrene, napthalene, pyrene, anthracene, and many others. Napthalene is used to make mothballs; anthracene goes into dyes, insecticides, and wood preservatives. Eleven of the chemicals found in PAHs are known to be carcinogenic, mutagenic, and teratogenic.

The Chevron operations burn more than fifty million cubic feet of gas a day without any controls. Thirty major oil spills have also discharged 16.8 million gallons of oil into the jungle, and a single 1989 spill fouled the Napo River with 294,000 gallons. It is estimated that a hundred major spills occur every year. The Cofan have been breathing, drinking, and eating these compounds for more than three decades.[52] They and adjacent communities suffer from extraordinarily high rates of cancer, including liver, larynx, leukemia, lymphoma, cervical, bile duct, and stomach. People in the region report skin lesions, breathing problems, and malnutrition. Their cattle die from polluted water, and rivers once rich in fish are now barren.[53]

During the time of the oil boom, Ecuador became a significant debtor nation, ending up owing more than it had gained from its oil riches. In response, the IMF has encouraged further investment to open up an additional two million hectares of virgin rain forest to exploration. In 1993 some thirty thousand members of indigenous tribes and local campesino communities brought a class action lawsuit against Texaco in the United States. The lawsuit was dismissed seven years later and remanded to the Ecuadorian courts, with the judge lecturing the plaintiffs that U.S. courts do not have "a general writ to right the world's wrongs."[54] In 2003 a lawsuit was filed in Ecuador claiming $6 billion in damages from the company. Amazon Watch is challenging Chevron to reveal to its shareholders and the SEC that it faces significant liabilities resulting from its contamination of an area the size of Rhode Island with a level of pollution so toxic that the land would qualify as uninhabitable under U.S. law. In letters to Chevron counsel, Amazon Watch has warned that its nondisclosure is similar to Enron's actions prior to its collapse. In the case of Chevron, a $6 billion loss would be material but not damaging; it represents sixty days' worth of operating income. To prevent the dumping of formation water and to conduct its operations safely, Chevron would have had to spend an additional $1.5 billion, or 5 percent of its total revenue from Ecuadorian oil, two weeks' income today.

If you drive up to the guardhouse at Chevron headquarters on Bollinger Canyon Road in San Ramon, California, you see an imperial white office building set a half-mile back from the road. At lunch, workers swarm across the road to Whole Foods to buy vegetable juice, organic salads, and freshly made soups. A small part of their headquarters is powered by a hydrogen fuel cell, and the company has expressed a commitment to conserve energy there. Pictures of African, Hispanic, and Asian Americans are featured every-where in Chevron's reports and literature, emphasizing its commitment to diversity. Its Web site includes a tab for social responsibility, where you can download the latest copy of the company's Corporate Responsibility Report. The 2005 edition discloses information on how it reports carbon emissions from refineries; it contains sections on business ethics, stakeholder involvement, climate change, community engagement, and, in a five-line paragraph on stockholder proposals, a single sentence referring to the situation in Ecuador: Chevron will "vigorously defend against litigation alleging environmental damage." In a peer-reviewed paper published in the *International Journal of Occupational Environmental Health* in 2004, the authors report significantly elevated levels of childhood leukemia between the ages of birth and four years in the oil-producing areas of the Amazon basin of Ecuador.[55] (Benzene is a well-known cause of leukemia.) What Chevron did not disclose in its Corporate Responsibility Report is that it has lobbied heavily to exclude Ecuador from the Andean Free Trade Agreement negotiations unless Ecuador moves to stop all legal proceedings against it. Ironically, Chevron's lawyers persuaded the U.S. federal judge to have the case tried in Ecuador, with the understanding that it would honor the jurisdiction and the eventual outcome.

The juggernaut of growing corporate power cannot last forever. In Bolivia, an Aymara Indian, President Juan Evo Morales Ayma, a coca leaf farmer without a high school diploma, made good on his campaign promises and took charge of all gas fields in May 2006, declaring that any country had the sovereign right to control its own natural resources. Oil and gas companies had six months either to leave the country or to sign new contracts more favorable to the Bolivian people. The proposed renegotiation with foreign oil companies comes twenty-five years after the World Bank and IMF forced Bolivia to privatize its gas fields. At that time natural resource revenues to the government provided 60 percent of Bolivia's national income. By the time of Morales's election, multinational corporations that controlled Bolivia's natural resources were contributing 12 percent of the national income, a critical drop of revenue for the eighth-poorest country in the world. During his political campaign, Morales and his running mate specifically referred to the 513-year-long period of colonization that preceded the election. In subsequent speeches, Morales often acknowledged the power of social organizations as the basis for re-creating Bolivia, "starting from scratch."

Two weeks after Morales's declaration in Bolivia, Ecuadorian energy minister Ivan Rodriguez canceled Occidental Petroleum's Block 15 oil concession because of numerous violations of its contract. The 494,000-acre holding in the northeast corner of Ecuador produces 20 percent of the country's oil, and was taken over by the national oil company, Petro-Ecuador. The move followed years of intense pressure by the Confederation of Indigenous Nationalities to remove Occidental because of its long history of abuse and harm to indigenous populations.[56]

The assumption that older and smaller cultures should either draw near to or get out of the way of modern ones, that homogeneity is a desirable end state, is the unwritten code of globalization. But globalization may also be having the unintended consequences of fostering diversity, as well. In New Zealand, the amounts of Maori-controlled lands and income are increasing rapidly. The Chiapas rebellion, although headed by nonindigenous leader Subcomandante Marcos, is a Mayan movement to reclaim autonomy. As Moises Naim, editor in chief of *Foreign Policy*, writes, "Global and local activism have transformed intolerance for human rights violations, for ecological abuses, and for discrimination of any kind into increasingly universal standards among governments, multilateral bodies, NGOs, and the international media. . . . Environmentalists and indigenous populations are thus obvious political allies. Environmentalists bring resources, the experience to organize political campaigns, and the ability to mobilize the support of governments and the media in rich countries. Indigenous groups bring their claims to lands on which they and their ancestors have always lived."[57]

Whether or not the earth itself is an organism, as some scientists believe, it is one system, and within that system all life as we know it coexists. If the erosion of indigenous ways of life represented only a cultural clash, it would qualify as a human rights issue, certainly, but one without much legal impetus, because nowhere is there a law that a culture has a right to survive. But it is now recognized as a salient issue for the whole of the globe, and it will only grow stronger as corporations face people who believe that the life of the earth, the mother of all nourishment in Mayan culture, is even more valuable than their individual lives. As resource values rise and living systems decline, the conflict will grow more pronounced, unless we change our understanding of what we value. If we want greater amounts of gas and gold, they will come at a severe cost to ourselves and all who follow.

In simple political terms, the aggregation of minorities on the planet is far greater than the aggregation of supposed majorities; they are what Jonathan Schell refers to as the unconquerable world. Darwin's encounters with the Fuegians were followed by journal entries of distaste and dismissal. Today, it is the indigenous people whose blogs and political action express distance from and disapproval of Western customs. Jesus was very clear about who will inherit the earth; so, too, are most indigenous people.

WE INTERRUPT THIS EMPIRE

The modern conservative . . . is engaged in one of man's oldest exercises in moral philosophy. That is the search for a superior moral justification for selfishness.

—John Kenneth Galbraith[1]

The Earth is not dying—it is being killed. And the people who are killing it have names and addresses.

—U. Utah Phillips[2]

November 30, 1999, was the longest day in Seattle's history, a day when hundreds of small citizen organizations came together to bell the cat of runaway undemocratic policies established by the World Trade Organization. More than seven hundred groups, and between forty thousand and sixty thousand individuals, took part in protests against WTO's Third Ministerial in Seattle, constituting one of the most disruptive demonstrations in modern history and, at that time, the most prominent expression of a global citizens' movement resisting what protesters saw as a corporate-driven trade agreement. The demonstrators and activists who took part were not against trade per se. They wanted proof, rather, that trade—at least as WTO envisions it—benefits the poor, the workers, and the environment in developing nations, as well as at home. That proof had yet to be offered, because it could not be offered. Because it does not exist, protesters came to Seattle to hold WTO accountable. Their frustration arose because one side held most of the cards; that side comprised heads of corporations, trade associations, government ministries, most media, stockholders, and WTO.

From the point of view of those on the streets, WTO was trying to put the finishing touches on a financial autobahn that would transfer income to a small portion of the population in wealthy nations under the guise of trade liberalization. The assumptions that undergird market fundamentalism are so pervasive that they have become conflated with fact. IMF, World Bank, and WTO are populated by many macroeconomists who believe that there is no such thing as involuntary unemployment, because in their economic models *demand always equals supply.* For true believers, markets are exquisitely calibrated mechanisms that always work perfectly; thus economic aberrations such as unemployment, poverty, or malnutrition must be caused by external factors. Because markets theoretically balance demand and supply, imbalances are caused by regulations or restrictions. According to this logic, it is unions and high wages that cause unemployment, while poverty is the result of high taxes imposed on people who aren't poor.[3] In this upside-down world, idealism harms society and greed benefits the needy. Those who question the inevitability of supranational corporations to supply most of our material and employment needs are seen as out of step, if not nostalgic. But even the free market's most articulate defender, *New York Times* columnist Thomas Friedman, knows better: "The hidden hand of the market will never work without a hidden fist. McDonald's cannot flourish without McDonnell Douglas. And the hidden fist that keeps the world safe for Silicon Valley's technologies to flourish is called the U.S. Army, Air Force, Navy and Marine Corps."[4]

Of course, globalization does have potentially positive effects. They include dissolution of exclusionary political borders, increased transparency of political actors, connectivity among people around the world, and in general a wealth of new opportunities in employment, education, and income. But these benefits obscure the liabilities: resource and worker exploitation, climate change, pollution, destruction of communities, and diminished biological diversity. What is lost in the concept of a globalized market system is economic resiliency, the ability of regional economies to withstand bust and boom cycles. Also forsaken is economic security. When communities depend almost entirely on sources of production thousands of miles if not continents away, they become spectral towns lined with fast-food outlets and big-box retailers.

An inordinate focus on wealth creation also obscures poverty creation. No country advocates trade liberalization more ardently than the United States as a means to improve any country's social welfare, a policy known as the Washington Consensus. The irony of America's overheated emphasis on free-market ideology is how miserably it has failed its most ardent proponent. The United States has the worst social record of any developed country in the world, and it is worse than that of many developing countries. By almost any measure of well-being, the United States brings up the rear: It is number one in prison population (726 prison inmates per 100,000 people versus 91 in France and 58 in Japan);[5] first in teen pregnancies, drug use, child hunger, poverty, illiteracy, obesity, diabetes, use of antidepressants, income disparity, violence, firearms death, military spending, hazardous waste production, recorded rapes, and the poor quality of its schools. (The United States is the only country in the world besides Iraq where schools need metal detectors.) It has the highest trade budget deficit as a proportion of national income and has seen more than 30 million workers laid off by corporations since 1984, most of whom were permanently consigned to lower-wage jobs. As a uniform trading system sweeps over the world, the monetary gains are called GDP, but the losses that are suffered, even in the industrialized West, much less in the Third World, are not tallied, as if one were recording sales at the cash register but ignoring thefts at the back of the warehouse.

The theory behind market liberalization is beguiling and, on the face of it, inarguable: If poor countries had more money and freedom, everyone would be better off; the greater flow of material goods would eventually improve everyone's life. To those who carp about low wages and poor working conditions in developing countries, free-market advocates argue that freedom and prosperity require time and sacrifice. But whose time and whose sacrifice? Critics see the further concentration of wealth and power, not the spread of freedom. The world's top two hundred companies have twice the assets of 80 percent of the world's people, and that asset base is growing fifty times faster than the income of the world's majority. Wealth flows uphill from the poor to the rich.

Do sixteen-year-old girls want to live in prisonlike concrete dormitories enclosed by gates, razor wire, and guards, a thousand miles from their villages, doing piecework that nets them $50 a month? Yes, in China they absolutely do. They can send nearly half of that sum back to their parents to help pay for medicine, books, and clothing for their younger siblings. Compare this description of workers toiling "day after day, for three hundred and thirteen days of the year, fourteen hours in each day, in an average heat of eighty-two degrees" with this account of "over four hundred men and women held by slavers in debt bondage, forced to work 10–12 hour days, 6 days per week . . . under the constant watch of armed guards. Those who attempted escape were assaulted, pistol-whipped, and even shot."[6,7] The first passage describes Lancashire textile workers in 1824; the second, Mexican and Guatemalan agricultural workers in South Carolina and Florida in 1997. Many believe that the world can do better after three hundred years than to retrace the Hobbesian dehumanization of the first Industrial Age.

Globalization has commoditized work into a series of fungible bits and pieces. Every aspect of production has been put up for bid; every developing country is an eager bidder. The rural poor provide a supply of labor for factories that make anything anytime for anyone anywhere, just as they did at the beginning of the Industrial Revolution. The anomie in these new factory towns, which typically lie near transportation corridors, is palpable. Prostitution, loss of identity, and feelings of powerlessness engulf you. The migration to factory towns today echoes the plight of commoners who in the early 1700s lost their homes after the Enclosure Movement and ended up working fourteen-hour days in Manchester textile mills. Inside modern contract-manufacturing facilities, there is no brand loyalty, pride, learning, or culture because the finished products have no intrinsic meaning to workers or management. They are abstractions: Guadalupe Virgins sold in Tijuana, athletic shoes for Kmart, a Ralph Lauren polo shirt, a pink baby bonnet for Kids "R" Us. The lives, culture, and skills of the workers are of no value because the work has been de-skilled to the point of mindlessness. Only speed, hand-eye coordination, and stamina are required.

From a consumer perspective, market globalization appears to have taken on a life of its own, as if lightning had ignited a wildfire in the sagebrush of the commercial world. One day your Levi's are made in San Antonio, the next in a country you can't easily locate on a map. In reality, however, the world trade system is heavily planned, controlled, and regimented. When you dissect the issue of globalization with respect to trade, it is less about integration, and primarily about the rights of business. WTO eliminates normal checks and balances because it performs all three roles of governance: executive, legislative, and judicial. The executive branch operates through the G 6 nations' ministerial meetings; the judicial functions through the Dispute Settlement Panel, which meets in secret and does not have to notify national legislative bodies of pending challenges to their laws; the legislative branch is the General Council, which sets policies, though in practice most decisions are made in private green room meetings in which the richer nations conduct the most important negotiations.

The purpose of the organization could not be simpler: the elimination of constraints on the flow of trade, including how a product is made, by whom it is made, or what happens after it is made. By doing so, WTO removes individual countries' and regions' ability to set standards, to express values, or to determine what they do or do not support if those standards conflict with WTO rules. From WTO's perspective, a fish is a fish is a fish, and it doesn't matter what happened to sea turtles, dolphins, or workers when that fish was caught. The underlying principle is to prevent protectionism and unfair discrimination against member countries, thus providing a form of civil rights for goods. What this means in practice is that child labor, prison labor, forced labor, substandard wages, and poor working conditions cannot be used as a basis to discriminate against imports. Environmental destruction, toxic waste production, and the presence of transgenic materials or synthetic hormones have likewise been rejected as a basis to screen or stop goods from entering a country.

In 1994, when the WTO's Uruguay Round of Multilateral Trade Negotiations was sent to Congress for ratification, Ralph Nader offered $10,000 to the charity designated by any senator or representative who signed an affidavit stating that he or she had read the 550-page document and could answer several questions about it. There had been no public hearings, dialogues, or education regarding this massive adventure in free trade, an agreement that gave WTO the ability to overrule or undermine international conventions, acts, treaties, and agreements. Many delegates to the negotiations, even heads of country delegations, were not aware of some provisions that had been drafted by subgroups of bureaucrats and lawyers, many of whom represented multinational corporations. In the United States, only Senator Hank Brown, a Colorado Republican, took Nader up on his offer. After reading the document, Brown changed his mind and voted against ratification.

What WTO seeks to protect is business and growth, not people and the environment, with an underlying assumption that the wealthier a country becomes, the better it is able to protect its people and its environment. It has not turned out that way. The people who were most articulate and vociferous in Seattle were not protesting against globalization per se but against what it actually (and, they believe, inevitably) entails, which is the corporatization of the commons. The commons that are being subsumed include the human genome, seeds, water, food, airwaves, media, and more. In a deeper sense, the commons include culture, place, self-determination, and democracy. The Seattle protests were about people's need for a standing and for a voice in their communities and factories. They do not constitute an anti-globalization movement but seek rather to bring about "globalization from below." Publicly held corporations promulgate an unspoken falsehood: that capital has a right to grow, a right greater than the rights of people, communities, and cultures. But corporations are extensions of us: shareholders, pension funds, endowments, and fiduciaries demand maximum returns. The community of protesters in Seattle argued that the world cannot achieve sustainability and equity if trade and corporate policies destroy local economies. The change in world economies has given us the means to disagree, but it is a lopsided discussion because of the realities of the balance of power. Most of the world's economies and governments are under the control of corporations, which seem to be successfully tightening their grasp; at the same time the world itself is increasingly out of control.

The corporatization of the world means the loss of economic and cultural diversity. Historian Arnold Toynbee cautioned that civilization is a movement, not a condition, and the rise of uniformity consistently marks its decline.

Most accounts of the Seattle demonstrations refer to them as "riots," even though they were 99.9 percent nonviolent. A person commits to a nonviolent path of resistance when avenues of dialogue have been closed. This is satyagraha. It is a deliberate choice and usually the last option, and anyone who has marched toward armored police, snarling dogs, or military units knows how gut-churning it is to face such physical threats with palms turned outward. Being beaten by another human being without resisting clashes with survival instincts and lights up the adrenal glands. To protect oneself but not respond, to see the person beating you as a friend, and not an enemy, is extraordinarily challenging. This response is not passive, nor is it weak. It was Gandhi who said he prized nonviolence only when he began to shed his cowardice. Seattle was not a riot. A riot occurs when people fight back.

During the afternoon and into the night, downtown Seattle became surreal. Ad hoc marchers imitated fife and drum corps of the American Revolution using empty five-gallon buckets for instruments. A few demonstrators danced on burning Dumpsters that had been ignited by pyrotechnic tear-gas grenades (the same ones used in Waco, Texas, against the Branch Davidians). Despite their steadily dwindling numbers, as many as 1,500 hardy protesters held their ground, seated passively in front of police, hands raised in peace signs, submitting to tear gas, pepper spray, and riot batons. As they retreated to the medics, new forces replaced them. The mandate for WTO vanished sometime that afternoon, because by then all media attention had been diverted to what was happening in the streets. By nine that night, the police order to clear the downtown area had been accomplished, but some police, perhaps fresh recruits from outlying towns, didn't want to stop there. They chased demonstrators into neighborhoods where the distinctions between protesters and citizens blurred (the majority of protesters were from Seattle) and began attacking bystanders, witnesses, residents, and commuters. They gassed commuters on passing Metro buses. They dragged a member of the Seattle City Council out of his car and tried to arrest him. When President Clinton sped from Boeing Field to the Westin

Hotel at 1:30 A.M. Wednesday, his limousines entered a police-ringed city of littered streets, patrolling helicopters, and boarded windows. Michael Meacher, environment minister of the United Kingdom, said afterward, "What we hadn't reckoned with was the Seattle Police Department, who single-handedly managed to turn a peaceful protest into a riot."

The day before the ministerial, Madeleine Bunting had filed a report in the *Guardian Weekly*: "Expect out of Seattle's World Trade Organization meeting this week lurid reports of multicolored-haired, body-pierced, tattooed anarchists. In gleeful detail, we will hear of the wilder shores of environmentalism and anarchism among the 150,000 protesters Seattle police and the FBI are bracing themselves for. . . . This is a colorful way of reporting a tediously difficult trade summit, but it is a gross distortion of a crucially important event. The protesters in Seattle cannot be dismissed as nutters; you could hardly describe the World Wildlife Fund, Oxfam or the Royal Society for the Protection of Birds or many of the other 1,200 environmentalist, development and human rights groups who signed a petition to the WTO in advance of this meeting as extremists. What is depressing is that this distortion serves only one interest. The wise chief executive of a global multinational will put up with a bit of tear gas floating over his lobster lunch this week. It gives him the perfect opportunity to dismiss his critics as fanatics. . . ."[8]

Bunting's prediction proved accurate: the unexpected action reverberated across the world and was portrayed as a threat to the nation-state itself. A surprised press corps went to work, expressing outrage and pointing fingers at brash, misguided white kids. Thomas Friedman, in his December 1 column, wrote that demonstrators were "a Noah's ark of flat-earth advocates, protectionist trade unions and yuppies looking for their 1960s fix." In fact, the protesters were hardly anarchic, but organized, well-educated, and determined. The vast majority were human rights activists, labor activists, nuns, indigenous people, people of faith, steelworkers, and farmers. They were forest activists, environmentalists, social justice workers, students, and teachers. They were citizens.

The business community believes that other cultures, once they have tasted the fruits of self-interest and economic freedom pioneered in America, will want to be like us. There are good reasons to think so. *Baywatch* is the most-watched TV program in the world, Nike the largest shoe company in the world,

Coca-Cola the biggest beverage company in the world, and McDonald's the biggest restaurant chain in the world. Given such commercial dominance, it is a short step to believing that the global market system is the perfect instrument to bring to less-developed countries "democratic capitalism." With a tin ear to the voice of the poor, however, no American corporate leader has ever expressed doubts that the future of every country in the world lies in adopting some version of Western values, or entertained the idea that cultural diversity may be more important than corporate profits. They and other ideological supporters of market fundamentalism do acknowledge that the rough and tumble of unrestrained markets can diminish the role of democracy in human affairs (the no-omelet-without-breaking-eggs analogy), but they insist that because people vote every day with their pocketbooks, the total impact of global markets is a true plebiscite in the end. *New York Times* columnist Thomas Friedman employs the term "economic democracy" for this principle, but the underlying idea is the same: markets are not merely a medium of exchange but a "medium of consent" that helps the poor, improves income, and protects citizens' interests.

Thomas Frank, author of *What's the Matter with Kansas?*, calls all such nostrums by yet another name: "Market populism is an idea riven by contradictions. It is the centerpiece of the new American consensus, but that consensus describes itself in terms of conflict, insurrection, and even class war. It is screechingly undemocratic, and the formal institutions of democracy have never seemed more distant and irrelevant than under its aegis. It speaks passionately of economic fairness, and yet in the nineties the American economy elevated the rich and forgot the poor with decisiveness we hadn't seen since the 1920s. Market populism decries 'elitism' while transforming CEOs as a class into one of the wealthiest elites of all time. It deplores hierarchy while making the corporation the most powerful institution on earth. It hails the empowerment of the individual and yet regards those who use that power to challenge markets as robotic stooges. It salutes choice and yet tells us this triumph of markets is inevitable."[9]

Many critics in other countries agree. For them, "democratic capitalism," "economic democracy," and "market populism" are all the same oxymoron. Dismissed by the market fundamentalists for wanting to turn back the clock, they make an excellent case for reversing that argument. By rapidly reducing tariffs and barriers to capital flows, governments and WTO are effectively re-creating the pitiful labor conditions of the nineteenth century:

deracination, urban blight, child and migrant labor, exploitation, mindless jobs, and on-the-job abuse, topped with laissez-faire indifference. Many petitioners to WTO ask whether the breakneck speed of corporate-led globalization might pause to solicit a second opinion from humanity. Must the economic mutations of rapid change replicate the same levels of "creative destruction" as seen in the past? To "globalize" literally means to make something round. Globalization began just over five hundred years ago when Western Europeans began to accept the idea that the earth is round, something Indian and Chinese civilizations already knew. Ever since then, in myriad ways, commerce, armies, travelers, and scholars have worked toward integrating human activity with geography, encircling the globe with development that arose from Western appetite.

The divisions in the world today have no better analog than the Green and Red Zones of Baghdad.[10] Armored fencing, earthen berms, sensors, Humvees, and machine guns defend the Green Zone, four square miles of a leafy green, irrigated desert surrounded by a twelve-foot concrete blast wall.[11] Within are villas, the Al-Rashid Hotel, bowling alleys, karaoke bars, fast-food restaurants, and the former presidential complex of Saddam Hussein. There, employees of the Coalition of the Willing do their work alongside American military commanders, Iraqi ministers, and American corporations such as Bechtel and Halliburton. They feed at cafeterias that offer pork morning, noon, and night—sausages, bacon, pork chops, hot dogs—a constant affront to the Muslim staff and Iraqi secretaries and translators.[12] Women jog down boulevards in tank tops and shorts while Iraqi children sell pornographic DVDs to soldiers at the bazaar. Bars are packed, administrative assistants are known to double up as hookers on their second shift, and armed military contractors lounge on cushions at the Green Zone Café sucking on hookahs.[13] Annual room, board, and office expenses run $300,000 per person, not counting six-figure salaries, travel, and the costs of military protection.[14]

Outside its perimeters is a traumatized city of 5.6 million residents, with open-air markets, mosques, neighborhoods, schools, tea stalls, and a roiling civil war fought with car bombs, beheadings, and executions. The Red Zone is noisy and crowded, pervaded by fear and anger. The two zones mirror the global split between those who rule and the majority, who don't. Throughout the world, at meetings such as that of the World Bank, corporate executives, political leaders, and heads of international agencies gather in secure resorts to map solutions to the world's problems, ringed by police, razor wire, attack dogs, and checkpoints. Outside is the majority world, and like Baghdad, its population feels stripped of power, security, and hope.

Tens of thousands of NGOs work toward amending the market policies of globalization because markets are not designed to be surrogates for ethics, values, and justice. So great a number of organizations have been founded because their function is the opposite of uniform trade rules: they try to deliver specific solutions tailored to the individuals and places they address. Several NGOs filed a lawsuit that brought tens of millions of dollars in back wages and repatriation costs to thousands of indentured workers in Saipan. These organizations were dismissed by free-market guru Jagdish Bhagwati as a "radical . . . fringe phenomenon," and in one retrospect, Bhagwati is correct: the organizations that argue, demonstrate, and litigate for human rights *are* on the fringe. Why must such groups operate at the margins of society simply if they believe that social justice and human rights should not be sacrificed when corporations shift their manufacturing to the lowest-wage countries? History may wonder why so few cared so little about so many for so long. Critics of NGOs sincerely believe that the brutality, slavery, and colonial exploitation of the previous five centuries have effectively been vanquished, and the modern corporate march to global markets represents a new page in economic history. In fact, that expansion is fundamentally a predictable stage in the march of market economics; what is new is the global coordination of resistance to it.

Patricia King, one of two *Newsweek* reporters in Seattle, called me after the ministerial to ask if I thought the protests represented a return of the 1960s. I replied that I didn't believe so, and I still don't. The sixties were primarily a Western event, whereas the protests against WTO are international. Who were the leaders? I said that there are no leaders in the traditional sense, as there were in the civil rights, women's, or antiwar movements. But there are thought leaders. Who are they? she asked. Martin Khor and Vandana Shiva of the Third World Network in Asia, Maude Barlow of the Council of Canadians, Tony Clarke of Polaris Institute, Jerry Mander of the International Forum on Globalization, Susan George of the Transnational Institute, John Cavanagh of the Institute for Policy Studies, Lori Wallach of Public Citizen, Anuradha Mittal, Owens Wiwa of the Movement for the Survival of the Ogoni People, Chakravarthi Raghavan of the Third World Network in Geneva, Debra Harry of the Indigenous Peoples Coalition Against Biopiracy, José Bové of the Confederation Paysanne, Tetteh

Hormoku of the Third World Network in Africa, Randy Hayes of Rainforest Action Network. "Stop, stop," she said. "I can't use these names in my article." I asked why not. "Because Americans have never heard of them." Instead of trying to explain the intellectual underpinnings of the Seattle protests, King's *Newsweek* editors, in an editorial decision that would have made William Randolph Hearst proud, prominently placed the picture of the Unabomber, Theodore Kaczynski, in their cover story on Seattle because he had once purchased a pamphlet by anarchist writer John Zerzan. A handful of Zerzan's followers did show up at Seattle to break windows, but his tangential essay could scarcely begin to account for what motivated the tens of thousands of other people who also took part in the protest.

In the end, it was not on the streets that the Seattle ministerial broke down, but within its conference rooms. Once the meeting finally got underway, it ended in a rancorous stalemate, with African, Caribbean, and some Asian countries refusing to support a draft agenda that had been negotiated behind closed doors without their participation. The outsiders see a difference between globalization and internationalization. (Former World Bank economist Herman Daly has long made the same distinction.) In internationalization, each nation sets its own trade standards and will do business with other nations that are willing to meet those standards. Do nations abuse this system? Always and constantly, and the United States is among the worst offenders in that regard. But where democracies prevail, internationalization does provide a means for people to set their own policy, influence decisions, and determine their own future. Globalization, in contrast, envisions standardized legislation for the entire world, with capital and goods moving at will superior to the rule of national laws. Globalization supersedes nation, state, region, and village. While diminishing the power of nationalism is a good idea, elimination of sovereignty may not be if it is replaced by a corporate boardroom.

An example of WTO's power is Chiquita Brands International, a $4 billion corporation that donated $500,000 to the Democratic Party in 1996. Two days before the contribution was made, the Clinton administration filed a complaint with WTO against the European Union, alleging European import and tariff policies favored bananas coming from small family-owned growers in the Caribbean who made a living wage, instead of from U.S.-owned multinational banana conglomerates that have a long history of worker strife, low wages, and toxic poisoning due to use of agrochemicals. The Europeans freely and proudly acknowledged the charge to be true, having set aside a portion of their market—7 percent—for Caribbean imports from Dominica, St. Lucia, and St. Vincent. For Europeans, the banana tariff was the decent thing to do; besides, everyone thought the bananas from the smaller growers tasted better. For the banana giants, this *attitude* was untenable, even though Chiquita already had 50 percent of E.U. market share.

The WTO banana hearings would have made Lewis Carroll proud. Trade representatives from St. Lucia and St. Vincent were not allowed to make presentations because they were not properly schooled in the arcana of WTO case law. They were asked to hire lawyers knowledgeable in the field of trade, but WTO then barred their attorneys from the hearing because it was a government-only body, which left both islands without representation at the banana hearings.[15] The Europeans ultimately made several concessions in 1998 but insisted that a two-tier quota regime remain in place. Demonstrating the customary political flexibility of U.S. corporations, Chiquita Brands donated $350,000 to the Republican Party. Two months later the Republican-controlled Congress introduced a retaliatory law against the E.U., which required the Clinton administration to impose sanctions on European products, including goat cheese, cashmere, and biscuits. In 1999 the E.U. was forced to rescind its preferential quotas for the small Caribbean growers. While the United States prevailed in this WTO-arbitrated case, who really won and who lost? Did the Central American employees at Chiquita Brands win? Ask the hundreds of workers in Honduras who were made infertile by use of dibromochloropropane on the banana plantations. Ask the mothers whose children have birth defects from pesticide poisoning. Did the shareholders of Chiquita win? By the end of 1999 Chiquita Brands was losing money. Its stock was at a thirteen-year low, shareholders were angry, and the company was up for sale, but the prices of bananas in Europe remained low. Who lost? Caribbean farmers who could formerly make a living and send their kids to school could no longer do so because of low prices and demand. Because bananas comprised nearly 50 percent of the islands' GDP, some displaced farmers have become growers for the drug cartels, and their governments have quietly pulled out of cooperative narcotics enforcement programs. This is America's loss. The Caribbean is one of the few places where the United States has a trade surplus. With lower export earnings, companies suffer—American companies.

No one much cares about WTO trade rules until they are directly affected by them, by which point it is often too late. In 1996 a WTO tribunal ruled that U.S. clean air standards regarding gasoline emissions violated Venezuela's right to export gasoline that did not meet those criteria, a ruling that came after American companies had invested $37 billion to bring their products into compliance. Lewis Carroll again: WTO member nations have a sovereign right to establish environmental standards and objectives, but they may implement environmental regulations only in a manner consistent with WTO rules. In other words, they are entitled to create meaningless rules. As it now stands, a member country cannot prevent substances, whether materials or food, from being imported unless they have WTO-approved proof of harmful consequences. Industry-led coalitions from the United States are contesting the fact that Europe allows food to be labeled to indicate whether its ingredients have been genetically modified. So far the United States has prevailed in eliminating European restrictions on hormone-treated beef, even though the vast majority of European citizens want to limit its sale. Monsanto hoped that Seattle would be its moment for eliminating E.U. restrictions on genetically modified corn and soy. When the E.U. tried to eliminate PVC teething rings containing phthalates, which cause liver and kidney damage, the U.S. toy industry tried to stop it because the science had not yet provided definitive proof. WTO secretary general Renato Reggiero declared that regulations and standards that are environmentally based are "doomed to fail and could only damage the global trading system."16

In all WTO rulings one common denominator prevails, and that denominator is money. As it happens, however, it is not the corporations that hold the biggest bag of cash, but the World Bank, which dominates the flow of money into the developing world. Like all institutions, the bank has a point of view, a modus operandi, and a culture to support it. What primarily guides it is the theory that unimpeded market-based systems deliver more goods to more poor people faster than any other method. Given the catastrophic failure of state-sponsored socialism in the former Soviet Union, it is a difficult point to argue. Buttressing that theory is the observation that corrupt governments do not efficiently provide equitable services to their people. Because the World Bank cannot directly meddle with the internal politics of a given country, it can only insist on privatization and monetary reforms to curtail irresponsibility. It can remove assets from the hands of the state and place them into the marketplace, where consumers vote every day with their wallets.

The theory behind the World Bank's operations is laudable. But there is scant evidence to support its claims in the real world. The situation is comparable to the difference between official history and the people's history. Columbus is generally credited with having discovered America. The inhabitants of the lands he "discovered" also made a discovery: that a race of barbaric men with developed navigational skills had no qualms about raping their women, eviscerating their elders, hunting down anyone who resisted, and packing them off naked like so many animals to Europe to be sold as slaves—a fate so horrific that masses of Arawak people committed infanticide and then suicide rather than face an uncertain future. The point here is to give dimension to how particular individuals actually experience the published narrative of history. What poor people experience when market reforms and other fiats sweep through their barrios or squatter settlements is, similarly, at a far remove from what World Bank bureaucrats imagine in their pristine offices in Washington, D.C. Asking the impoverished to make a structural adjustment is like asking a toddler to cross a freeway. People at the bottom of the economic pyramid don't have the latitude to be able to pay more for water, give up free health care, lose jobs, compete with subsidized Iowa corn farmers, or begin funding schooling.

In the 1960s, the World Bank financed the fish-processing industry on Lake Victoria in Tanzania. Nile perch were introduced, a ravenous predator that eliminated 350 varieties of the lake's native fish. Today, flash-frozen perch fillets are exported to Europe via massive Ilyushin cargo planes. These aging jumbo jets return, sometimes carrying dried peas and flour to feed refugees, and other times Kalashnikovs and munitions that support Central Africa's numerous refugee-creating wars. The native population does not benefit from this $400 million industry. Having lost most of their traditional fisheries, they rely for food on the fish heads and scraps thrown away by the fish processors. Because the oily fish heads require wood-fire for drying, shorelines are becoming rapidly deforested, causing erosion, runoff, and eutrophication, all of which threaten the Nile perch. Children are protein-deficient, HIV-infected prostitutes ring the fishing camps, and midges that were once controlled by the now-absent native fish swarm near the shore.[17] The 2004 documentary *Darwin's Nightmare* depicts the carnage of Lake Victoria in terrifying detail. It should be mandatory viewing for every World Bank manager, every market fundamentalist, every op-ed globalizer.

In 1990 fewer than 50 million people in the world bought their water from private companies. Ten years later, that number has risen to 460 million. Why? Countries borrowing from the World Bank must have a water privatization plan as a precondition for all loans.[18] European companies, dominated by Suez and Veolia, control 70 percent of the private water market. Water and sanitation are estimated to be a $200–$400 billion business, which for corporations is an opportunity comparable to discovering oil. The recipient of World Bank financing for water privatization in Ghana was Enron.[19] After several years of coercive threats, the World Bank forced Bolivia, South America's poorest country, to sign a forty-year privatization lease in September 1999 with Aguas del Tunari, a company owned partly by Bechtel Corporation, to privatize Cochabamba's water, a deal that guaranteed the company a 16 percent return on investment over the life of the contract. What this meant was that people in the eighth-poorest country in the world were paying more for their water than Bechtel executives living in the wealthy suburbs of San Francisco. When water rates doubled and tripled for the Bolivian poor to an overall average of one-fourth of their monthly income, demonstrations began and continued until the contract was repealed. Eduardo Galleano has described the electoral victory banning water privatization in Uruguay, writing that Uruguayans, who are so melancholic that they could be described as "Argentineans on Valium, are dancing on air. . . . A few days before the election of the President of the planet in North America, in South America elections and a plebiscite were held in a little-known, almost secret country called Uruguay. . . . [F]or the first time in world history, the privatization of water was rejected by popular vote . . . [and] the people asserted that water, a scarce and finite natural resource, must be a right of all people and not a privilege for those who can pay for it."[20]

You can try to determine the future, or you can try to create conditions for a healthy future. To do the former, you must presume to know what the future should be. To do the latter, you learn to have faith in social outcomes in which citizens feel secure, valued, and honored. It is difficult to estimate precisely how many NGOs in the world are trying to rectify and heal the effects, some would say fiascos, of World Bank's development policies, but they number in the thousands. In borrower countries—including Bolivia, Argentina, Haiti, India, Brazil, Mexico, Ecuador, and South Africa—the indigenous and the poor have initiated countless protests against the corruption, political cronyism, privatization schemes, job losses, environmental destruction, human rights violations, deracination, mega-dams, and loss of sovereignty brought about by the World Bank. The bank's response has been to increase spending on public relations, which now exceeds its research budget.[21]

One of the legacies of the World Bank is the misery of unpayable debt, in the form of money owed for tractors, dams, and power plants that were supplied by big corporations from the developed world. That debt, which can be crushing for developing countries, especially when development schemes have failed to deliver the promised jobs and growth, can drag a country backward into an austerity that truncates formerly affordable education and health care. Many new World Bank loans are not for development but for debt service on deficits created by previous loans. World Bank money comes from wealthy countries, and throughout the institution's history, the flow of money from the southern hemisphere's less developed countries to the northern hemisphere's more developed ones has been greater than the reverse. World Bank headquarters in Washington, D.C., are lavish to the point of embarrassment. U.N. headquarters in New York are in disrepair; part of the building collapsed in 2003.

Just as democracies require an informed and active citizenry to prevent abuse, markets require constant tending to prevent them from being diverted or exploited. A free market, so lovely in theory, is no more feasible in practice than a society without laws. Democracies can sustain freedom because their citizens and representatives continually adjust, maintain, and as necessary enforce standards, rules, and laws. Markets are unequaled in providing feedback, fostering innovation, and allocating resources. Market competition is ultimately a matter of financial capital: those activities that most efficiently accrete and concentrate money gain market advantage; those that don't are marginalized. But there is no comparable competition to improve social or natural capital, because markets for such commodities simply don't exist. The only way those issues are dealt with is through legislation, regulation, citizen activity, and consumer pressure. Removing the laws and regulations that create market constraints leaves the body politic with very few means to promote economic democracy. The localized poor, primary forests, the stratosphere, and ecosystem viability, which are the source of life for every economy in the world, have no voice at all in market systems. That voice comes from citizen organizations, although when it does, it is often ignored or patronized.

To raise the decibel level where they can be heard, outcries and protests from civil society may at times appear to be narrowly focused and shrilly expressed. Any resistance to economic liberalization strikes some critics as pathetic, even absurd. Ironically, what citizen organizations, NGOs, and trade unions are calling for is exactly the same thing as the glossy PowerPoint presentations that emanate from World Bank and IMF: economic integration. From the citizens' viewpoint, integration of policy needs to be bottom-up, not just top-down.[22] It needs to combine public and private needs, the poor and the middle class, owners and the landless. Today, a growing number of economists believe that the exclusion of civil society as a valid, necessary factor in economic planning has undermined growth and equitable development throughout the world. These economists are now calling for the same policy changes that NGOs have been promoting. Nancy Birdsall of the Carnegie Endowment for International Peace and Augusto de la Torre of the World Bank have written a paper entitled "Washington Contentious: Economic Policies for Social Equity in Latin America," which puts forth eleven critical development prerequisites, including promoting services and schools for the poor, protecting workers' rights, enacting land reform, giving small businesses more stable footing to compete, and removing regressive consumption levies and taxing the rich. All of these would have found a place on the placards of the demonstrators in Seattle.

Arguments for and against globalization of trade fall into the symmetric patterns seen in debates between progressives and conservatives on just about any issue. "You cannot switch off these forces except at great cost to our own economic well-being," writes Fareed Zakariah in the *New York Times.* "Over the last century, those countries that tried to preserve their systems, jobs, culture or traditions by keeping the rest of the world out stagnated. Those that opened themselves up to the world prospered." That argument sounds good, but it is not true. Japan, the second-largest economy in the world, continues to maintain rigid barriers and tariffs for many goods and products, as do Korea and China. Zakariah is referring to countries such as Burma, North Korea, and the former Soviet Union, which were effectively strangled by totalitarian economic controls. He overlooks, however, a country like Bhutan, which has done a remarkable job in preserving its culture, restraining corruption, promoting the economy, and preserving the environment, using an approach it calls the Gross National Happiness.

Generalities, whether from the right or left, aren't helpful when discussing free trade, because a plethora of factors determines the first- and second-order effects of liberalized markets. In his book *The Clock of the Long Now: Time and Responsibility*, Stewart Brand discusses what makes a civilization resilient and adaptive. Scientists have applied that very question to ecosystems, and it is worthwhile to ask it of economic and market systems as well. How does a system, be it cultural or natural, manage change, absorb shocks, and survive, especially when the forces of change are rapid and accelerating? The answer has much to do with time, both our use of it and our respect for it. Biological diversity in ecosystems helps buffer against disasters due to sudden environmental shifts because different organisms operate on different time scales—flowers, fungi, spiders, trees, and foxes all have unique life cycles and rates of change. Some respond quickly to their environment, others slowly, so that the system, when subjected to stress, can move, sway, and give, and then return and be restored to balance.

WTO policies engage and affect four chronologies or time frames, but WTO considers only one of them. The dominant time frame of our age is commerce. Businesses are responsive, welcome innovation in general, and have a bias for change. They need to grow more quickly than ever before, due to the integration of capital markets and globalization; they are punished, even bankrupted, if they do not. With the efficiency of worldwide capital mobility, companies and investments are rewarded or penalized instantly by a network of technocrats and money managers who move $2 trillion a day, seeking the highest return on capital.

The second time frame is culture. It moves more slowly, as cultural revolutions are resisted by deeper, established beliefs. The first institution to blossom under perestroika was the Russian Orthodox Church. I walked into a church near Boris Pasternak's dacha in 1989 and heard priests and babushkas reciting the litany with perfect recall, as if seventy-two years of repression had never happened. Culture provides the slow template of change within which family, community, and religion prosper. Culture stabilizes identity, and in a fast-changing world of displacement and rootlessness, becomes an ever more important anchor.

Between culture and business is a third time frame: governance, which moves faster than culture, slower than commerce. The fourth and slowest chronology is earth, nature, and the web of life. As ephemeral as it may seem, it is the slowest clock, always present, responding to long, ancient evolutionary cycles that extend beyond any one civilization's reckoning. Nature has the greatest inertia but the most resilience.[23]

These chronologies often come into conflict. What makes life worthwhile and enables civilizations to endure are all the elements and qualities that have poor returns under commercial metrics: universities, temples, poetry, choirs, parks, literature, language, museums, terraced fields, long marriages, line dancing, and art. Nearly everything humans hold valuable is slow to develop and slow to change. Healthy commerce requires the governance of politics, art, culture, civil society, and nature, to slow its pace, to make it heedful, to make it pay attention to people and place. As Brand points out, business unchecked becomes criminal. Consider Russia. Look at Enron, Tyco, Unocal, WorldCom. The extermination of languages, cultures, forests, and fisheries is occurring worldwide in the interests of speeding up business, even while business itself is stressed by increasingly rapid change. The rate of change is unnerving to all, even to those who benefit. To those who are not benefiting, it is devastating.

When there was an abundant earth supporting relatively few people, it was not necessary for markets to allocate resources with an eye toward the future. On a crowded earth with failing ecosystems, that lapse will be fatal. The continuity of the human species requires a fundamental change in market structures so that they include and harmonize with longer, slower time frames. The challenge of civilization has changed, and markets must change accordingly. As effective as markets are, they are tools, not reality. Markets make great servants, but bad leaders and ridiculous religions. To impose on contemporary global trade nineteenth-century laissez-faire ideology—an economic fundamentalism that was practiced in rhetoric only—in hopes of alleviating poverty and addressing environmental degradation is like slitting an artery to reduce high blood pressure. Trade is not the salient issue; the critical question is: Who sets the rules and who enforces them? There can be no sustainability when institutions whose primary purpose is to create money are dictating the standards.

One of the failures of the arguments opposing market globalization is the visible lack of an alternative economic model that might address the plight of the world's poor. The failure of those making the case for globalized free trade is their inability to adequately address the results of rapid economic change in human and ecological terms, how it creates prosperity *and* misery *and* ecological degradation, roughly in equal measure, incomparable though they may seem. The worldwide diaspora of immigrants, refugees, and peasants to urban slums is growing faster than even the most optimistic forecasts of the benefits of free trade. No institution stands more solidly behind free trade than the World Bank, yet few institutions are more pessimistic about the plight of humanity. It has predicted that more than 5 billion people will receive less than $2 a day in income by 2030, and 2 billion of them will live in slums in dozens of cities with populations greater than 10 million.[24] The future of the world is being cultivated in the despair, anger, and bleakness of the *chawls* of Mumbai, the *favelas* of Rio, the *kampongs* of Jakarta, the *shammasas* of Khartoum, the *pueblos jóvenes* in Lima, the *villa miseria* of Buenos Aires, and the *umjundolos* of Durban, not in the Pilates studios of the Hamptons and Santa Monica.[25] In Darwinian terms, it is a rapid breeding pool of human evolution as 200,000 new migrants join the urban poor every day.[26] Citing theories that crime burgeons when the population of youth rises, military analysts predict that control of the slums will fall into the hands of psychopaths, lunatics, demagogic firebrands, and clan-based militias.[27] Mike Davis writes in *Planet of Slums* that "the cities of the future, rather than being made of glass and steel as envisioned by earlier generations of urbanists, are instead largely constructed out of crude brick, straw, recycled plastic, cement blocks, and scrap wood. Instead of cities of light soaring toward heaven, much of the twenty-first century urban world squats in squalor, surrounded by pollution, excrement, and decay. Indeed, the one billion city dwellers who inhabit postmodern slums might well look back with envy at the ruins of the sturdy mud homes of Çatal Hüyük in Anatolia, erected at the dawn of city life nine thousand years ago."[28]

In his book *The End of Suffering*, Pankaj Mishra describes attending a conference of radical Islamists in Pakistan near the Afghanistan border. There 200,000 men, mostly in their teens and twenties, attended a "medieval desert fair" for the disenfranchised, where fiery speakers described centuries of humiliation at the hands of Western powers. It was a recruiting event for jihad:

> It took me some time to sort out my own responses to all this. I knew about the corruptions of jihad; of the leaders grown fat on generous donations from foreign and local patrons, sending young men to poorly paid *shahadat* (martyrdom) in Kashmir and Afghanistan. But I hadn't expected to be moved by the casual sight in one *madrasa* of sixty young men sleeping on tattered sheets on the floor. I hadn't thought I would be saddened to think of the human waste they represented—the young men, whose ancestors had once built one of the greatest civilizations of the world, and who now lived in dysfunctional societies under governments beholden to, or in fear of, America, and who had little to look forward to, except possibly the short career of a suicide bomber. . . . The other kind of future once laid out for them had failed. This was the future in which everyone in the

world would wear a tie, work in an office or factory, practice birth control, raise a nuclear family, drive a car and pay taxes. . . . The forward march of history was to include only a few of them. For the rest, there would be only the elaborate illusion of progress, maintained by a thousand "aid" programmes, IMF and World Bank loans, by the talk of underdevelopment, economic liberalization and democracy. But the fantasy of modernity, held up by their state, and supported by the international political and economic system, had been powerful enough to expel and uproot them from their native villages . . . hundreds of millions of stupefied and powerless individuals, lured by the promise of equality and justice into a world which they had no means of understanding, whose already overstrained and partially available resources they were expected to exploit in order to hoist themselves to the level of affluence enjoyed by a small minority of middle-class people around the world.[29]

Inuit mythology tells the story of Skeleton Woman. The tale begins with a fisherman trolling an inlet for his dinner. Suddenly he feels a heavy pull on his line, something so strong that it drags his kayak out to sea. He believes he has caught a fish so great he can eat for weeks, one so fat that he will prosper ever after, a fish so amazing that the whole village will marvel at his prowess. As the fisherman deliriously imagines fame and material ease, he reels in his line. Instead of a fish, he pulls up the decomposed, flesh-eaten carcass of a young woman who had been flung into the sea by her angry father years before. The frightened fisherman tries to free himself of Skeleton Woman, but she is so snarled in his fishing line that she is dragged behind his kayak wherever it goes. He paddles hurriedly back to shore, horrified by the trailing bones and flesh. In his haste to escape, he drags his lines over the beach and brings Skeleton Woman into his cabin, where he collapses in terror. In the retelling of this story by Clarissa Pinkola Estes, Skeleton Woman, who is still alive but cadaverous and half-eaten, represents both life and death, a specter reminding us that every beginning brings with it an end, that for all that is taken, something must be given in return. The fisherman eventually calms down and, cowering in the corner, dares to look at his "catch," asleep in the corner of the hut. He begins to feel pity for her, so he quietly creeps over and carefully disentangles the fishing line and seaweed from her hair so as not to wake her. He straightens and rearranges her bony carcass, puts a blanket over her, and finally, no longer afraid, falls asleep. In his sleep, tears of sadness fall from his eyes. Skeleton Woman awakens and crawls across the floor, drinks the tears of the dreaming fisherman, grows a new body, and is transformed into a young woman again.[30]

Like all fishermen, WTO's proponents want to catch the big one. They see few downsides to unleashing the benefits of untrammeled corporate growth. And yet death is always attached to life; all growth comes with an accounting, a reckoning. Birth and death are each other's consorts, inseparable and fast. The expansive dreams of the world's future wealth were reflected perfectly by Bill Gates III, cochair of the Seattle host committee, the world's richest man, hosting delegates at his $97 million, 66,000-square-foot house. Skeleton Woman showed up in Seattle as the uninvited guest. Dancing, drumming, ululating, marching in black alongside symbolic coffins, she wove through the sulfurous rainy streets of the night. She couldn't be killed or destroyed, no matter how much gas or pepper spray, or how many rubber bullets were used. She kept coming back and sitting in front of the police and raised her hands in peace, and was kicked, and trod upon. Skeleton Woman told the corporate delegates and the rich nations that they could not have the world, that it was not for sale, that if business was going to trade with the world, it had to recognize and honor the world, her life, and her people. She told WTO that it had to be brave enough to listen, strong enough to yield, courageous enough to give. Skeleton Woman was brought up from the depths and regained her eyes, voice, and spirit. She is about in the world, released that day and night, November 30 in Seattle, and her dreams are different. She believes that the right to self-sufficiency is a human right; she imagines a world where the means to kill people are described as not a business but a crime, where all crimes against women are crimes against the earth, and all crimes against nature are crimes against humanity, where families do not starve, fathers can work, children are never sold, and women cannot be impoverished because they choose to be mothers. Skeleton Woman does not see a time when a man holds a patent to any living thing, or where animals are factories, or where rivers belong to stockholders. Hers are deep and fearless dreams from slow time. She will not be quiet or be thrown back to sea anytime soon.

IMMUNITY

Our immune systems, and only our immune systems, prevent us from becoming everyone else all at once. We are who we are only because we defend ourselves every moment of every day. And who we are is everything. We are pieces of others. Portraits painted somewhere between our brains and thymuses. We are the dirt we've eaten and the songs we've sung. We are the light of stars and darknesses old beyond imagining. We are at once spontaneous fires and sacred water. We are faith and forgiveness. We are our own deaths and we are the eternal thought of others.

—Gerald Callahan, *Faith, Madness, and Spontaneous Human Combustion*[1]

One of the beauties of biology is that its facts become our metaphors.

—Kenny Ausubel, *Nature's Operating Instructions*[2]

I n the 1960s, Sir James Lovelock began examining the possibility that
the earth might be a single living entity. The Gaia hypothesis, as he later
named it, is ecology writ large. It asserts that the earth, in creating conditions favorable to life, exhibits qualities of self-organization and self-regulation
that are similar to those of a living organism. Two centuries earlier, Immanuel Kant and French economist Jacques Turgot imagined humanity itself as a similar entity, a system with some of the attributes of an organism.
They were not alone. From Spinoza to Gandhi, from Lewis Thomas to Teil-
hard de Chardin, philosophers, religious teachers, and scientists have all
wondered if the entire human race might be integrated in mysterious and
inexplicable ways. "Joined together, the great mass of human minds around
the earth seems to behave like a coherent, living system," writes Thomas.[3]

One of the differences between the bottom-up movement now erupting
around the world and established ideologies is that the movement develops
its ideas based on observation, whereas ideologies act on the basis of belief
or theory—the same distinction that separated evolution from creationism
in the time of Charles Darwin and William Paley, the same distinction that
George Soros draws between open and closed societies. Darwin did not try
to disprove creationism. He was a field scientist who tried instead to make
sense of the evidence he found on the voyage of HMS *Beagle*. Likewise the
movement doesn't attempt to disprove capitalism, globalization, or religious
fundamentalism, but tries to make sense of what it discovers in forests, favelas, farms, rivers, and cities. Are there ideologues in the movement? To be sure,
but fundamentally the movement *is* that part of humanity which has assumed
the task of protecting and saving itself. If we accept that the metaphor of an organism can be applied to humankind, we can imagine a collective movement
that would protect, repair, and restore that organism's capacity to endure when
threatened. If so, that capacity to respond would function like an immune system, which operates independently of an individual person's intent. Specifically, the shared activity of hundreds of thousands of nonprofit organizations

can be seen as humanity's immune response to toxins like political corruption, economic disease, and ecological degradation.

Just as the immune system recognizes self and non-self, the movement identifies what is humane and not humane. Just as the immune system is the line of internal defense that allows an organism to persist over time, sustainability is a strategy for humanity to continue to exist over time. The word *immunity* comes from the Latin *im munis,* meaning ready to serve.[4] The immune system is usually portrayed in militaristic terms: a biological defense department armed to fight off invading organisms.[5] In the textbook case, antibodies attach themselves to molecular invaders, which are then neutralized and destroyed by white blood cells. Simple and elegant, but the process of fending off invaders and disease is more complex and interesting.

The immune system is the most diverse system in the body, consisting of an array of proteins, immunoglobulins, monocytes, macrophages, and more, a microbestiary of cells working in sync with one another, without which we would perish in a matter of days, like a rotten piece of fruit, devoured by billions of viruses, bacilli, fungi, and parasites, to whom we are a juicy lunch wrapped in jeans and T-shirt. The immune system is everywhere, dispersed in lymphatic fluid, which courses through the thymus, spleen, and thousands of lymph nodes scattered like little peanuts throughout the body. The thymus organ is a blueprint of who you are, a physiological inventory of your genes, past diseases, and current condition, and it floods the lymphatic fluid with helpers to prevent anything that is not you from taking over. Those thymus helper cells (T-cells) are a type of white blood cell called lymphocytes, and hundreds of billions of them meander throughout the body, able to identify current and past infections and diseases, an itinerant Alexandrian library of pathological memorabilia moving from the bloodstream to tissues and back again. When an outsider is spotted, a message is passed from the T-cells to the B-cells, another cellular cohort, which then make the antibodies that attach to receptor sites on the invading cell and neutralize it. Another group of antibodies stakes out the future by remembering the past. They pick up odd bits of pathogens, a protein here and some antigens there, even live diseases, and attach them to follicular dendritic cells that look like a cross between a burdock burr and spiky punk hairdo. Decorated with bits of virus, the follicular dendritic cells remain in the lymph nodes to maintain vigilance, creating an imperceptible low-grade response, pathology practice drills in the corners of the body. As I write this paragraph at a cottage near a beach, a one-minute siren blasts across town warning of an incoming tidal wave, except that there is no tidal wave: the alarm serves as a monthly drill, verifying that the siren works and townspeople don't forget. Just as the drill reminds people, the captives attached to dendritic cells serve as reminders, talismans that maintain readiness, and together with the antibodies they form a network of *immunological memory*, a store of knowledge that rivals the capacities of our brain. It is because long-lived memory cells created after a primary infection remember previously encountered antigens that we can vaccinate ourselves with dead or attenuated cells from toxic organisms.[6]

At the core of immunity is a miracle of recovery and restoration, for there are times when our immune system is taken down. Stress, chemicals, infections, lack of sleep, and poor diets can overwhelm the immune system and send it into a tailspin. When that happens, old diseases can resurface while protection from new ones breaks down. Pathogens burgeon and seem to hold sway, and a moment comes when death lurks at the threshold. At that point, given the odds and circumstances, something extraordinary can happen that really shouldn't: the immunological descent slows and halts, our life hangs in the balance, and we begin to heal as if stumbling upon Ariadne's thread, a comeback that rivals the climax of a Hollywood plot. How the disoriented and muddled immune system reverses course and recovers is not well understood; some would say it is a mystery.[7]

The immune system sounds orderly and precise, but it is not. Antibodies bind not just to pathogens but to many types of cells, even themselves, as if the lymphatic system were a chamber of commerce mixer of locals feverishly exchanging business cards. In *The Web of Life*, Fritjof Capra writes, "The entire system looks much more like a network, more like the Internet than soldiers looking out for an enemy. Gradually, immunologists have been forced to shift their perception from an immune system to an immune network."[8] Francesco Varela and Antonio Coutinho describe an immune system that can best be understood as intelligence, a living, learning, self-regulating system—almost another mind. Its function does not depend on its firepower but on the quality of its connectedness. Rather than "inside cells" automatically destroying "outside cells," there is a mediatory response to pathogens, as if the immune system learned millions of years ago that détente and getting to know potential adversaries was wiser than first-strike responses, that achieving balance was more appropriate than eradication. The immune system depends on its diversity to maintain resiliency, with which it can maintain homeostasis, respond to surprises, learn from pathogens, and adapt to sudden changes. The implication for medicine is clear: to fend off cancer and infection, we may need to understand how to increase the immune network's connectivity rather than the intensity of its response.[9]

Similarly, the widely diverse network of organizations proliferating in the world today may be a better defense against injustice than F-16 fighter jets. Connectivity allows these organizations to be task-specific and focus their resources precisely and frugally. Incremental success is achieved by consensus operating within handmade democracies, where no one person has all or much power. The force that such groups exert is in the form of dialogue and truthfulness. Computers, cell phones, broadband, and the Internet have created perfect conditions for the margins to unify. According to Kevin Kelly, author of *Out of Control*, the Internet already consists of a quintillion transistors, a trillion links, and a million e-mails per second. Moore's Law, which predicts that processing power will double in power and halve in price every eighteen months, is meeting Metcalfe's Law, which states that the usefulness of a network grows exponentially with arithmetic increases in numbers of users. These laws enable big corporations as much as small NGOs, but the latter gain greater advantage because they amplify smallness more effectively than largeness. Large organizations don't need networks; small ones thrive on them. Webs are complex systems of interconnected elements that link individual actions to larger grids of knowledge and movement. Web sites link to other sites with more links to other sites ad infinitum, creating a critical, fluid mass of information that evolves and grows as needed—very much like an immune response. At the heart of all of this is not technology but relationships, tens of millions of people working toward restoration and social justice.

The state of the world today suggests that, given the number of organizations and people dedicated to fighting injustice, the movement has not been particularly effective. The counterargument to this claim is that globalization's depredations have had a nearly five-hundred-year head start on humanity's immune system. The exponential assault on resources and the production of waste, coupled with the extirpation of cultures and the exploitation of workers, is a disease as surely as hepatitis or cancer. It is sponsored by a political-economic system of which we are all a part, and any finger-pointing is inevitably directed back to ourselves. There may be no particular *they* there, but the system is still a disease, even if we created and contracted it. Because a lot of people know we are sick and want to treat the cause, not just the symptoms, the environmental movement can be seen as humanity's response to contagious policies killing the earth, while the social justice movement addresses economic and legislated pathogens that destroy families, bodies, cultures, and communities. They are two sides of the same coin, because when you harm one you harm the other. They address what Dr. Paul Farmer calls the "pathologies of power," the "rising tide of inequalities" that breed violence, whether it be to people, places, or other forms of life. No culture has ever honored its environment but disgraced its people, and conversely, no government can say it cares for its citizens while allowing the environment to be trashed. Farmer writes, "More guns and repression may well be the time-honored prescription for policing poverty, but violence and chaos will not go away if the hunger, illness and racism that are the lot of so many are not addressed in a meaningful and durable fashion."[10] The pathologies that Farmer refers to are multitudinous, and the carrier is most often a corporation, government agency, or bank, rhetorically accompanied by the rationale that they are trying to help the very people they harm. Vectors of social and environmental disease also can be the military and governments, or they can just be bad thinking, bureaucratic rigidity, and institutional hubris.

The ultimate purpose of a global immune system is to identify what is not life affirming and to contain, neutralize, or eliminate it. Where communities, cultures, and ecosystems have been damaged, it seeks to prevent additional harm and then heal and restore the damage. Most social-change organizations are understaffed and underfunded, and nearly all are negotiating steep learning curves. It is not easy to create a system that has no antecedent, and if you study the taxonomy of the movement you will see a new curriculum for humankind emerging, some of it corrective, some of it restorative, and some of it highly imaginative. In many countries participation in the movement can be dangerous. We memorialize the well-known murders of South African activist Stephen Biko and rubber tapper and environmentalist Chico Mendes, yet people in the movement are killed and intimidated every day. When you see images of Amazon Indians marching in full regalia in São Paulo to protest Brazilian government policies, they are individuals who are as courageous as they are terrified. I have a photograph of a small Mayan girl holding her mother's hand looking up in wide-eyed disbelief at a phalanx of black polycarbonate shields and masked police gripping their batons in Guatemala. When the Revolutionary Association of the Women of Afghanistan march for women's rights without their burqas, they display an extraordinary valor, because they *know* there will be reprisals. When the Wild Yak Brigade was formed in Zhidou, China, to protect the endangered Tibetan antelope, poachers murdered its first two leaders. Most movement activists start like Chico Mendes, believing they are fighting for a specific cause, in his case rubber trees, and realize later they are fighting for a greater purpose: ". . . then I thought I was trying to save the Amazon rainforest. Now I realize I was fighting for humanity."[11]

To deal with the pathogens, the movement has had to become an array of different types of organizations. There are community development agencies, village- and citizen-based groups, corporations, research institutes, associations, networks, faith-based groups, trusts, and foundations. Within each of these categories are dozens of types of organizations defined by their activity; within these different activities, groups have a specific focus: rights of the child, pinnipeds, cultural diversity, coral reef conservation, democratic reform, energy security, literacy, and so on. (To fully appreciate the range of organizations involved, please turn to the Appendix for a taxonomy of the movement.) Because people typically hear about the same larger, established organizations frequently in the media, the underlying diversity of the movement is often hidden, or reduced to a stereotype of demonstrators, placards, and protests. These activities do represent an important effort but constitute the tiniest sliver of the work being undertaken by the movement.

What follows below are descriptions of the major types of organizations and some specific examples of each. The energy and achievements of them all are astonishing, but just as astonishing are the categories and groups that have not yet been accounted for. There is simply no way to adequately present the range of undertakings of this global movement.

Keeper groups are organizations inspired by Robert Kennedy Jr.'s representation of the Hudson River Fisherman's Association against GE and other polluters of the river in 1984. The association was later renamed Hudson Riverkeepers and so far has spawned another 150 organizations under the Waterkeeper Alliance, an international umbrella organization of Riverkeepers, Lakekeepers, Baykeepers, Soundkeepers, Streamkeepers, Inletkeepers, Reefkeepers, and Coastkeepers. According to Kennedy, keeper organizations are part scientist, part lawyer, part lobbyist, part investigator, and part public relations agent. Their client is water and all its users, from fish to fishermen, from mothers to boaters—a perfect illustration of the immune response.

Watch organizations monitor corporations, projects, institutions, and places: Gunpowder Creek Water Watch, Passumpic River Watch, Kurdish Human Rights Watch, Vietnam Labor Watch, Vermont Wal-Mart Watch, and Australian Paper Watch, which monitors logging in the forests of Victoria and how the resulting pulp is used. *Friends organizations* are almost the opposite of watch organizations: They go out into the field and improve, clean up, assist, and support. For example, Friends of Coyote Creek in Santa Clara Valley, California, and Friends of North Kent Marshes in the United Kingdom. The grandmother of all *Defender* groups is Defenders of Wildlife. There are also Defenders of the Rainforest and the Association of Human Rights and Tortured Defenders in Cameroon. *Coalitions* include the Coalition Against Trafficking in Women, the Indiana Recycling Coalition, and the East Tennessee Clean Fuels Coalition. *Alliances* include the India Alliance for Child Rights, the Alliance of Religions and Conservation, and the Mendocino Wildlife and Watershed Alliance.

Except for the multibillion-dollar Nature Conservancy, based in Virginia, even the largest organizations in the movement would not qualify in financial terms as small enterprises on Wall Street. The vast majority of organizations are so small as to be almost laughable, but they are far from amusing to the pathogens they target. More than seven thousand Roots and Shoots youth groups organized by Jane Goodall involve children in community-based projects to help animals, the environment, and human beings. Earth Island Institute is not an institute so much as it is an *incubator NGO* that supports thirty-six different organizations, from Baikal Watch, which promotes activities to preserve the greatest body of freshwater in the world, to the Tibetan Plateau Project, which works on conservation and biodiversity issues in Tibet. There are *networks* such as the African Parliamentary Poverty Reduction Network and Youth for Social Justice Network in the Maritime Provinces of Canada. There are *workers' rights organizations* such as Jobs with Justice and United Students Against Sweatshops.

The environmental message has until recently had difficulty getting traction with conservative Christians because it was branded early and often as nature worship, economically threatening, and godless. It became an easy target for right-wing politicians, but all of that may be in the past. Climate change has become a triggering point for Christians to reexamine notional beliefs about the environment, and an important group of leaders did precisely that by turning to a fellow Christian. He is Sir John Houghton, cochair of the Intergovernmental Panel on Climate Change's working group, lead editor of the first three IPCC reports, professor of atmospheric physics at Oxford, and founder of the Hadley Centre for Climate Prediction and Research. And he is an evangelical Christian. When a group of evangelical leaders (including naysayers and doubters) met privately in Maryland to discuss climate change, it was Houghton who was the primary speaker. By the meeting's end, the participants had agreed to gather signatures for the Evangelical Climate Initiative document, which was later issued with eighty-six prominent signatories. The rhetoric is heating up, and Christians are now taking on once unthinkable positions: "We've spiritualized the devil," says one prominent Christian leader. "But when Exxon is funding think tanks to basically confuse the lessons that we're getting from this great book of creation, that's devilish work. We find ourselves praying to God to protect us from the wiles of the devil, but we can't see him when he's staring us in the face."[12] This is a perfect illustration of how the movement grows: Groups do not subsume their identities to the movement. Rather, the number of groups that address environmental and social issues grows, each keeping its unique character and focus while adding to the richness of the movement as a whole.

Street theater groups parody, satirize, or tease the object of protest. The Deconstructionist Institute for Surreal Topology worked with the progressive wing of the Society for Creative Anachronism during the Free Trade Agreement of America protests in Quebec City in 2001. The two groups created a full-scale, rough-hewn replica of a wooden catapult that lobbed teddy bears, velveteen rabbits, and acrylic dragons across the two-mile chain link and K-rail barrier separating corporate lobbyists and political leaders from demonstrators. Like a scene from the movie *Road Warrior*, the catapult was manned by medieval yeomen with tin pots on their heads, and drawn across the cobbled streets by ten slaves clad in burlap and hemp. As police launched tear gas canisters over the fence onto protesters, fuzzy animals arced back across the barrier onto police. The barrier, erected by police, was eventually taken down by a coalition of skinny vegans, well-muscled steelworkers, and Mohawk warriors wielding bolt cutters and grappling hooks.[13] Police seized what they later called a "dangerous weapon," a pink rabbit, and arrested twenty-nine-year-old activist Jaggi Singh and held him for seventeen days without bail for attacking police with stuffed toys. Singh, who had nothing to do with the catapult, was a favorite target of police because he had been an effective organizer in Quebec and previous Canadian demonstrations. (When Singh showed up at the Republican National

Convention in 2004, the New York *Daily News*—"POLICE ON GUARD VS. VIOLENT TACTICS"—reported that he had shot gasoline-soaked teddy bears at Quebec City police, implying they were terrorist weapons.[14]

There are public activists known as *culture jammers,* exemplified by Kalle Lasn at *Adbusters* magazine in Canada. Culture jamming subverts the blandishments of corporate advertising, public relations firms, and image-makers to expose the realities behind the myth, oftentimes by appropriating their slogans, icons, or artwork. There are guerrilla shoppers who will push empty shopping carts up and down aisles in a daze at big-box retailers such as Toys "R" Us and Sainsbury's in the United Kingdom, never buying anything. According to its tongue-in-cheek Web site, Billionaires for Bush is "a grassroots network of corporate lobbyists, decadent heiresses, Halliburton CEOs, and other winners under George W. Bush's economic policies." Headquartered on Wall Street and with more than sixty chapters nationwide, the group said in 2004 it would "give whatever it takes to ensure four more years of putting profit over people. After all, we know a good president when we buy one." Founded by Chuck Collins and Andrew Boyd (aka Phil T. Rich), Billionaires for Bush debuted at presidential candidate Steve Forbes's campaign kickoff in Concord, New Hampshire, in 1999. Young, clean-shaven, and conservatively attired in dark suits and ties, members filled a third of the pressroom and were given front-row seats. As Forbes began his announcement and national television cameras rolled, the well-camouflaged troublemakers unfurled a banner reading "Billionaires for Steve Forbes, Because Inequality Isn't Growing Fast Enough" and held up signs reading "Tax cuts for me, not my maid!" During the 2004 Bush campaign, they dressed in tuxedos or ball gowns, adorned themselves with bling, held up champagne glasses, and chanted "Four More Wars!" Five years of national attention in the media, including every network and major newspaper, cost them a total of $100,000.

There are also *real billionaires* in the movement. Rich and powerful men who are setting up NGOs and are committed to saving the planet and its people. This group includes George Soros, Bill Gates, Warren Buffett, Gordon Moore, and the collection of private sector funds orchestrated by President Clinton under the auspices of the Clinton Global Initiative. Some of these fortunes were, certainly, accumulated in controversial ways. Warren Buffett continues to be a director and large shareholder of Coca-Cola; Soros speculated in the fall of the British pound to the tune of a $1 billion gain, and Microsoft was hounded by attorneys general for antitrust activities. But these and other hyperwealthy individuals are collectively moving into the vacuum left by the collapsing legitimacy of government. Soros has already donated $5 billion to closely directed projects that promote the concept of an open society, a term originated by Henri Bergson and popularized by Karl Popper at the London School of Economics, where Soros studied. It calls for transparent democratic practices that are based on pluralism and multiculturalism—a politics that is flexible, tolerant, and open to new information, the opposite of an ideological or totalitarian state.[15] The Open Society Institute now operates in sixty countries and ironically has turned its attention from the former dictatorships of Eastern Europe to the Bush administration, whose rhetoric mirrors that of the closed society that Popper and Soros warn about.[16] Warren Buffett, having accumulated spectacular amounts of money in his holding company, Berkshire Hathaway (he paid $8 for his first shares in 1962; today they are worth $90,000, a 21 percent annual rate of return for over forty years), announced in June 2006 the largest bequest in the history of philanthropy, a gift of $31.6 billion to the Gates Foundation, which already had assets of $29 billion. Buffett believes that giving money away is far more difficult than aggregating it. The Gates Foundation, which focuses primarily on the eradication of disease in the developing world, now has an annual budget twice that of the World Health Organization.[17] In a three-day meeting in September 2006, President Clinton received pledges of $7.3 billion to address poverty, AIDS, and climate change, including a promise by Sir Richard Branson to reinvest $3 billion, all of his Virgin Air and other transport company profits for the next ten years, into projects that would address climate change. Between 2001 and 2005, the most generous act of philanthropy was Intel cofounder Gordon Moore and his wife Betty's gift to their own foundation, totaling $7 billion, most of which is dedicated to conservation and the environment.[18]

Perhaps the most original philanthropic idea goes to Mohammed Ibrahim, the founder of Celtel, Africa's largest cell phone network, with 78 million subscribers. Ibrahim believes that one of Africa's main problems is its leaders, a deficiency that is exacerbated by poorly drawn colonial boundaries whose legacy is weak national identity. NGOs and aid agencies have likewise long decried the issue of corrupt and incompetent governance in Africa. Ibrahim is creating a foundation that will grant an award to Africa's democratically elected leaders who deliver security, health, education, and development to their citizens and who leave office when their term is up. Ibrahim is offering those who step down in a timely fashion $5 million over ten years, plus another $200,000 a year for life.[19] It will be the world's richest prize.

Pierre Omidyar, founder of eBay, and his first employee, Jeff Skoll, have created organizations that aggressively promote social values. Along with the Google Foundation, they have instituted ways to fund both profit-making and nonprofit social enterprises, erasing the age-old distinction of civil society as separate from commerce. Omidyar Network has divided $400 million into two for- and nonprofit funds, focusing primarily on microfinance, technology, and community development ideas. Skoll believes his mission is social change, and he tries to achieve it in a variety of ways. His film company, Participant Productions, produces movies squarely aimed at social justice: its releases include *Syriana*, about the dark intrigues of petro-empires; *Good Night, and Good Luck*, which reminded us how the McCarthy era is not so far from where we are today; and Al Gore's *An Inconvenient Truth*. Asked if he was worried about losing his money in Hollywood, Skoll replied that his film company was philanthropy.[20] His eponymous foundation is the leading funder of social entrepreneurship in the world, and has endowed the Skoll Centre for Social Entrepreneurship at Oxford University. The $1 billion Google Foundation addresses poverty, disease, and climate change but has forsaken nonprofit tax status under IRS regulations in order to put money anywhere it wishes, whether in for- or nonprofit enterprises. The Google Foundation can make money, lose money, donate money, or invest money, whatever it takes to accomplish its social mission.

In the past decade, critics of the environmental movement have taken it to task, citing the worldwide collapse of ecosystems as proof of its ineffectiveness. What may be happening is the opposite. Although the momentum of damage and exploitation continues to accelerate in both the social and environmental arenas, the activity addressing it is increasing exponentially and has broken out of its traditional institutional boundaries. Emblematic of this shift is the growth of social entrepreneurship, which refers to activists who use entrepreneurial methods to address systemic social problems. Social entrepreneurs are innovative risk takers who use ideas, resources, and opportunities to tackle problems and produce social benefit. The scope of social entrepreneurship encompasses every category and subcategory listed in the appendix, from war orphans to animal trafficking. Although they can work in both the for-profit and nonprofit realms, their success is measured by social profit; monetary criteria are used where applicable to gauge the sustainability of their programs.

The term *social entrepreneur* is relatively new, dating back to the 1950s and the work of Michael Young of the UK, who created sixty different social benefit organizations in the world, including the School of Social Entrepreneurs in 1997. The person who deserves the credit for promoting the concept and practice tirelessly is Bill Drayton. Beginning in 1980 in the United States, Drayton's Ashoka organization has funded over eighteen hundred fellows from around the world who create practical and replicable solutions to pressing issues. Starting with a meager $50,000, Ashoka had a $30 million budget in 2006. Although the term is relatively new, the practice of social entrepreneurship extends back to the public health movement during the Industrial Revolution and would include such notables as Florence Nightingale, Susan B. Anthony, and M. K. Gandhi. The best-known practitioner of social entrepreneurship is Muhammad Yunus, the creator of microfinance and microcredit, the founder of Grameen Bank in Bangladesh, and the 2006 Nobel Peace Prize winner. Despite a modicum of publicity, the work of most social entrepreneurs remains largely unnoticed compared to business entrepreneurs. Even if they are unnoticed, their influence is not. Managers and executives in large corporations, from GE to Wal-Mart, understand issues concerning the environment in a way that would have been radical in the nonprofit world not even ten years ago. Essentially, the nonprofit and social entrepreneurship sector is a source of memes that are moving into the governmental and for-profit world.

This hybridization of business, philanthropy, technology, and nonprofit activity is exemplified in the work of Daniel Ben-Horin, who became involved in the effort through Larry Brilliant, now head of the Google Foundation, and Stewart Brand, who together in 1985 became founders of the WELL, the Whole Earth 'Lectronic Link, one of the first online virtual communities and one of the first dial-up ISPs in the world. One year later, Ben-Horin posted a question on the WELL about a problem he was having with a printer, and was overwhelmed by the number of helpful, timely responses. From that kindness of strangers emerged his idea for Compumentors (or the Emergency Nerd Response Team, as it was sometimes known), which worked with nonprofits on all their IT needs.[21] Twenty years later, it is a 110-person organization with a $13.5 million budget that not only offers technical assistance, but through its subsidiary, TechSoup, has distributed free or highly discounted software and hardware to over 50,000 nonprofits at a savings to them of $400 million.[22]

Another example of how the movement morphs and grows is the US Green Building Council (USGBC), which addresses a topic that has the greatest environmental impact in the world: buildings, which use 40 percent of all material and 48 percent of the energy in the United States alone. In 1993 David Gottfried, a successful but disillusioned developer, and Rick Fedrizzi, an executive at the Carrier corporation, gathered a small group of architects, suppliers, builders, and designers in order to create a rigorous set of green building standards. Today, the USGBC comprises 6,200 institutional members and 85,000 active participants, and green building councils exist in Japan, Spain, Canada, India, and Mexico. No one has done the metrics, but in its short life USGBC may have had a greater impact than any other single organization in the world on materials saved, toxins eliminated, greenhouse gases avoided, and human health enhanced. It collaborates with designers, architects, and businesses—not always easy, because their movement means the company's products must change—in order to define and incrementally raise the environmental standards of green buildings by means of a rating system called Leadership in Energy and Environmental Design (LEED). LEED has been adopted by cities, government agencies, and universities, and is being applied to the new World Trade Center. USGBC has helped spawn many other organizations, including the Building Materials Reuse Association, which includes over 1,100 companies and nonprofits that recycle and reuse building materials from deconstruction. Architect Edward Mazria has created a nonprofit called Architecture 2030, which has proposed the 2030 Challenge, a bid to make all buildings carbon neutral by that year. It is supported by the American Institute of Architects with its 78,000 members and by the U.S. Conference of Mayors.[23] The challenge begins with requiring all new buildings to use one-half as much energy as the national average of similar buildings, a goal that can be accomplished immediately at little or no incremental cost by employing existing technologies, according to Mazria. By 2015, the goal rises to 60 percent and then increases 10 percent every five years until 2030, when no fossil fuel energy is consumed by a new building. There are three paths to successfully meeting the Challenge: innovative design, new technology, and renewable energy.

Stewart Brand, the author of *How Buildings Learn*, has preoccupied himself for the past decade with how civilizations learn. The purpose of the

Long Now Foundation, cofounded by him and Danny Hillis, one of the inventors of parallel processing, is to foster responsibility by engendering long-term thinking—*really* long-term. The centerpiece of this novel idea is the construction of the world's slowest computer, a clock with a 32-bit mechanical processor designed by Hillis that will keep accurate time for ten thousand years without power and with minimal assistance from humans. The first functioning prototype is on display in London at the Science Museum; the finished product will ultimately be housed in a remote limestone cave in the Great Basin National Park in Nevada. Brand and Hillis call it the Clock of the Long Now. It will chime once a millennium.

Brand and Hillis (and just about everyone else who has considered the matter) believe that shorter attention spans are causing decisions to be made within quicker time horizons; fast results demand fast thinking, which is making for a foreshortened and badly imagined future. This is especially true in business, but also permeates politics, development, economics, buildings, and planning. Founded in 01996, the Long Now Foundation aims to extend time out in iconic and massive forms in order to create the mental continuum in which a future can be imagined. Indeed, it is an attempt to restore human imagination to the very idea of a future. When Pulitzer Prize–winning author Michael Chabon asked his eight-year-old about what would happen in the future, he replied that there probably wouldn't be one, that humans probably didn't deserve a future. His guess was that we would perish by water or fire due to climate change. When Chabon "told my son about the 10,000-year Clock of the Long Now, he listened very carefully, and we looked at the pictures on the Long Now Foundation's website. 'Will there really be people then, Dad?' he said. 'Yes,' I told him without hesitation, 'there will.' I don't know if that's true, any more than do Danny Hillis and his colleagues, with the beating clocks of their hopefulness and the orreries of their imaginations. But in having children—in engendering them, in loving them, in teaching them to love and care about the world—parents are betting, whether they know it or not, on the Clock of the Long Now. They are betting on their children, and their children after them, and theirs beyond them, all the way down the line from now to 12,006. If you don't believe in the Future, unreservedly and dreamingly, if you aren't willing to bet that somebody will be there to cry when the Clock finally runs down, ten thousand years from now, then I don't see how you can have children. If you have children, I don't see how you can fail to do everything in your power to ensure that you win your bet, and that they, and their grandchildren, and their grandchildren's grandchildren, will inherit a world whose perfection can never be accomplished by creatures whose imagination for perfecting it is limitless and free."[24]

What Brand proposes complements and connects to a movement based around the concept of "slow"—Slow Food, Slow Cities, Slow Fish, slow just about everything. What began as a protest against the opening of a McDonald's in Rome's Piazza di Spagna has bloomed into a booming international organization that defends small farmers, local markets, agricultural biodiversity, artisanal producers, the environment, human dignity, small business, and human health. Founded by Carlo Petrini in Orvieto, Italy, in 1986, Slow Food (*alimento lento*) is the long overdue response to dead food, processed food, fast food, agribusiness, Unilever, Nestlé, and General Foods. Like Brand and Hillis, Petrini questions every aspect of the "fast life," an industrial system of commerce and globalization that is causing untold environmental and social damage. The logo for the movement is the snail, an "amulet against exasperation."[25] Slow Food is about pleasure, the delight of taste, place, and conviviality. Petrini believes we have forgotten the simple satisfaction of eating, that sharing food is communion with friends and the earth, and that hosting is more "art than philanthropy."[26] The story of fast food traces a path of wreckage that starts with chemical factory shipments to the farm, proceeds through inhumane slaughterhouses to portion-control factories churning out uniform buns and patties to numbed minimum-wage workers, and ends up in the hospital in the form of obesity, diabetes, and heart attacks—an allegory of modernity. Slow Food supports the re-creation of networks of traditional food producers with customers so that both may thrive. It is about conserving the heritage of the exquisite variety of tastes humankind has created, which means organizing farmer's markets and ensuring both that varieties of fruits and vegetables and rare breeds of animals do not become extinct, and that the people who are artisans of food are supported and can pass on their craft to future generations. Slow Food is spreading throughout the world, with 83,000 members in Italy, Germany, Switzerland, the United States, France, Japan, and Great Britain. At their Terra Madre gathering in Turin in 2006, Senegalese cereal farmers met Peruvian potato growers and Hmong long bean farmers from California.[27] To those who argue that gastronomy is a privilege of the affluent and hardly a suitable environmental cause, Petrini replies that food lovers who are not environmentalists are naive, and an ecologist who does not take time to savor his food and culture leads a deprived and sad life.

Food has always been at the heart of cultural identity. The loss of its traditional foods is just as devastating to a culture as the loss of its language. Although globalization has caused havoc in all areas in every country, slow movements are not anti-globalization; they are pro-localization. Savoring something—a spice, a radish, a piece of cheese—brings us back home to the world in which we walk and breathe. It slows us down. Taste is social. We come together, sit and talk together around food; we clink glasses and laugh and engage in small gossips and whispers in the presence of local beers or wines, tisanes and small cakes with gooseberry preserves and clotted creams, or thin wafers bearing full-fatted cheeses daubed with slices of purple figs. It is how we share being alive. We can engage in the virtual world of iPod music and TV drama, but there is no virtual world of taste. It is in our mouth, and every day our mouth connects us to place.

Every second we chew, our tongue and its ten thousand taste buds evaluate hundreds of millions of molecules—sorting, testing, probing, like a doorman making sure that every bit of food is on the guest list. Our mouth is a direct extension of billions of years of knowledge and learning connected to every cell in our body. It is the way our body detects toxins, the first and most powerful expression of the immune system's deciding what can become part of you and what should not. This ability enabled us to become the only form of life that can say we are a form of life, the only form of life that is self-aware through consciousness, and the first form of life that consciously destroys its habitat and consciously knows that there are biological limits to its desires and appetites. The mouth can tell us so much, but only if we remember to take the time to taste. As Petrini points out, when we lose a flavor, we lose a recipe, and when a recipe is lost, the use of a natural food is lost, and when the use of a food is lost, the cultivation and source of that food is lost, and when production of a food is lost, the seeds of rare animal breeds are lost, and when local food production is lost, people are forced to become consumers of food produced far away by multinational companies.[28] To counter this hollowing out of culinary and agricultural diversity, Petrini has created food arks to save endangered varieties (and recipes), and established tasting schools, especially for children. The taste buds in our mouths are not to be toyed with; they are evolution itself, a teacher, a kindness, a guide. They have guided us here and they can heal us, and heal the earth.

Thousands of NGOs around the world are organizing around localization of the economy, particularly where cities or entire regions have been gutted by globalization, and where people are preparing for a different future. By rejecting the deterioration of the quality and variety of food, localization creates food webs that produce fresher, higher quality food, and provides food security, because it lessens dependence on distant sources. It reduces shipping, energy, and packaging and engenders farmer's markets, festivals, and engagement. Localization strengthens the economy, as money circulates when spent on locally produced items. It also functions as a response to climate change. A growing post-carbon movement is trying to organize communities to reduce their energy use and, as with food, reduce their dependency on imported energy. To do this means rethinking the entire system of a community, from transport to food to housing. Proponents do not dispute that globalization is a fact, but are simply going in another direction. The main business localization group in the United States, Business Alliance for Local Living Economies, has spread to twenty-nine cities. In Bellingham, Washington, five hundred local, independently owned and operated companies have formed Sustainable Connections, which is working together to green their businesses, educate their customers, and support environmental building, farming, energy, and transportation alternatives. If you are envisioning hippies, think again: these are engineers, contractors, technology integrators, homebuilders, printers, Realtors, orchardists, Laundromat owners, health clinicians, and more. In other words, these individuals are creating a community, or in the case of Vermont, a local eatery called Farmer's Diner, which purchases and serves the unique foods of the region, which is what Petrini started doing twenty-five years ago.

The localization of food has a corresponding initiative in the world of information and knowledge. Wikipedia, operated by a nonprofit foundation, has a simple goal: to provide every person free access to all human knowledge. Founder Jimmy Wales was influenced by Friedrich Hayek's arguments about combining individual knowledge to create a better approximation of truth, as well as by the emergence of the open-source software movement in the late nineties, which gave users free access to code, allowing them to change a program in any way they wished.[29] The users of Wikipedia create encyclopedia entries, organize communities, set the rules, joust about content, and comb entries for errors or mistakes. Wikipedia has just five employees, operates on a small budget, is almost entirely supported by user contributions, and is one of the top twenty Internet sites in the world, getting 14,000 page views a minute. In January 2007 Wikipedia included 1,556,317 English articles, fifteen times the number of articles in the *Encyclopædia Britannica* online, and it is adding 45,000 articles a month. There were 518,681 articles in German, 416,550 in French, and more than 100,000 in Polish, Spanish, Italian, Japanese, Chinese, Dutch, Swedish, and Portuguese. Forty-one other languages feature more than 10,000 articles, and nascent Wikipedias are sprouting in 150 more languages, including Asturian, Tatar, Cornish, Diné (Navajo), Avar, and even Gothic (which is extinct). It is blocked in China, but Chinese citizens circumvent the blockade through secondary links to Taiwan.

Wikipedia is the work of a global community and it is not copyrighted. All information can be freely used, modified, and repurposed on other online sites or publications. In a study published on December 14, 2005, comparing the accuracy of the online *Encyclopædia Britannica* with Wikipedia, the editors of *Nature* evaluated them to be roughly equal, although Wikipedia had a few more errors, albeit minor.[30] The for-profit *Britannica*, whose sales have been falling for over a decade, attacked the study in press releases and a half-page advertisement in the London *Times*.[31] One of the key differences between the two institutions is that within five days, the Wikipedia entry for *Encyclopædia Britannica* contained the story of the *Nature* study written in neutral and objective prose, a bedrock Wikipedia principle. The *Encyclopædia Britannica* online site discussing Wikipedia makes no mention of the *Nature* article. The *EB* editors complain that no one is in charge at Wikipedia and that no expert has overall responsibility for any given article, criticisms Wikipedia considers to be badges of honor. Unlike at *Encyclopædia Britannica*, no one can censor content. What may seem like a minor difference is a major shift in how a global social order is organizing itself. Wikipedia is the product of a bottom-up world, the wisdom of the many, also known as crowdsourcing, the rise of the amateur.[32]

I recently received three annual reports from movement organizations in the same week: the Audubon Society, Friends of the Earth, and the India Resource Center. The centennial Audubon annual report was a graphic masterpiece, printed in soft sepia tones on recycled paper. Inside were photographs from the Centennial Gala, featuring the Rockefellers and other black-tie attendees. At the back of the report was the list of the society's 44 vice presidents; 34 directors; 127 state offices, centers, and sanctuaries; and 492 chapters in fifty states, from the Arctic Audubon Society in Alaska to the Florida Keys Audubon Society, 3,457 miles away. One can only imagine how many flyways, species, sinkholes, inlets, forests, lakes, riparian corridors, and marine estuaries Audubon is able to keep an eye on, supported as it is by 254 corporations and 528 foundations, and more than half a million members.

The Friends of the Earth report was issued on a CD, to save paper, and listed sister organizations in seventy countries, which involved over a million activists. Its assets (U.S. only) are a half percent of those of the Audubon Society. In fact, Friends of the Earth has fewer staff in the United States than Audubon has directors, yet in the prior year it managed to stop snowmobiles in Rocky Mountain National Park, pressure JP Morgan to adopt environmental standards in lending, block logging in the Green Mountain National Forest, help obtain a ban on jet skis to protect manatees in the Florida Keys, help push through new rules in the International Maritime Organization on ship emissions, block the planting of biopharmaceutical rice in Missouri, lobby against Newmont Mining's plans in Peru, win a clean-air lawsuit against the EPA in the federal court of appeals, run advertisements in the *New York Times* against Ford Motor Company to improve its fleet's fuel economy, urge the World Bank to increase its funding of renewable energy projects, and more.

The third report I received was from the one-person India Resource Center, and it was e-mailed. IRC's total budget is equivalent to the salary of one staff person at Audubon. It is part of a larger network of NGOs taking on the world's biggest beverage company, Coca-Cola, over concerns about water pollution, toxicity, product safety, and worker rights. The movement against Coke is emblematic of how the smallest organizations confront large institutions. A coalition of tiny organizations comprising farmers, indigenous people, students, and Dalits (formerly untouchables) is using sit-ins, protests, Web sites, and phone cards to create a formidable international grassroots campaign against Coca-Cola that is singeing its global reputation and costing it tens of millions of dollars.[33]

The original protest began in Plachimada, Kerala, instigated by an Adavasi (indigenous) widow named Mylamma, who had farmed land adjacent to a new Coca-Cola bottling plant all her life. Two years after the factory went into production in March 2000, the water table dropped from 150 to 500 feet, drying up the farmers' wells, and the water that Mylamma and the villagers were able to obtain was polluted and undrinkable. In 2003 the district medical officer declared the water hazardous to human health. Mylamma believed that Coca-Cola's pumping of groundwater was responsible for the water situation and led a series of protests that culminated in the local village council's canceling the company's license to operate. Coca-Cola sued, claiming the local council had no jurisdiction, but was turned down by the High Court, which ruled that the company had to obtain its water in some other manner because groundwater belonged to all of the community, and that it could draw no more groundwater from its thirty-four-acre site than could a farmer using it for agriculture.[34]

In 2003 the Center for Science and the Environment (CSE), a nonprofit institute in New Delhi, published analyses of popular soft drinks (including Coke and Pepsi) showing that they contained levels of pesticides, including DDT, lindane, and malathion, eleven to seventy times higher than those established by E.U. drinking water standards.[35] (CSE's analysis of American Coke and Pepsi showed zero pesticide residues.) The Indian parliament banned Coke and Pepsi from its cafeterias, and ten thousand schools and colleges on the subcontinent became Coke/Pepsi-Free Zones. The U.S. companies fought back. Allegedly, Secretary of State Colin Powell intervened at the highest levels. Lawyers swept in from the United States to lobby Indian officials. A massive ad campaign was mounted featuring Aamir Khan, one of the most popular film stars in India, assuring his countrymen that Coke was safe. In 2006 the tests were repeated, and Pepsi and Coke again contained the highest pesticide residues of all soft drinks analyzed.[36] Adding to the controversy, Coca-Cola had been distributing wastes from its bottling plants containing lead, chromium, and cadmium to local farmers for use as fertilizer.

The India Coca-Cola protest movement runs on very little money, whereas Coca-Cola has a market capitalization exceeding $100 billion. The protests have spread to soft-drink plants throughout India and all the way to college and university campuses in South America, the United States, and Europe, spawning an organization called the Student Coalition to Cut Contracts with Cola-Cola. Michigan, Swarthmore, Bard, and other colleges have banned the sale of Coca-Cola products because of the company's environmental practices and violations of human rights. Despite such annoyances, the company's worst enemy has proven to be itself, not global activists. Responding to questions from a *Wall Street Journal* reporter, the company's spokesperson was caught dissembling, if not lying, about analyses of the heavy metal content of processed sludge distributed to farmers.[37] There is a perverse naivete in Coca-Cola's response, the credulity that comes with having read its own press releases for so long, that prohibits it from recognizing the obvious: the company is harming the environment and peo ple. Coke sees no contradiction in promoting sixty-four-ounce plastic bot- tles of Coke containing more than a half-pound of sugar to American teenagers who drink sixty gallons of soft drinks every year, despite the rapid increase of obesity and diabetes in the United States, or in having Olympic athletes promote the product. Coke's so-called natural products, such as Fruitopia, contain 5 percent juice and even more sugar than a Classic Coke. Coke may not notice the 44 billion soft drink cans tossed into landfills and apparently looked in the opposite direction when offering farmers cadmium- tainted wastes from its bottling plant. It has no problem with its India Web site asserting, "Sugar does not cause heart disease, cancer, diabetes or obe- sity," a statement of farcical disregard for nutrition and medical science. The problems it faces are pathologies of intent and power.

The Coca-Cola conflict concerns the rights of a community versus the rights of a corporation, and in that respect is emblematic of NGO campaigns throughout the world. Coca-Cola has appealed to the Indian Supreme Court, hoping to overturn the High Court ruling, and in so doing it is pursu- ing a strategy universally employed by corporations: resorting to higher realms of governance if local bodies impede corporate goals. In 2005 Monsanto, one of the two largest companies in the world selling genetically modified seeds, pressured fourteen state legislatures to pass laws that would prohibit lo- cal restrictions on genetically engineered crops after they lost county elections that called for their ban. In 2005 factory farm and food lobbyists, including Coca-Cola, Pepsi, and Unilever, began to push legislation through the U.S. Congress that would overturn more than seventy-six different state laws regu- lating labels, toxins, and other issues of food safety. The federal government has always been able to set minimum safety standards, but under the National Food Uniformity Act of 2005, the logic is reversed. This legislation—which could also be called the Faster Food Bill—would prevent a state from setting different standards from the FDA's, and has passed the House at the time of this writing. Under California state law, for example, fish containing danger- ous levels of mercury must be so labeled, a vital health issue with respect to food choices for pregnant women and their unborn children. Under the new federal law, such labeling would be forbidden. New legislation being passed and proposed increases the rights of business at the expense of consumers, much as WTO rulings can override national standards, as discussed earlier.

Karl Marx had one goal: to change the world. He was dismissive of those who spent their lives in an ongoing effort to interpret the world because he had the luxury of believing there was only one correct interpretation. Such a narrow view is common to ideologies, and it is why many diverse groups are arising in opposition to this sort of rigid thinking. Ideologies exclude openness, diversity, resiliency, and multiplicity, the very qualities that nourish life in any system, be it ecosystem, immune system, or social system.[38] Hundreds of thousands of small groups are trying to ignite an array of ideas in the world, fanning them like embers. Ideas are living things; they can be changed and adapted, and can grow. Ideas do not belong to anyone, and require no approval. This may sound ethereal but it is in fact the essence of praxis, the application of grassroots democracy in a violent and exploitative age.

Some would argue that it is counterproductive to conflate all the different organizations and types of organizations into a single movement, that it is self-evident that such divergent aims cannot create an effective, unified body. It's true that pluralism, the de facto tactic of a million small organizations, functions best in a society that cultivates diversity, dialogue, and collaboration; in a you-are-either-with-us-or-against-us society, small, single-issue organizations are effectively marginalized. In the United States, the environmental and social justice movements emerged in what was a pluralistic society. Because that is increasingly not the case, the stratagems and goals of the movement may be inadequate to the increasing centralization of power. Volunteer organizations that focus on the health of a stream or riparian corridor are still needed, but their work may be futile if the larger issue of climate change is not addressed. Addressing climate change is futile, in turn, if political corruption is not eliminated. The corrupting influence of large corporations cannot be addressed unless campaign finance is addressed. Similarly, vaccinating the children of the developing world, an extraordinarily helpful act that has cascading benefits, including population reduction (families have fewer children if they are confident their children will survive), could be undermined by shifts in climate that could cause mass starvation.

Can myriad organizations work together to challenge deeper systemic issues? Do organizations step back and see where there is overlap? Do they cooperate sufficiently? Do they try to create synergies, maximize funding, encourage efficiencies, and sublimate their identities to a larger whole? Not as much as is possible, and is necessary. The fact that the movement is made up of pieces does not mean it can only work piecemeal. The Garfield Foundation, a small New York group, spent a year working with twelve nonprofit organizations and eight foundations in the Midwest to create and implement a plan whereby clean energy would rapidly increase as a proportion of overall consumption. They determined four leverage points, all of which were being addressed in some ways by a number of the groups, but not with the systematic approach that would produce results: retiring old coal-fired generating plants that produced the greatest pollution, stopping the licensing of new ones, increasing the production of renewable energy, and raising energy efficiency in the region. Then they went to work and spent months in meetings, with the group eventually growing to include thirty NGOs. From that collective came a unifying vision, ReAmp, with an agenda that was accessible to citizens, government, and business: to create a vibrant Midwest clean energy economy that would increase jobs, investment, and prosperity while addressing climate change, culminating in an 80 percent reduction of greenhouse gases by 2030.

If anything can offer us hope for the future it will be an assembly of humanity that is representative but not centralized, because no single ideology can ever heal the wounds of this world. History demonstrates all too eloquently that no ideology has ever amounted to more than a palliative for any dire condition. The immune system is the most complex system in the body, just as the body is the most complex organism on earth, and the most complicated assembly of organisms is human civilization.

The movement, for its part, is the most complex coalition of human organizations the world has ever seen. The incongruity of anarchists, billionaire funders, street clowns, scientists, youthful activists, indigenous and native people, diplomats, computer geeks, writers, strategists, peasants, and students all working toward common goals is a testament to human impulses that are unstoppable and eternal. The founder of Earth First, Dave Foreman, and the chair of the New York Council of the Alaska Conservation Foundation, David Rockefeller Jr., want the same things for Alaska: no drilling in the Arctic National Wildlife Refuge, moratoriums on indiscriminate game hunting, wildlife corridors for migratory species, permanent protection for the roadless areas of the Tongass and Chugach old-growth forests, elimination of all clear-cutting in the national forests, challenges to all timber sales and concessions by the Department of the Interior, and banishment of destructive bottom trawling in the fisheries. The list goes on. The two Davids do not know each other. They do not have to hoist a pint or exchange e-mails to work together, because their goals are the same, however different their politics, backgrounds, wealth, and education. This is the promise of the movement: that the margins link up, that we discover through our actions and shared concerns that we are a global family.

The ability to respond to the endless injustices and hurts endured by the earth and its people requires concerted action and hinges in part on understanding both our function and potential as individuals and where we fit into a larger whole. Antigens dot the surface of our body's cells like lapel pins that proudly proclaim, "It's me, don't hurt me, I am you." Viruses and invasive diseases have their own antigens that warn the body that a "not me" has arrived. Millions of different kinds of antigens tag the different microorganisms and cells that find their way into the body, especially detrimental ones. With almost perfect symmetry, millions of different antibodies, proteins that can lock on to antigens as neatly as a key to a hasp, neutralize these invaders while simultaneously signaling for help. This is the beginning of the immune response, the ability of the body to maintain the self, to be a human rather than a petri dish for opportunistic microorganisms. The hundreds of thousands of organizations that make up the movement are social antibodies attaching themselves to pathologies of power. Many will fail, for at present it is often a highly imperfect and sometimes clumsy movement. It can flail, overreach, and flounder; it has much to learn about how to work together, but it is what the earth is producing to protect itself.

For much of medical history, the immune system and brain were considered two completely separate entities. Over the past two decades, science has mapped the many interactions between the two, demonstrating that each affects the other, right down to what we are thinking. Gerald Callahan, associate professor of immunology at Colorado State University, has upped the ante, stating what may be obvious from an evolutionary view: *the brain is part of the immune system.*[39] The immune system predates the brain by a good billion years. While the immune system responds to microscopic threats, the brain defends against risk that is too big for our natural immunity to handle. "The mind is for bears, coral snakes, sharks, snapping turtles, wife beaters, and Buicks," expains Callahan. The immune system addresses organisms that have been around for billions of years; the brain confronts relatively newer dangers.

The massive growth of citizen-based organizations responds to threats that are new, immense, and, in some cases, game-ending. These groups defend against corrupt politics and climate change, corporate predation and the death of oceans, governmental indifference and pandemic poverty, industrial forestry and farming, and depletion of soil and water. Five hundred years of ecological mayhem and social tyranny is a relatively short time for humanity to have learned to understand its self-created patterns of systematic pillage. What has changed recently, and has offered evidence that hope may be a rational act despite the onslaught of countervailing data, is the use of connectivity. Individuals are associating, hooking up, and identifying with one another. From that meeting and experience they are forming units, inventing again and again pieces of a larger organism, enjoining associations and volunteers and committees and groups, and assembling these into a mosaic of activity as if they were solving a jigsaw puzzle without ever having seen the picture on its box. The insanity of human destructiveness may be matched by an older grace and intelligence that is fastening us together in ways we have never before seen or imagined.

There is fierceness at work here. There is no other explanation for the raw courage and heart displayed over and again in the people who march, speak, create, resist, and build. It is the fierceness of knowing we are human and intend to survive. To witness the worldwide breakdown of civility into camps, ideologies, and wars, to watch the accelerating breakdown of our environmental systems, is harrowing and dispiriting.

But immune systems do fail; this movement most certainly could fail as well. What can help preserve it is the gift of self-perception, the gift of seeing who we truly are. We will either come together as one, globalized people, or we will disappear as a civilization. To come together we must know our place in a biological and cultural sense, and reclaim our role as engaged agents of our continued existence. Our minds were made to defend ourselves, born of an immune system that brought us to this stage in our development and evolution. We are surfeited with metaphors of war, such that when we hear the word *defense,* we think *attack,* but the defense of the world can truly be accomplished only by cooperation and compassion. Science now knows that while still in diapers, virtually all children exhibit altruistic behavior. Concern for the well-being of others is bred in the bone, endemic and hardwired. We became human by working together and helping one another. According to immunologist Gerald Callahan, faith and love are literally buried in our genes and lymphocytes, and what it takes to arrest our descent into chaos is one person after another remembering who and where they really are.

RESTORATION

Although we carry the ocean within ourselves, in our blood and in our eyes, so that we essentially see through seawater, we appear blind to its fate. Many scientists speak only to each other and studiously avoid educating the press. The media seems unwilling to report environmental news, and caters to a public stalled by sloth, fear, or greed and generally confused by science. Overall, we seem unable to recognize that the proofs so many politicians demand already exist in the form of hindsight. Written into the long history of our planet, in one form or another, is the record of what is coming our way.

—Julia Whitty, "The Fate of the Oceans"[1]

What is abundantly clear is that all *life—from bacterium to elephant—shares common characteristics at the level of molecules. There is a common thread that runs through the whole of biological existence. . . . These molecules run through life in the same way as the musical theme runs through the last movement of Brahms's Fourth Symphony. There is a set of variations which superficially sound very different but which are underpinned by a deeper similarity that binds the whole. The beauty of the structure depends upon the individuality of the passing music, and also upon the coherence of the construction. That vital spark from inanimate matter to animate life happened once and only once, and all living existence depends on that moment. We are one tribe with bacteria that live in hot springs, parasitic barnacles, vampire bats and cauliflowers. We all share a common ancestor.*

—Richard Fortey,
Life: A Natural History of the First Four Billion Years of Life on Earth[2]

Every cell's dream is to become two cells.   —François Jacob[3]

One quadrillion cells make up a human being, and 90 percent of them are bacteria, fungi, yeasts, and other microbes, without which we could not survive. Therein lies a paradox: what makes us fully human is, well, not human. The prospect that we descended from lower primates, anathema to Christian fundamentalists, is a relatively minor phenomenon in the larger picture of science. Within our body is the backstory of the earth four billion years ago, the molecular chains, elemental compounds, simple bacteria, and salty fluids that wash our eyes and surround our cells, forming a compendium of life that has preceded us. We have always been a work in progress, a cumulative animal, a chimeric fusion of different organisms from the beginning of life "bound together by the elastic string of time."[4] It is thought that our microbial ancestors came into being when a soupçon of carbon polymers, nucleotides, and amino acids combined over a sulfurous oceanic vent. It was a preternatural event, if ever there was one: a living cell made of inanimate compounds. In truth, we do not know precisely how life started, and as one biologist noted, only fools and knaves would venture such a claim. The leap from chemical soup to microbe seems astounding, as if Gutenberg's first printing press were a laser printer attached to a laptop—and not just any laptop, but one that could generate new laptops endlessly from simple compounds. No doubt, preceding the creation of the first cell were tens of millions of years of chemical experiments in which precursor forms of life existed in different combinations. Yet until recently it was thought that the odds of life arising in such a dilute broth were so remote that no statistical analysis could approximate the eventuality—one analogy suggested a hurricane assembling a Boeing 747 after sweeping over a wrecking yard full of parts.[5] Science, which once bought into the junkyard odds, has flipped the logic: we now regard the miracle of life not as its impossibility but as its inevitability.[6] Given conditions similar to those that existed on earth when life began, life would emerge again and again.

That first form of life was a microscopic single-celled organism called a prokaryotes, which roughly translates as "pre-kernels," cells without true nuclei. They are graphically depicted in textbooks as capsules with wiggly material inside, which is a gross simplification. A single bacterium cell, *Escherichia coli,* contains 2.4 million protein molecules of nearly 4,000 different types, 280 million small metabolite and ion molecules, 22 million lipids, a genome consisting of 4.6 million base pairs of nucleotides, and 40 billion water molecules, all packed into a cell whose diameter is one-hundredth the width of a strand of hair.[7] Those first cells, in Robinson Jeffers's words, "had echoes of the future" in them[8] and essentially took over the planet. They are in every ditch, on every leaf, in the sky, at the South Pole, on our tongues, three miles deep in the ocean, and throughout the deserts of the world. They created photosynthesis, respiration, and fermentation, and eventually mitochondria and chloroplasts, the organelles that digest, breathe, and circulate nutrients in our cells. Although we have identified the molecules in a single *E. coli* cell, we do not understand how they work together to create shape, reproduction, mentation, and purposeful behavior. When we take apart a cell, life disappears, because molecules are *all* we find.

Prokaryotes pioneered the different pathways through which life can metabolize energy, from sunshine to sulfur, and then started a grassroots movement, combining into compartmentalized colonies to create a new form of life, eukaryotes, cells with nuclei that could assemble themselves into skinks, honeysuckle, and humans. Eukaryotes took the molecular palette devised by prokaryotes and became artists, assembling into millions of different life forms, creating praying mantises, chanterelles, night-blooming jasmine, and limbic systems, a great stew of life spiced with shyness, coffee beans, caribou herds, and Mahler's Second Symphony. The difference between a prokaryote and a eukaryotic cell is like an igloo compared to the city of Paris. With 30,000 genes and 400 billion molecules, a single animal cell exceeds any Intel microprocessor in its prodigious computational breadth. Each cell simultaneously conducts millions of molecular processes involving trillions of atoms. Multiply that activity by the trillions of cells in the quintillions of creatures on the planet, and it raises a fair question: Who, exactly, is in charge? The total amount of intra- and intercellular activity in just one human body is staggering: one septillion actions at any one moment—a one with twenty-four zeroes after it. In one second, our body has undergone ten times more processes than there are stars in the universe, exactly what Charles Darwin foretold when he said that science would discover that each living creature was a "little universe, formed of a host of self-propagating organisms, inconceivably minute and as numerous as the stars in heaven."[9]

Complex, multitrillion-celled replicating organisms called *Homo sapiens* can argue with one another about the environment and how serious climate change is. But we cannot sit down with a cell and discuss our personal aspirations or the flaws of free-market capitalism. Life is life, and nothing politicians have said or voted for can influence primary biological principles. As author Bill McKibben wryly noted, given a choice between the laws of Congress and the laws of physics, he was pretty sure the laws of physics would prevail. It is an understandable vanity for humans to believe that their cells are privileged or unique, but the distinction between human cells and those of a sunflower is shockingly narrow, while between primates and humans, the difference is slender as a thread. To distinguish between human cells and those of newts, seals, or coyotes, one has to descend to the molecular level of the cell to find the odd dissimilarity. We *are* nature, a realization that stopped Emerson dead in his tracks in Paris, and may it stop us in ours. We live in community, not alone, and any sense of separateness that we harbor is illusion. Humans are animals, albeit extraordinary ones, and have no special immunity conferred upon them. Given the present rate of planetary pollution and destruction, we need to negotiate a détente with nature and ourselves.

The collective functions of the septillion concurrent activities in our body have a name: resilience. This extraordinary redundancy is why we can be callous about our physical needs, bolting fast food, poisoning ourselves with alcohol and drugs, living in polluted air, and still surviving. Likewise, we can insult nature in myriad ways and still have cornflakes, an SUV, and a functioning planet each morning. Resilience is one of the secrets of life, and functions as the opposite of the domino theory. Cells mutate and fail constantly, but do not take down the cells around them. Healthy organisms and ecosystems are diverse, unpredictable, redundant, and adaptive. Life is astonishingly connected, but it refuses to march in lockstep or synchronize its watches. Any living system is a dialectic of harmony and autonomy, persistence and flux, predictability and instability. We don't have one brain but three—reptilian, limbic, and neocortical—and each has different functions and capabilities. A human can lose his eyesight, limbs, a frontal lobe, a kidney, and half his lung capacity, and still survive. By definition, evolution produces creatures and systems that have the greatest ability to persist

over time, and resilience allows an organism to withstand the greatest range
of disturbances. This is as true for social systems as it is for environmental
ones, for governments and corporations as it is for fisheries and reefs. The
more resilient a system, the more shocks and impacts it can withstand and
still recover.[10] Conversely, as systems lose diversity and thus functional redundancy, they become vulnerable to disruption or collapse.

Ecology is about how living organisms interact with one another and
their environment. Sustainability is about stabilizing the currently disruptive relationship between earth's two most complex systems—human culture and the living world. The interrelation between these two systems
marks every person's existence and is responsible for the rise and fall of every
civilization. Although the concept of sustainability is relatively new, every
culture has confronted this relationship, for better or ill. For thousands of
years civilizations have not been able to reverse their tracks with respect to
environmental damage but rather have declined and disappeared because
they forfeited their own habitat. Today, for the first time in history, an entire
civilization—its people, companies, and governments—is trying to arrest
the downspin and understand how to live on earth, an effort that represents
a watershed in human existence. Life is either increasing or decreasing; there
is no Goldilocks happy medium where everything is just right. At this point
in our environmental freefall, we need to preserve what remains and dedi-
cate ourselves to restoring what we have lost.

Some would argue that the task is futile. Robert Kaplan's grim auguries
of a world in which crime, violence, and anarchy are fed by environmental
scarcity and inequities seem uncomfortably present. Jared Diamond's *Collapse* chronicles the ways human beings have repeatedly ignored environmental feedback and spiraled into oblivion. Despite the popularity of such
books, the general public is, at best, only dimly aware of the extent to which
problems are rapidly multiplying. Nevertheless, the world is fast reaching a
we're-not-in-Kansas-anymore moment whether it realizes it or not. Although the scale of environmental and social breakdown is so vast it isn't
possible for any one individual or institution to be fully informed about it,
the warning signs are omnipresent.

Recently two largely unnoticed events occurring within a span of twelve
months marked our future on earth. On March 30, 2005, the *Millennium
Ecosystem Assessment* report was released by organizations representing 1,360
scientists in 95 countries. The $24 million *Assessment* was the largest such
scientific study ever undertaken of the planet's carrying capacity. It was the

first time global civilization surveyed the world's biological resources and assessed how the increasing losses would affect our future. Although detailed in its analysis, the final diagnosis was straightforward: the earth is wearing out and will soon become exhausted, incapable of supporting life as we know it. The report included little data that hadn't been said, published, or ignored before. The novelty of the study was its breadth and the deliberateness of tone. It acknowledged what scientists have been alarmed about for more than a decade: ecosystems, like all nonlinear systems, do not necessarily wind down gradually when under assault but may reach triggering thresholds, ecological heart attacks, where they suddenly collapse and die. Resilience can only protect a system to a point before failure can take down the whole.[11] What the scientists couldn't say, journalists around the world could and did: We are on the brink of disaster.[12]

What that report did not address was climate change, a parallel brink. In the past year prognostications about the timing and effects of climate change have become even more dire. The transformation of the earth's climate has typically been portrayed as a gradual warming that will cause a series of adverse effects: rising sea levels, loss of some species, the movement of tropical disease into temperate zones, more droughts, more intense rainfall and floods, more powerful hurricanes, and shifts in agricultural productivity. In other words, tough stuff, but something we can adapt to and live with. New science, however, suggests that current climate models greatly understate the rate and magnitude of change.[13] We may have already broken the thermostat, and if we do not get it repaired, all other ameliorative activities will dim in relevance. The greatest warming today is occurring at the poles, not the equator, and rising temperatures there are releasing another gas, methane, from permafrost, where it has been locked up for millions of years. Methane is twenty-four times more powerful than carbon dioxide as a greenhouse gas. A rapid rise in its release into the atmosphere would create a dramatic increase in warming, a positive feedback loop that would accelerate additional methane release. If such a runaway event were to take place, it could occur within forty years or less, and would transform the planet into a biological desert. Once that is precipitated, no change in our energy use could affect the outcome. As the long-term predictions became more dire, the daily weather becomes more biblical, which perhaps explains why 2006 was the year that many die-hard skeptics dropped their objections and urged action on climate. *The Economist*, which a year earlier was praising Michael Crichton's book comparing climate scientists to Nazi eugenics professors, reversed its course in a special issue addressing climate change.[14] Rupert Murdoch's right-wing London tabloid *The Sun*, which had heretofore ridiculed environmentalist and climate issues, printed ten things readers could do to reduce their energy use. And as previously noted, it was the year Christian evangelical leaders called for rapid and massive efforts to address climate change. In short, the movement to restore the earth heated up.

How could something so important as this movement grow so much and be largely unseen? Describing the breadth of the movement is like trying to hold the ocean in your hand. It is that large. When an iceberg rises above the waterline, the massive ice beneath is unseen. When Wangari Maathai won the Nobel Peace Prize, the wire service stories didn't mention the network of six thousand different women's groups in Africa that were planting trees. When we hear about a chemical spill in a river, it is never reported that more than four thousand organizations in North America have adopted a river, creek, or stream. We read that organic agriculture is the fastest-growing sector of farming in America, Japan, Mexico, and Europe, but no connection is made to the more than three thousand organizations that educate farmers, customers, and legislators about sustainable and biological agriculture.

The distinctive bent of the movement is to tackle problems directly. We live in a faith-based economy, and by that I do not refer to religious practice. People are asked to place their faith in economic and political systems that have polluted water, air, and sea; that have despoiled communities, sacked workforces, reduced incomes for most people in the world for the past three decades, and created a stratosphere sufficiently permeated with industrial gases that we are, in effect, playing dice with the planet. One does not have to demonize the corporate system to recognize that it has no means to account for its negative impacts, except as a charitable footnote to its annual reports if it is inclined to donate a small part of its earnings. As that faith begins to seem more and more misplaced, the way to change the world is to change one's own practices, including one's home, source of energy, method of agriculture, diet, transport patterns, and communities. Not that Kyoto Protocols shouldn't be signed or adopted—symbols are ever important—but you can't get there from here by any mechanism that depends on support from institutions that benefit from the status quo. Efforts must continue to be directed to bring about institutional change, but such efforts cannot succeed unless people reexamine how they behave and consume in their own lives. The movement can be seen as weak when measured against large institutions, but its goals are more important. The goal is to create a more resilient social and economic understory in what is basically an oligarchic world, a powerful act that restores a measure of autonomy and power to citizens.

To understand what the movement is, you need to ask what it does.[15] Molecular biologist Mahlon Hoagland wrote a primer entitled *The Way Life Works* that identifies sixteen qualities common to all living organisms, and most apply to social movements.[16] The first trait: *Life builds from the bottom up.* Just as complex organisms are built of cooperating communities of cells, the movement to address environmental and social issues has been built up by small, cooperating groups of people. Just as cell communities in the body attend to different functions, from taste buds to kidneys, groups organize around specific causes, missions, and objectives. Because the movement's growth rises from the grassroots, it can at first appear powerless; powerful people, after all, have the means to express themselves and satisfy their needs. For those who feel excluded by governments, corporations, or institutional processes, voluntary and nonprofit associations are the sanctioned and sometimes nonsanctioned way they can express their needs. Together these groups form a different body politic, and create different social building blocks. There is no better example of building from the bottom up than the thousands of organizations that focus on microfinance, providing loans to disenfranchised or poor people who do not qualify for loans from traditional financial institutions because they lack assets or have low income, or because of discrimination, such as that against women. The intention of microcredit institutions is to alleviate poverty by funding income-generating activities for self-employment wherever the poor live, from Apoyo Para el Campesino Indígena del Oriente in Bolivia to the Zimbabwe Association of Microfinance.

Just as *life assembles itself into chains,* nonprofits aggregate either by linking up interests, people, or communities, or by linking to related organizations. The building blocks of all life forms are polymers, long chains of smaller units called monomers. We have many names for polymers, depending on their composition: leather, starch, protein, DNA, cashmere, cellulose, egg whites, spider silk, cotton, toenails, rubber, crab shells, and enzymes. The basic function of the movement is linking, and there are many names for the constituent subsectors, depending on the units being joined. The social polymers are profuse and include women's rights, wetlands, wildlife corridors, water, waste reduction, wealth disparity reduction, wind power, workers' rights, and women's health—and these, as you may have noticed, are just some of the areas that begin with the letter "w."

What works in nature persists, and it works because it evolved through unceasing invention and experimentation. *From a few themes, life generates many variations.* Those variations are generated endlessly and relentlessly. To continuously take advantage of new possibilities requires constant change and adaptation, which results in many diverse organizations. I have heard it said many times that the movement is inefficient, that there are too many groups and too much overlap. I have no doubt that is true, but I suspect the opposite is also true: namely, that citizen-based organisms are the most efficient social entities on earth, outstripping corporations and institutions manyfold in how effectively they deploy resources. Rather than being the lowest common denominator of social organization, they should be regarded as the fundamental unit of social change. Democracies aggregate votes into centralized bureaucratic institutions. The movement morphs social intention into agile, responsive organizations. They are more effective precisely because they make the most efficient use out of limited resources. I doubt there is a health organization in the world as beneficial *and* efficient as Paul Farmer's Partners in Health, a community-based nonprofit in Haiti's Central Plateau. Patients who live near the clinic that the Partners operate with their sister organization, Zanmi Lasante, get better attention and health care than people in the South Bronx or South Central Los Angeles. No money-center bank in New York or London has a loan portfolio as diverse, cost-effective, and helpful to its depositors and society as the Grameen Bank in Bangladesh, the pioneer in microlending. The Rainforest Action Network has done as much to instigate corporate reform (Citibank, Goldman Sachs, JP Morgan) as Eliot Spitzer did as attorney general of New York at a fraction of the cost. Very few of these organizations have sustained resources or savings; they live year to year, and it is difficult to waste money when you don't have it. Governments and corporations spend money to solve problems because they can. The NGO community has no such capability. It must find social niches within which to survive, make itself immediately useful, and constantly adapt to the needs of its stakeholders. Feedback loops are short, learning is accelerated.

Because the movement is not an ideology, there can be no concision of goals, no succinct slogans representative of the whole. It is a body of thought that coheres into a values system but not a belief system; it is a confluence of evolving ideas that never ceases; a creator of choice, actions, and solutions that confront suffering and degradations visited upon people and the earth. Its atomization may prevent it from ever becoming coherent enough to challenge marauding institutions as large as the U.S. government or ExxonMobil; it could succeed on many of the smaller issues and still lose the day. In response to the charge that the movement represents a gloomy assessment of the state of the world, may it be said that every person working in the movement has a parallel goal: to prove all doomsayers wrong. It is not burdened with a syndrome of trying to *save* the world; it is trying to *remake* the world. Nonprofits find partners wherever they can, and address key issues by joining with other institutions, whether they are corporations, governments, universities, or religions. What distinguishes the movement from more venerable institutions is that it does not require an overriding structure, central authority, or dominance to function. Just as *life organizes with information,* the most powerful instrument wielded by the movement is an unimpeded flow of information, for that directionless communication is the only way the whole of humanity can reorganize itself.

Just as the human body cannot be explained or managed by conventional means, neither can humanity. Hoagland estimates it would require 1,500 encyclopedias to create an owner's manual for one person. The exquisite integration of movement, thought, physiology, sight, touch, and metabolism supersedes the complexity of any other system we can imagine. *Something* operates us, but what? Is it not the free flow of brilliant and ancient information, an involuntary and endemic ·intelligence freely exchanged on the cellular and intracellular level? This is the system in which we should place our faith, because it is the only one that has ever worked eternally. If this enlightening, enlivening pulse is God, then may we get on our knees and give thanks night and day. If it is Allah, may we face the east five times between sunup and sundown and humble ourselves. If it is Yahweh, may we touch the Holy Wall and shed tears of gratitude. If it is biology, may science touch the sacred. I believe it is all of these, but whatever it may be to each person, and however we name it, it is not knowable.

For similar reasons we may deduce that it is reasonable to conclude that we cannot embrace or manage the plethora of problems that confront us. The world simply appears to be out of control. Too often, however, such problems seem insoluble because of how they are managed—with ideological, top-down, oligarchic, militaristic management styles. If we tried to consciously control our bodies, we would die, just as the planet is dying. We don't manage our bodies because we cannot. We can, however, protect, nurture, listen to, and tend to them with food, sleep, prayer, friendship, laughter, and exercise. And that is all the planet asks from us: allies, rest, nurturance, respect, celebration, collaboration, and engagement. Can a global system of citizen-based organizations with simple, clear values turn the world away from war, climatic chaos, social devolution, and environmental collapse? If history is a guide, the answer is no. In the past two centuries, we have seen the struggles for fundamental rights of freedom, democracy, and human dignity repeatedly overtaken by chronic and endemic poverty. On top of that ongoing effort, we now face another task, a campaign to surmount our legacy of environmental neglect. To succeed requires ubiquity, a network of informants, a conspiracy of social imaginaries, groups that cultivate new knowledge, share it, seek information elsewhere, and provide it to agencies and citizens who need it. Without question, each individual part of the movement is not up to the task, for it will inevitably be outgunned by larger institutional forces. But as each organization attends to its mission, it need not attempt to match the sheer firepower arrayed against it. These groups do not have to dominate the world with a new order; they need only take their rightful place in a multicentric planet in which no institution is dominant. Rather than having megasolutions, they need to solve for pattern.[17]

The term *solving for pattern* was coined by Wendell Berry, and refers to a solution that addresses multiple problems instead of one. Solving for pattern arises naturally when one perceives problems as symptoms of systemic failure, rather than as random errors requiring anodynes. For example, sustainable agriculture addresses a number of issues simultaneously: It reduces agricultural runoff, which is the main cause of eutrophication and dead zones in lakes, estuaries, and oceans; it reduces use of energy-intensive nitrogen-based fertilizers; it ameliorates climate change, because organic soil sequesters carbon, whereas industrial farming releases carbon dioxide to the atmosphere, and is the second-greatest cause of climate change after fossil fuel combustion; it improves worker health because of the absence of toxic pesticides; it enables soil to retain more moisture and is thus less reliant on irrigation and outside sources of water; it is more productive than conventional agriculture; it is less susceptible to erosion; and it provides habitat for pollinators, birds, and beneficial insects, which promotes biodiversity. On top of all that, the resulting food commands a premium in the market, making small farms economically more viable. Solving for pattern is the de facto approach of the movement because it is resource constrained. It cannot afford "fixes," only solutions.

Nature works in cycles, and so does a healthy society. A self-correcting system thrives because of feedback. The movement is composed of small organizations because it is on the ground, with its people at the scene—a scale at which information can be generated and acted upon. At this level, organizations quickly adapt. Mistakes are hidden treasures, Joycean "portals of discovery," because we learn from our failures. The opposite of learning is a runaway system where mistakes are relegated to file cabinets and ignored. When a government, corporation, financial institution, or religious organization insulates itself, its initiatives, however well intended, create uncontrolled outcomes and second-order effects that generate newer problems. The current state of the world reflects a problem-solving methodology never seen in nature: remedies from above imposed upon the excluded. The movement offers a solution-creating methodology from below that is inclusive, a process that mimics biological adaptation and evolution. Every physical activity the human body sustains is part of a cyclical, biological system with a self-correcting bias. The same should be true of every social activity, with a system called democracy.

Nature recycles not only information, *nature recycles everything;* nothing is wasted, nothing is thrown away because there is no "away." All natural processes are cyclical, and every scrap of matter, atom, and molecule is reused and repurposed into new flows of life. Industrial society behaves like a spoiled child casting away its unwanted toys in every direction, the only creature that leaves a wake that cannot be recycled by nature *or* industry. The movement doesn't merely advocate recycling, it actively imagines a system of human production that is as elegant, frugal, and abundant as what we observe in nature. One of the first people to have discussed human production in biological terms was economist Kenneth Boulding, a native of Liverpool who became a brilliant academician on two continents. In 1965 Boulding introduced the concept of "spaceship earth" in a lecture as a trope to help people understand that our prowess in development and subduing nature was changing our perception of a limitless earth into one that was a "tiny sphere, closed, limited, crowded, and hurtling through space to unknown destinations."[18]

In his book *Operating Manual for Spaceship Earth,* published four years after Boulding introduced the trope, Buckminster Fuller commented that spaceship earth had been so extraordinarily designed that human beings, who had been traveling on it for at least two million years, had yet to recognize they were on a spaceship. And indeed, how would one design a spaceship to support biological life for two million years, or four billion?

This is a question I have sometimes posed to corporate managers who could not see the practicality or necessity of transforming their business practices into ecological ones. One event at a large company that specialized in agricultural chemicals was particularly instructive, because it was precipitated by a vice president's sharp retort to a colleague's statement that there needed to be equitable distribution of resources as a prerequisite for moving toward a more sustainable world. His exact reply: "That is communism, socialism— it has nothing to do with ecology or the environment." Sixty of the company's chemical engineers were then divided into four teams, each with the same task: in two hours, design a spaceship that could leave earth and return in one hundred years with its crew alive, healthy, and happy. A biome was called for—an ecosystem that would provide food, clean water, medicinal plants, and fiber for a century. Each team also had to design the entire culture of this society—who would be on the ship, what they would do, the lines of authority, and all the messy details of creating and maintaining a society. The spaceship could be as big as necessary, and it could receive light from outside. But it had no escape hatches, and what happened on the spaceship stayed on the spaceship for a century.

All four proposals were sophisticated, but one stood out as the preferred ship for the long voyage out and back. The winning designers set up some unusual features. Instead of bringing caches of DVDs and display screens for onboard entertainment, they decided that a significant proportion of the passengers should be artists, musicians, actors, and storytellers. To endure for one century, the passengers needed to *create* a culture rather than simply consume one. They brought onboard a large variety of weeds, not just useful seeds, to enliven the soils and bring minerals to the surface. They brought mycorrhizae and other fungi, bacteria, insects, and small animals— everything their company poisoned on earth for a profit. (The company's number-one product was a pesticide.) Of the several thousand products this company made, none were invited along on the trip. The designers realized they were too toxic to be released in a small environment, that being a spaceship five miles in diameter. Essentially, the winning team created a diverse ecosystem within which a socially just and equitable society practiced organic agriculture and designed all objects for disassembly, reuse, and recycling. When the participants were asked if it was fair that 20 percent of the passengers received 80 percent of the fruits, vegetables, and medicines produced onboard, all of them, including the vice president who had been disgusted with the idea of equity, shouted the idea down and agreed that it would be unacceptable. Then the VP realized what he had said. After the exercise, a group of employees began an organic garden at headquarters, and several engineers quit their jobs.

The power of the spaceship model is not only metaphorical but also pedagogical. It teaches systems thinking, a holistic approach to the interaction and interdependence of constituent parts and how they function together over time. How we came to believe that the earth could support disposables, heavy-metal contamination, Superfund sites, and nuclear testing is a question I leave to cultural historians. Despite centuries-long practices of despoliation and pollution, almost every responsible corporation in the world is moving away from destructive practices and trying to institute more sustainable ones, and all of them have turned to NGOs to assist, teach, inspire, and urge them on. The stereotype of civil society is groups resisting corporations, and that is true as outlined in previous chapters. What is also true, however, is that nonprofit groups have formed productive relationships with corporations to help them develop in more benign ways. Wal-Mart, which has been in the crosshairs of nonprofits for just about every possible issue for more than a decade, has made a commitment to sustainable practices in every aspect of its business. These include tripling the efficiency from 6 to 18 mpg in what is the biggest truck fleet in the world, converting to 100 percent renewable energy, and going to a zero-waste system in which nothing is thrown away. To achieve those goals, Wal-Mart actively consults with dozens of NGOs on topics that include seafood, organic food and farming, textiles, climate change, China, electronics and waste, jewelry, chemicals, green chemistry, logistics, forest products and certification, green buildings, transportation, packaging, and renewable energy. (It is important to note that an equally large group of NGOs continues to oppose Wal-Mart's siting, labor, and business practices.)

A wasteful society is a relatively new phenomenon. I spent part of my childhood on a farm belonging to my Swedish grandmother and Scottish grandfather, where nothing much was thrown away. The barn was full of used washers, bolts, wire, and doodads. In the kitchen every other plate was chipped, but the china never left the dinner table with food on it. Gravy and juices were mopped up by homemade bread, vegetable peelings went to the chickens, the shells from the eggs eaten at breakfast were put into the coffee grounds, the coffee grounds were placed into the compost, the compost was tilled into the garden, the tomatoes and corn from the garden were sealed in glass jars that joined the jams and jellies lining cool, dark basement walls. Paper lunch bags were brought back from school and neatly folded for use the next day. Our idea of play included capturing horned toads and pretending they were dinosaurs or lying faceup in irrigation ditches, the tiny fry tickling our toes, and imagining we were floating down a great Amazonian river. Our notion of a toy was a bald tire swinging from the sycamore. Had my grandparents been from Chile, Korea, or Kerala, life essentially would have been the same. Nothing would have been wasted.

Today, the creation of a zero-waste society is a global movement carried out by thousands of organizations. The ostensible purpose is to institute cyclical systems that eliminate waste by design, not by management at end-of-pipe. Zero-waste strategies have intellectual origins in the thinking of Walter Stahel of the Product Life Institute in Switzerland and biologist John Todd of the New Alchemy Institute in Cape Cod, Massachusetts. Stahel coined the term "cradle to cradle" in 1985 in connection with the cyclical use of materials. About three-fourths of all energy is used to extract and produce basic materials such as steel, and the balance is used to finish raw materials into products. Conversely, three-fourths of industrial employment is deployed in manufacturing goods, and only one-fourth is in raw materials extraction. By closing the loop so that recycled instead of virgin materials are employed in manufacturing, energy use plummets and employment increases. John Todd is a genius in the design of aquatic systems, and based on his work Yale architecture student Paul Bierman-Lytle coined the term "waste equals food" to express the concept that waste from one system should provide food for another, whether industrial system or ecosystem. Todd envisaged industrial and municipal effluent as a potential nutrient source rather than as water pollution, and proceeded to design a water treatment process using living organisms and plants to transform them into safe, nontoxic nutrients. Buckminster Fuller perceived what all thermodynamicists know, that spaceship earth is powered by a mother ship, the sun, and that to sustain the earth we need to run off current solar income. Dipping into the carbon bank of the past is not only tantamount to going into debt, it overwhelms the waste-absorption capacity of the earth (which is exactly what global greenhouse gases do). That trilogy of concepts—cradle to cradle, waste equals food, and stay within current solar income—lays out the basic tenets of the greening of industry and elimination of pollution, waste, and toxins. It could not be more clear.

Beyond elimination of material waste resides a deeper issue: the elimination of waste on a social level. We are the only species without full employment,

again defying the nature of nature. No academic yet has satisfactorily explained the wisdom of an economic system that marginalizes human beings.
A zero-waste society means wasting nothing, and foremost among these resources are people, especially children. If we are to care for our children,
then we must address the needs of their mothers and fathers. Tens of thousands of organizations in the movement create dignified, living-wage jobs
for impoverished men and women. They work in villages, communities,
and rural areas, attacking the challenge job by job. At the same time, NGOs
will mass or swarm to protest treaties such as NAFTA that put small farmers
out of business in Chiapas.

   Life tends to optimize rather than maximize. Maximization is another
word for addiction. "Humans exhibit addictive tendencies when trying to
maximize such values as wealth, pleasure, security, and power. . . . Too
much of a good thing is not a good thing,"[19] writes Hoagland. Critics of the
movement complain that it is against free markets, expanding wealth, and
security, which is not true. What is missing in that critique is a discussion of
how we gauge sufficiency. A sense of balance—of knowing what is too
much wealth, what is too much power, what constitutes license instead of
freedom—is not easy to achieve, but it raises crucial questions. In *Lyrical
and Critical Essays*, Camus wrote of how beauty has been exiled in Western
culture and replaced by the cult of reason that constantly seeks to overcome
limits. "But limits nonetheless exist and we know it. In our wildest madness
we dream of an equilibrium we have lost, and which in our simplicity we
think we shall discover once again when our errors cease—an infantile presumption, which justifies the fact that childish peoples, inheriting our madness, are managing our history today. . . . We turn our back on nature, we
are ashamed of beauty. Our miserable tragedies have the smell of an office,
and their blood is the color of dirty ink."[20]

While so much is going wrong, so much is going right. Over the years the
ingenuity of organizations, engineers, designers, social entrepreneurs, and
individuals has created a powerful arsenal of alternatives. The financial and
technical means are in place to address and restore the needs of the biosphere and society. Poverty, hunger, and preventable childhood diseases can
be eliminated in a single generation. Energy use can be reduced 80 percent
in developed countries within thirty years with an improvement in the
quality of life, and the remaining 20 percent can be replaced by renewable

sources. Living-wage jobs can be created for every man and woman who wants one. The toxins and poisons that permeate our daily lives can be completely eliminated through green chemistry. Biological agriculture can increase yields and reduce petroleum-based pollution into soil and water. Green, safe, livable cities are at the fingertips of architects and designers. Inexpensive technologies can decrease usage and improve purity so that every person on earth has clean drinking water. So what is stopping us from accomplishing these tasks?

It has been said that we cannot save our planet unless humankind undergoes a widespread spiritual and religious awakening. In other words, fixes won't fix unless we fix our souls as well. So let's ask ourselves this question: Would we recognize a worldwide spiritual awakening if we saw one? Or let me put the question another way: What if there is already in place a large-scale spiritual awakening and we are simply not recognizing it?

In a seminal work, *The Great Transformation*, Karen Armstrong details the origins of our religious traditions during what is called the Axial Age, a seven-hundred-year period dating from 900 to 200 BCE, during which much of the world turned away from violence, cruelty, and barbarity. The upwelling of philosophy, insight, and intellect from that era lives today in the works of Socrates, Plato, Lao-tzu, Confucius, Mencius, Buddha, Jeremiah, Isaiah, and others. Rather than establishing doctrinaire religious institutions, these teachers created social movements that addressed human suffering. These movements were later called Buddhism, Hinduism, Confucianism, monotheistic Judaism, democracy, and philosophical rationalism; the second flowering of the Axial Age brought forth Christianity, Islam, and Rabbinical Judaism. The point Armstrong strongly emphasizes is that the early expressions of religiosity during the Axial Age were not theocratic systems requiring belief, but instructional practices requiring action. The arthritic catechisms and rituals that we now accept as religion had no place in the precepts of these sages, prophets, and mystics. Their goal was to foster a compassionate society, and the question of whether there was an omnipotent God was irrelevant to how one might lead a moral life. They asked their students to question and challenge and, as opposed to modern religion, to take nothing on faith. They did not proselytize, sell, urge people to succeed, give motivational sermons, or harangue sinners. They urged their followers to change how they behaved in the world. All relied on a common principle, the Golden Rule: Never do to anyone what you would not have done to yourself.[21]

The Axial sages were not interested in providing their disci-
ples with a little edifying uplift, after which they could return
with renewed vigor to their ordinary self-centered lives. Their
objective was to create an entirely different kind of human being. All the sages preached a spirituality of empathy and compassion; they insisted that people must abandon their egotism
and greed, their violence and unkindness. Not only was it wrong
to kill another human being; you must not even speak a hostile
word or make an irritable gesture. Further, nearly all of the Axial
sages realized that you could not confine your benevolence to
your own people: your concern must somehow extend to the entire world. . . . If people behaved with kindness and generosity
to their fellows, they could save the world.[22]

No one in the Axial Age imagined that he was living in an age of spiritual awakening. It was a difficult time, riddled with betrayals, misunder-
standings, and petty jealousies. But the philosophy and spirituality of these
centuries constituted a movement nevertheless, a movement we can recognize in hindsight. Just as today, the Axial sages lived in a time of war. Their
aim was to understand the source of violence, not to combat it. All roads led
to self, psyche, thought, and mind. The spiritual practices that evolved were
varied, but all concentrated on focusing and guiding the mind with simple
precepts and practices whose repetition in daily life would gradually and
truly change the heart. Enlightenment was not an end—equanimity, kindness, and compassion were.

These teachings were the original source of charities in the ancient
world, and they are the true source of NGOs, volunteerism, trusts, foundations, and faith-based charities in the modern world. I suggest that the contemporary movement is unknowingly returning the favor to the Axial Age,
and is collectively forming the basis of an awakening. But it is a very different awakening, because it encompasses a refined understanding of biology,
ecology, physiology, quantum physics, and cosmology. Unlike the massive
failing of the Axial Age, it sees the feminine as sacred and holy, and it recognizes the wisdom of indigenous peoples all over the world, from Africa to
Nunavut.

I have friends who would vigorously protest this assertion, pointing
out the small-mindedness, competition, and selfishness of a number of
NGOs and the people who lead them. But I am not questioning whether

the human condition permeates the movement. It does so, most surely. Clay feet march in all protests. My question is whether the underlying values of the movement are beginning to permeate global society. And there is even a larger issue, the matter of intent. What is the intention of the movement? If you examine its values, missions, goals, and principles, and I urge you to do so, you will see that at the core of all organizations are two principles, albeit unstated: first is the Golden Rule; second is the sacredness of all life, whether it be a creature, child, or culture. The prophets we now enshrine were ridiculed in their day. Amos was constantly in trouble with the authorities. Jeremiah became the root of the word *jeremiad,* which means a recitation of woes, but like Cassandra, he was right. David Suzuki has been prescient for forty years. Donella Meadows was prophetic about biological limits to growth and was scorned by fellow scientists. Bill McKibben has been unwavering and unerring in his cautions about climate change. Martin Luther King was killed one year after he delivered his "Beyond Vietnam" address opposing the Vietnam War and berating the American military for "taking the young black men who have been crippled by our society and sending them 8,000 miles away to guarantee liberties in Southeast Asia which they had not found in southwest Georgia and East Harlem."[23] Jane Goodall travels three hundred days a year on behalf of the earth, speaking, teaching, supporting, and urging others to act. Wangari Maathai was denounced in Parliament, publicly mocked for divorcing her husband, and beaten unconscious for her work on behalf of women and the African environment. It matters not how these six and other leaders will be seen in the future; for now, they are teachers who try or have tried to address the suffering they witness on earth.

I once watched a large demonstration while waiting to meet a friend. Tens of thousands of people carrying a variety of handmade placards strolled down a wide boulevard accompanied by chants, slogans, and song. The signs referred to politicians, different species, prisoners of conscience, corporate campaigns, wars, agriculture, water, workers' rights, dissidents, and more. Standing near me a policeman was trying to understand what appeared to be a political Tower of Babel. The broad-shouldered Irishman shook his head and asked rhetorically, "What do these people want?" Fair question.

There are two kinds of games—games that end, and games that don't. In the first game the rules are fixed and rigid. In the second, the rules change whenever necessary to keep the game going. James Carse called these, respectively, finite and infinite games.[24] We play finite games to compete and win. They always have losers and are called business, banking, war, NBA, Wall Street, and politics. We play infinite games to play; they have no losers because the object of the game is to keep playing. Infinite games pay it forward and fill future coffers. They are called potlatch, family, samba, prayer, culture, tree planting, storytelling, and gospel singing. Sustainability, ensuring the future of life on earth, is an infinite game, the endless expression of generosity on behalf of all. Any action that threatens sustainability can end the game, which is why groups dedicated to keeping the game going assiduously address *any* harmful policy, law, or endeavor. With no invitation, they invade and take charge of the finite games of the world, not to win but to transform finite games into infinite ones. They want to keep the fish game going, so they go after polluters of rivers. They want to keep the culture game going, so they confront oil exploration in Ecuador. They want to keep the hope game alive in the world, so they go after the roots of poverty. They want to keep the species game happening, so they buy swaths of habitat and undeveloped land. They want to keep the child game going; consequently, when the United States violated the Geneva Conventions and bombed the 1,400 Iraqi water and sewage treatment plants in the first Gulf War, creating sewage-, cholera-, and typhus-laden water, they condemned it as morally repugnant. When the same country that dropped the bombs persuaded the United Nations to prevent shipments of chlorine and medicine to treat the resulting diseases, the infinite-game players thought it hideous and traveled to the heart of that darkness to start NGOs to serve the abandoned. People trying to keep the game going are activists, conservationists, biophiles, nuns, immigrants, outsiders, puppeteers, protesters, Christians, biologists, permaculturists, refugees, green architects, doctors without borders, engineers without borders, reformers, healers, poets, environmental educators, organic farmers, Buddhists, rainwater harvesters, meddlers, meditators, mediators, agitators, schoolchildren, ecofeminists, biomimics, Muslims, and social entrepreneurs.[25]

David James Duncan penned a response to the hostile takeover of Christianity by fundamentalists, with advice that applies to all fundamentalisms: the people of the world do not need religious fanatics to save them any more than they need oleaginous free-trade hucksters to do so; they need *us* for their salvation, and *us* stands for the crazy-quilt assemblage of global humanity that is willing to stand up to the raw, cancerous insults that come from the mouths, guns, checkbooks, and policies of ideologues, because the movement is not merely trying to prevent wrongs but actively seeks to love this world. Compassion and love of others are at the heart of all religions, and at the heart of this movement. "When small things are done with love it's not a flawed you or me who does them: it's love. I have no faith in any political party, left, right, or centrist. I have boundless faith in love. In keeping with this faith, the only spiritually responsible way I know to be a citizen, artist, or activist in these strange times is by giving little or no thought to 'great things' such as saving the planet, achieving world peace, or stopping neocon greed. Great things tend to be undoable things. Whereas small things, lovingly done, are always within our reach."[26] Some people think the movement is defined primarily by what it is against, but the language of the movement is first and foremost about keeping the conversation going, because ideas that inform it never end: growth without inequality, wealth without plunder, work without exploitation, a future without fear.[27] To answer the policeman's question, "these people" are reimagining the world.

To salve the world's wounds demands a response from the heart. There is a world of hurt out there, and to heal the past requires apologies, reconciliation, reparation, and forgiveness. A viable future isn't possible until the past is faced objectively and communion is made with our errant history. I suspect that just about everyone owes an apology and merits one, but there are races, cultures, and people that are particularly deserving. The idea that we cannot apologize to former enslaved and First Peoples for past iniquities because we are not the ones who perpetuated the evil misses the point. By receiving sorrow, hearing admissions, allowing reparation, and participating in reconciliation, people and tribes whose ancestors were abused give new life to all of us in the world we share. Making amends is the beginning of the healing of the world. These spiritual deeds and acts of moral imagination lay the groundwork for the great work ahead.[28]

The movement is not coercive, but it is relentless and unafraid. It cannot be mollified, pacified, or suppressed. There can be no Berlin Wall moment, no treaty to sign, no morning to awaken to when the superpowers agree to stand down. This is a movement away from the maximization of anything that is not conducive to life. It will continue to take myriad forms. It will not rest. There will be no Marx, Alexander, or Kennedy to lead it. No book can explain it, no person can represent it, no group can stand at its forefront, no words can encompass it, because the movement is the breathing, sentient testament of the living world. The movement is an outgrowth of apostasies and it is now self-generating. The first cells that assembled and metabolized under the most difficult of circumstances deep in the ocean nearly forty million centuries ago are in our bodies now, and we are, in Mary Oliver's words, determined, as they were then, to save the only life we can.[29] Life can occur only in a cell, and a cell is where all disease starts, as well. In Franklin Harold's book *The Way of the Cell*, he points out that for all its hard-bitten rationalism, molecular science asks us to accept a "real humdinger . . . that all organisms have descended . . . from a single ancestral cell."[30] This quivering, gelatinous sensate mote is the core of everything we cherish, and places us in direct relation to every other form of life. That primordial connection, so incomprehensible to some yet so manifest and sacred and incontestable to others, links us inseparably to our common fate. The first gene was the password to all subsequent forms of life, and the word *gene* has the same etymological root as the words *kin, kind, genus, generous,* and *nature.* It is our nature to cultivate life, and this movement is a collective kindness produced over the course of four million millennia.

I believe this movement will prevail. I don't mean it will defeat, conquer, or create harm to someone else. Quite the opposite. I don't tender the claim in an oracular sense. I mean that the thinking that informs the movement's goals will reign. It will soon suffuse most institutions, but before then, it will change a sufficient number of people so as to begin the reversal of centuries of frenzied self-destructive behavior. Some say it is too late, but people never change when they are comfortable. Helen Keller threw aside the gnawing fears of chronic bad news when she declared, "I rejoice to live in such a splendidly disturbing time!" In such a time, history is suspended and thus unfinished. It will be the stroke of midnight for the rest of our lives.

My hopefulness about the resilience of human nature is matched by the gravity of our environmental and social condition. If we squander all our attention on what is wrong, we will miss the prize: In the chaos engulfing the world, a hopeful future resides because the past is disintegrating before us. If that is difficult to believe, take a winter off and calculate what it requires to create a single springtime. It's not too late for the world's largest institutions and corporations to join in saving the planet, but cooperation must be on the planet's terms. The "Help Wanted" signs are everywhere. All people and institutions, including commerce, governments, schools, churches, and cities, need to learn from life and reimagine the world from the bottom up, based on first principles of justice and ecology. Ecological restoration is extraordinarily simple: You remove whatever prevents the system from healing itself. Social restoration is no different. We have the heart, knowledge, money, and sense to optimize our social and ecological fabric. It is time for all that is harmful to leave. One million escorts are here to transform the nightmares of empire and the disgrace of war on people and place. We are the transgressors and we are the forgivers. "We" means all of us, everyone. There can be no green movement unless there is also a black, brown, and copper movement. What is most harmful resides within us, the accumulated wounds of the past, the sorrow, shame, deceit, and ignominy shared by every culture, passed down to every person, as surely as DNA, a history of violence, and greed. There is no question that the environmental movement is critical to our survival. Our house is literally burning, and it is only logical that environmentalists expect the social justice movement to get on the environmental bus. But it is the other way around; the only way we are going to put out the fire is to get on the social justice bus and heal our wounds, because in the end, there is only one bus. Armed with that growing realization, we can address all that is harmful externally. What will guide us is a living intelligence that creates miracles every second, carried forth by a movement with no name.

It is perhaps not too much to say that, in the first decade of the new millennium, humanity has entered into a condition that is in some sense more globally united and interconnected, more sensitized to the experiences and suffering of others, in certain respects more spiritually awakened, more conscious of alternative future possibilities and ideals, more capable of collective healing and compassion, and, aided by technological advances in communication media, more able to think, feel, and respond together in a spiritually evolved manner to the world's swiftly changing realities than has ever before been possible.

—Richard Tarnas*

No one knows how many nonprofit organizations exist in the world, and it is unlikely anyone ever will. They come and go, not as quickly as some suggest, but in sufficient numbers to defy a meaningful count. There are organizations and institutes that track this information, but the data presented are scant and concentrate on established urban organizations. Smaller nonprofit organizations can exist for years in many countries before they are recognized or counted, if ever. The Internet has created virtual organizations that may only have one full-time employee but a network of correspondents, volunteers, and part-time activists. Developed countries count nonprofits through tax rolls, but some organizations can be excluded as several may function under one roof. Some social entrepreneurs operate out of their laptop and never form an organization. Data about nonprofits vary widely in different parts of the world and are never definitive, due to the lack of accurate, up-to-date lists. Where government regimes are oppressive, nonprofits will hide within informal networks and shadow economies that mask their presence to authorities. After reading books, reports, government registries, and monographs, I believe researchers underestimate the extent of civil society; there may be as many as ten million nonprofits in the world. Of that number, there are at least one million nonprofit organizations addressed by this book, groups that address the environment, indigenous rights, and social justice.

In order to write *Blessed Unrest* I needed to do more than assert the existence of these organizations, and I had to define what is meant by the words "the environment, indigenous rights, and social justice." Those catchphrases mask the depth and breadth of this humanitarian movement. To begin to solve the problem, my colleagues and I at the Natural Capital Institute, a California-based nonprofit research institution, created a database of civil society organizations from around the world directed to the issues of the environment and justice. It is now the largest database of its kind, listing organizations in 243 countries, territories, and sovereign islands. As impressive as the number of organizations is, equally important is the classification scheme that emerged

---

*Richard Tarnas, *Cosmos and Psyche, Intimations of a New World View* (New York: Viking, 2005), p. 483.

from mapping this social landscape. There are literally thousands of disciplines or practices that engage these NGOs. One could say that environmental or social justice are convenient labels but not truly descriptive. What we did find, for example, are organizations that address photochemical oxidants, perennial polyculture, perissodactyls, microinsurance, abandonment prevention, and food autonomy. These will seem obscure to many but fall under the categories of air pollution, agriculture, biodiversity, green business, rights of the child, and hunger, respectively. In the pages that follow is a taxonomy of the movement that describes it in granular detail. It can be scanned or read, but either way I think you will see another world described.

The taxonomy is, as far as I know, the most complete listing of issues that need to be addressed to achieve a just and sustainable world. It is a curriculum and course catalog for a university that should but does not exist. It was created by staff who carefully examined the records and Web sites of existing organizations. Rather than begin with a predetermined classification system, the taxonomy accreted over a span of two years, one organization at a time. We would read an organization's mission, purpose, or goal, and locate it in the system. If it wasn't there, we added it. As we added categories and subcategories, we added keywords and discussed placement. As with any system, it is subject to different opinions as to the proper interpretations and relationships. Here's how you can change and improve what you are about to read:

Along with the database of organizations the Natural Capital Institute created a Web site to accelerate public awareness and to increase the ability of the movement to connect and collaborate. This is called WiserEarth, with WISER standing for World Index of Social and Environmental Responsibility (www.wiserearth.org). It is the first online database that can be edited by the community it serves. Web sites that can be edited by users, such as Wikipedia, are not databases; they are alphabetical lists of text. Structured content makes searches and connectivity more accurate and effective. You are invited to list your own organization, add one you know of that is not listed, or modify and improve existing listings. You can become a creator, editor, or author of the site by joining in and participating with others. We concluded early on that we would be unable to enter the data about every organization, and even if we could it would be outdated before we got to the end of the list, which is why the site is community-driven and -operated.

The WISER platform contains the technology that makes awareness, support, and communication possible. The software employed by WISER is created under open-source license as governed by the standards of the Open Source Initiative. The code and the application are freely available for the good of the community. It grants new users the right to modify and redistribute the software without payment, but restricts a user from selling it. The philosophy of open source is simple: when code can be modified and changed by many programmers, it evolves and improves. It is the same principle that informs all of WISER. If it can provide a means to exchange information and communicate ideas freely, it can accelerate understanding, social evolution, and adaptation.

Besides extensive listing of organizations, following are some of the features provided by the site. It:

- provides a relational and editable database of organizations and individuals searchable by areas of interest, geography, type of organization, profession or pursuit (for individuals), and scope of activity;

- offers advanced search tools, enabling users to find out quickly and easily who is working on what and where so that organizations can leverage their experience, knowledge, and resources;
- provides funders with an information landscape of the organizations engaged in program activity in their fields of interest, a helpful tool to better evaluate proposals and dockets;
- provides an instant and effective means for many people to give small amounts of money to organizations all over the world, thereby broadening the global philanthropic base;
- offers free listings of jobs, positions, and resources for organizations, prospective employees, interns, volunteers, and students;
- supports individuated calendars that notify users of any and all events in their specific geographical area regarding their areas of interest;
- establishes the means for bioregional hubs to empower local economies;
- facilitates free or low-cost VOiP communication between listed organizations in the world;
- contains the first detailed taxonomy of the organizations within civil society;
- provides lists of resources, including books, conferences, events, other databases, definitions, magazines, articles, podcasts, streaming audio and video, maps, research reports, and educational opportunities.

The success of this movement will be defined by how rapidly it becomes a part of all other sectors of society. If it remains singular and isolated, it will fail. If it is absorbed and integrated into religion, education, business, and government, there is a chance that humanity can reverse the trends that beset the earth.

Except for quotations, I wrote what you have read to this point in the book. Everything you read after this page was written by a community, which includes me; that is also true of WiserEarth. It is axiomatic that we are at a threshold in human existence, a fundamental change in understanding about our relationship to nature and each other. We are moving from a world created by privilege to a world created by community. The current thrust of history is too supple to be labeled, but global themes are emerging in response to cascading ecological crises and human suffering. These ideas include the need for radical social change, the reinvention of market-based economies, the empowerment of women, activism on all levels, and the need for localized economic control. There are insistent calls for autonomy, appeals for a new resource ethic based on the tradition of the commons, demands for the reinstatement of cultural primacy over corporate hegemony, and a rising demand for radical transparency in politics and corporate decision making. It has been said that environmentalism failed as a movement, or worse yet, died. It is the other way around. Everyone on earth will be an environmentalist in the not too distant future, driven there by necessity and experience.* The idea that we can solve our collective problems by narrow, single-issue, partisan politics is moribund. The world is a system, and it will soon be a very

---

*Schmitt, Mark, "We're All Environmentalists Now," *American Prospect Online*, September 18, 2005, www.prospect.org/web/page.ww?section=root&name=ViewPrint&articleId=10311.

different world, driven by millions of communities who believe that democracy and restoration are grassroots movements that connect us to values that we hold in common. The following appendix pages contain some of the vocabulary of that new world.

See you online at these URLs (one will take you to all the others):
  wiserearth.org
  naturalcapital.org

I can be reached at:
  wiserearth.org/user/paul or hawken@wiserearth.org

Thank you.
Paul Hawken

———————

WiserEarth Areas of Focus

Note: The number following each entry beginning on the following page indicates the number of organizations listed at WiserEarth.org for each category at the time this book was written. Although these numbers will be outdated because the database grows daily, they do provide a sense of the depth and breadth of organizations addressing specific interests. A list of keywords, in italics, follows each definition.

AGRICULTURE AND FARMING

Agriculture and farming is the science, art, or practice of cultivating the soil, producing crops, and raising livestock, and in varying degrees the preparation and marketing of the resulting products.

organic, farm, bioagriculture, pesticides, fertilizers, biological control, composting, gardening, livestock methane emissions, sustainable agriculture, permaculture, agricultural policy, soil management, cultivation, crops, livestock

Agricultural Policy (238): *Agricultural policy* refers to the practice of creating new regulations and procedures for the agricultural market chain-of-custody on local, regional, national, and international levels. The market chain includes on-farm use of seeds, agrochemicals, farm machinery, practices impacting soil and water; grading and quality separation of produce; processing; truthful branding and advertising; and the structure of the wholesale and retail markets. It includes changing existing policies through campaigning, lobbying, advocacy, direct action as well as reducing perverse subsidies, loans, taxes, tariffs, and price supports; and implementing incentives to encourage a transition to sustainable agriculture.

lobbying, advocacy, campaigning, food production, livestock, crops, environmental impact, farming methods, agrarian reform, farmland preservation, government farm policy, farm agency, agricultural agency, pollution standards and organic farm production standards, agriculture-related regulations, grazing regulations, agricultural economics, agricultural tariffs, duties, quotas, nation-state agricultural policy, food security, agricultural regulatory controls, biotechnology, farm location and transportation costs, boycotts, buycotts, food health, culturally significant foods, agricultural environmental impacts, water policy, biodiversity policy, soil conservation policy, subsidies, protectionism, currency values and commodity production, trade, watershed management, agricultural wastes and by-products, price supports, farm insurance, farm legislation, farm tax policy, agrarian reform, land concentration, credit extension, technical extension, participation in nation-state decisions, grassroots decision-making, food production infrastructure, agroecological research funding, farmer's markets

Agricultural Water Conservation and Management (133): On farms, soils can contain too much or too little water, or water of the wrong quality to produce the desired crops. Water also can cause erosion problems, leaching of nutrients and toxics, and salinization, and can require huge inputs of energy to pump irrigation water from wells or distant rivers. Water conservation can help reduce costs as well as leave water for other users, including the natural environment. On-farm water conservation and management attempts good conservation and soil moisture practices.

pumps, energy input, windmills, groundwater, water table, waterlogged soils, drainage, wetland loss, tailwaters, water pollution, rill erosion, sheet erosion, gully erosion, dryland farming, irrigation, wilting point, consumptive use, return flows, leaching losses, evapotranspiration, drip irrigation, agriculture, farming, water and agriculture

Agroecology (152): Agroecology is the science of applying ecological concepts and principles to the design, development, and management of sustainable agricultural systems.

sustainable agriculture, management, food systems, study, ecology, farming, research, whole-systems approach, feed, fiber production, minimal artificial inputs, biological pest control, diversification, ecological relationships, conserve resources, minimize toxics use, agroecological indicators, agrarian reform

Biological Control (143): Biological control is the practice of using naturally occurring predators, parasites, or disease-causing bacteria to control the population of unwanted pest and weed organisms, rather than using chemical pesticides and herbicides to achieve the eradication of pests and weeds.

integrated pest management, IPM, insects, mites, biocontrol, natural predators, parasites, pathogens, garden, fauna, flora, holistic, pest, weed, pest control, pesticides, herbicides, agriculture, farming

Composting (103): Composting is the practice of biologically decomposing organic waste, such as kitchen scraps and animal manure, in aerobic conditions to produce a soil-like product that can be reused as mulch or organic fertilizer. Composting is the recycling of organic waste and nutrients to improve the condition and fertility of the soil.

decomposition, aerobic, biological, plant, organic waste, garbage, recycle, reuse, compost, vermiculture, worm composting, sustainable agriculture, manure, sustainable farming

Farm Ecosystem Management (62): Sustainable farm management tries to be environmentally friendly while providing a fair profit to the farmer. Farm management addresses the farm's watershed (including runoff, sedimentation, and off-farm impacts on water and air quality); inputs (seeds, fertilizers, pesticides, biocontrols, weeds); soil health (including crop rotations, manures, plowing, compaction); water and soil moisture balance; as well as the technology and machinery used. The goal is long-term improvement of farming practices and soil fertility as well as least harm to biodiversity. Farm ecosystem management is the management of all living species and their interactions with a farm and the farm's watershed—in the largest sense, management of inputs to the farm so as not to harm the ecosystems in which the inputs were extracted or produced. Ecosystem management includes control of pests, pathogens, and invasive species. It includes nurturing the "living soil" of the farm, protecting in-stream flows and water quality required by aquatic plants and animals, and encouraging the continued existence of rare and sensitive species. Use of biological control agents and soil solarization with plastic sheeting as opposed to agrochemicals are typical tools for farm ecosystem management.

seeds, agrochemicals, organic fertilizers, inorganic fertilizers, manures, cover crops, herbicides, pesticides, fungicides, farm machinery, crop rotations, nutrient cycles, energy inputs, biocontrols, crop diversity, off-farm environmental impacts, tillage, no-till farming, humus, genetically modified organisms (GMO), weeds, invasives, agricultural wastes and by-products, habitat conversion, water pollution, air pollution, filter strips, field borders, hedgerows, agriculture, farming, biocontrol, Bacillus thuringensis, integrated pest management (IPM), pest control, invasives, in-stream flows, set-asides, safe harbor planning, soil solarization, riparian vegetation, buffer strips, ecosystem stability, apiculture, field borders, sustainable agriculture, sustainable farming, healthy soil, soil conservation, integrated green systems

Gardening (768): Gardening is the science and practice of cultivating plants and vegetables, usually on a small scale, for individual or community enjoyment and consumption.

horticulture, urban gardens, small farms, plant cultivation, lawn, turf, pesticide, organic gardens, community gardening, community gardens, school gardens, French intensive method, compost, mulches, green manure, farmer's markets, trees in gardens, food security, companion planting, water conservation, apiculture, bees in gardens, pollination, nutrition, seeds, horticultural, pollinator appreciation

Global Beef Industry (6): The beef industry has thoroughly globalized. About 25 percent of all beef is obtained by trade. Developing nations continue to increase beef consumption per person. The global system emphasizes grain-fed cattle in feedlots, special breeds that can graze more marginal and fragile habitats, and the use of growth hormones and antibiotics. A major sustainability issue: should grain go to livestock or directly to humans? Environmental and social impacts include: land conversion into pastures, pollution from pasture development, grain cropland, hide tanning and manure from feedlots, greenhouse gas emissions, and long-term health impacts of antibiotics and growth hormones. In addition, government subsidies for pasture conversion and grain production, and the concentration of the chain-of-custody in the hands of a few multinationals, create difficult barrier to sustainable beef production operations.

globalization, beef, trade, developing nations, grain-fed cattle, feedlots, bovine growth hormones, bovine antibiotics, land conversion, rangelands, pasture, leather industry, subsidies, multinationals, agribusiness

Livestock in Developing Nations (33): Livestock raising in Africa and similar ecoregions cannot be compared with the beef industry in more developed nations. Livestock raising in developing nations focuses on self-sufficiency and regional markets. The livestock are not specialized for meat production. The same cattle, for instance, may provide milk, meat, manure for fires, traction, transport, manure for fertilizer, skins, a major form of savings, assets, old-age and health insurance (cash for drugs), as well as intergenerational equity (inheritance). Livestock may be necessary for marriage dowries and other ceremonies. In developing nations, it is difficult to manage the rangelands. Pastoralists manage herd size, a mix of livestock species (e.g. sheep, camels, cattle), sex and age structure, and livestock management (grazing time). Desertification has been the result of very limited management choices in times of drought. Sustainable livestock raising includes access to supplement feeds (e.g. shrub and tree nurseries for browse), market reorganization during droughts for rapid destocking, secure access to reserved pastures and watering locales, early warning systems, and access to alternative sources of income.

pastoralism, transhumance, livestock investment, livestock sales, agro-sylvo-pastoralism, cattle, one-humped camels, hair sheep, wool sheep, goats, donkeys, horses, herd size, desertification, over-stocking, over-grazing, browse, war, drought, famine, locusts, urbanization, coping strategies, land tenure, early warning systems

Organic Farming (670): Organic farming is the practice of producing crops and livestock without the use of synthetic fertilizers, pesticides, and other chemicals, and is an approach to agriculture where the aim is to create integrated, humane, environmentally sustainable agricultural production systems. Organic farming systems rely on crop rotations, crop residues, animal manures, legumes, off-farm organic wastes, mechanical cultivation, and biological pest control to maintain soil productivity, to supply nutrients to plants, and to control weeds and pests.

food production, produce, ecological health, standards, labels, hormones, pesticides, synthetic fertilizer, food diversity, whole system agriculture, plant cultivation, low-impact, gardening, horticulture, farmer's markets, safe processing and packaging, markets, farm income, crop yield, organic production standards, soil conservation, small organic farming

Permaculture (186): Permaculture emphasizes high-yielding, small plot, and greenhouse agriculture and aquaculture ponds as well as forage agriculture, forest agriculture, and the development of self-sufficiency, including on-site alternative energy sources.

permanent agriculture, agricultural ethics, holistic management, perennial polyculture, aquaculture, agro-sylvo-pastoral farming, gardening, farm communities, crop diversity, crop rotation

Precision Farming (14): Precision farming attempts to combine the latest in info-technology with farming in order to increase efficiency of farming management. It includes, for instance, laser leveling of fields in order to best distribute irrigation water, and infrared satellite photography of farms to see where nutrient deficiencies and/or pathogens are most severe.

laser leveling, time domain reflectometry (TDR), infrared leaf reading, information technology, precision farming, agriculture

Rural Farming Communities (817): *Rural farming communities* refers to the activity of preserving and enabling rural farming communities to develop sustainably, protecting the environment while thriving socially and developing economically.

economic development, social development, rural development, erosion control, integrated rural development, renewable energy production, rural income generation, rural women, neglected crops, underexploited crops, landowners, agriculture, co-operative farming, agrarian reform, farmland preservation, farmer's markets, farmland trusts, zoning, urban sprawl, water costs, watershed management, fuel and transportation costs, agriculture, farmer's rights

Soil Conservation and Management (426): On farms, soil conservation and management attempts to maintain an artificial soil to maximize food production. Sustainable farm managers cultivate the soil to maximize its long-term fertility and its soil moisture holding capacity for the particular crop rotation without harming the off-farm environment. Sustainable soils, ideally, keep the soil in place (nonerosive), retain its organic content, maintain its structure of pore spaces for air, water, and microbes, and give life support to all the organisms that grow within it. Sustainable soils are difficult. Some are stripped for development and covered in asphalt or concrete. Some are drained for improved agriculture. Some of the greatest disasters of human history (e.g. the end of the Fertile Crescent, the Dust Bowl, the denuded landscapes of the Yangtze) come from abuse of soil integrity. There is no program to maintain legacy or historical soils as baselines for comparison. Soil is the medium for production of food, fiber, and forage. Soil is a form of natural capital whose financial value can be measured by a yield of crops. The natural capital portfolio increases in value with increased soil organic matter, soil-moisture holding capacity, specific structure and textures, as well as the soil's genetic resources. Sustainable fertility eliminates petrochemical inputs of fertilizers, pesticides, and herbicides as well as the heavy metals found in some sewage sludges. Soil provides numerous goods and services to society. Goods include paint pigments, ceramics, adobe, gravel, medicines, sand for cement and glass, clay for adobe and filters, silicon for microchips. Services include purifying water, humidifying the air, anchoring plants, storing carbon, exchanging and transforming nutrients, reducing toxic pollution, fertilizing crops, recharging groundwater, mitigating floods and heat islands, and more. Of all the crucial components of sustainability (soil, air, water, and energy), soils have received the least attention, even though many civilizations have collapsed from ignoring human impacts on soils.

topsoil, tilth, soil structure, soil texture, water holding capacity, soil fertility, humus, organic matter, soil erosion, runoff, gully erosion, rill erosion, sheet erosion, sedimentation, strip cropping, grassed waterways, contour plowing, leaching losses, soil nutrients, micronutrients, mycorrhizal fungi, soil compaction, no-till, mulch-till, tillage plowing, agrochemicals, windbreaks, soil stewardship, sustainable agriculture and farming, soil as a malleable material, soil as an organizer for life, soil as fertile life-support, soil as an organizer of society, sustainable soils, healthy soil

Sustainable Agriculture (3,349): Sustainable agriculture and farming is the science, art, or practice of cultivating the soil, producing crops, and raising livestock that is economically viable, socially responsible, and ecologically sound, renewing the land for continued agricultural use in the long term. In practice, agricultural sustainability may also include budgetary and politically acceptable policies of governmental aid to help transition farming to more sustainable practices, control prices of food in poorer nations, and prevent environmental harms. On a global level, two trends are on a crash course: there is a steady increase in the consumption of food and fiber, and a steady decline in the quality and productivity of agricultural soils. In addition, farming is the single greatest threat to the planet's biodiversity and ecosystem services. Sustainable agriculture requires the reorganization of producers, buyers, investors, regulators, traders, wholesalers, retailers, and consumers as well as commodity financial agents. Sustainable world agriculture works toward breaking perverse incentives and practices that encourage soil depletion and agribusiness favoritism, promotes water conservation, higher producer incomes, and protection of biodiversity.

sustainable world agriculture, sustainable agriculture education, alternative agriculture, organic agriculture, permanent agriculture, permaculture, biodynamic agriculture, ecological agriculture, ecosystem agriculture, perennial polyculture, agroecology, bioagriculture diversified farming, low-input farming, precision farming, agro-livestock farms, agroforestry, agro-sylvo-pastoral farming, diversified land use, crop diversity, crop rotations, time-of-planting practices, organic manures, closed nutrient cycles, crop rotations, low energy input, no agrochemical inputs, biological disease and pest control, renewable resources, holistic farming, ecosystem stability, soil stewardship, energy efficiency, food quality, animal welfare, no GMOs, green market products, certification, nonindustrial agriculture, agroecological zones, apiculture, shifting agriculture, swidden agriculture, land tenure, staple crop, terracing, usufruct right, agricultural tariffs, duties, quotas, nation-state agricultural policy, international commodity agreements, food security, agricultural regulatory controls, farm labor costs and incomes, biotechnology, farm location and transportation costs, boycotts, buycotts, food health, culturally significant foods, better management practices, agricultural soil conservation

Sustainable Livestock Husbandry (33): Sustainable livestock raising is custom-designed to region, has low inputs, maintains or improves soil conditions and forage, balances biodiversity with forage production, custom-designs predator control to minimize impacts on ecosystem, minimizes invasive exotics, shares watering areas with wildlife, and minimizes impacts on stream waterflows and riparian communities. Sustainable ranching takes responsibility for animal welfare during the ranch or farmyard stage as well as during transport, fattening, slaughtering, processing, packaging, wholesaling, and retailing. No growth hormones and minimal antibiotics are used in feedlots.

cattle, beef, dairy cow, goat, sheep, pigs, poultry, chickens, camels, water buffalo, livestock breeds, draft animals, ranch, farmyard, land tenure, grassland, grass, riparian, grazing, browse, rangeland, fire, climate change, urbanization, endangered species, overgrazing, trampling, desertification, paddock, fencing, stocking rate, salt, free-ranging, veterinarian supplies, fodder, agrochemicals, cattle producer, stocker, breeder, antibiotics, growth hormones, pesticides, feed grains, animal feed, commodity, feedlot, anti-foulants, biological oxygen demand (BOD), biosolids, BSE, livestock diseases, pathogens, slaughterhouse, packing house, processing plant, retailer, consumer, meat, milk, vegetarianism, cattle associations, animal welfare, food health, debt structure, global trade conventions, subsidies, organic beef, holistic resource management, predator-friendly beef, conservation ranching, natural beef, trade

AIR

Air is the mix of gases (predominantly nitrogen and oxygen, with small amounts of argon, carbon dioxide, and water vapor) that make up the earth's atmosphere that enable living creatures to breathe. The atmosphere is the gas layer that surrounds the planet, mixes with the planet's water, and penetrates into the soil.

outdoor, indoor, air quality, ozone, contamination, pollutants, smog, acid rain, greenhouse effect, particles, purity, particulates, concentration, environmental, health, pollution, standards

Acid Rain (19): Acid "rain" is any form of precipitation (rain, snow, hail, or fog) that contains harmful amounts of nitrogen and sulfur oxides. The major human sources are from industry, transportation, and power plants. These industrial pollutants from the air increase the acidity of the precipitation and harm plants, buildings, and humans.

sulfur dioxide, nitrogen oxides, runoff, atmosphere, pH, precipitation, rainfall, air pollution, atmospheric pollution, sulfuric acid, nitric acid, acidic precipitation, industrial pollutants, acid deposition, power plants, transportation, anthropogenic emissions, pollution source, plant damage, materials damage, building damage, emissions trading, forest decline, Waldsterben, *air quality*

Air Quality and Pollution (1,055): Air is the mix of gases that make up the earth's atmosphere. The atmosphere is the gas layer that surrounds the planet, mixes with the planet's water, and penetrates into the soil. Air pollution comes from a change in the mix of gases, particulates, or water-based vapors that interferes directly or indirectly with human health, comfort, or safety, and can eat away at metal and stone. Air pollution can contribute to a change in the planet's climate and impact all living creatures.

pollutants, particulates, ozone, radon, asthma, asbestos, temperature, concentrations, visibility, emissions, pollution sources, clean air, environmental health, solid waste disposal, transportation, fuel combustion, climate change, air quality standards, industrial air pollution chemical, carbon monoxide, nitrogen oxides, sulfur oxides, hydrocarbons, photochemical oxidants, greenhouse gases, aerosols

Indoor Air Quality (37): Indoor air quality (IAQ) describes the attributes and the pollution level of the air inside a confined space, such as a building. IAQ is determined by characteristics such as gaseous composition, temperature, relative humidity, and airborne contaminant levels (e.g. of mold, bacteria, chemicals, and allergens), which can exist in the air and that affect human and nonhuman health.

IAQ, ventilation, air conditioning, HVAC, standards, contaminant level, contamination, mold, airborne particulates, humidity, temperature, circulation, allergens, bacteria, airborne contaminants, clean air, indoor environmental quality

Ozone Layer (21): The ozone layer is a protective layer in the atmosphere, about fifteen miles above the ground, that absorbs some of the sun's ultraviolet rays, thereby reducing the amount of potentially harmful radiation that reaches the earth's surface. The destruction of the ozone layer is caused by the breakdown of certain chlorine- and/or bromine-containing compounds (chlorofluorocarbons or halons), which break down when they reach the stratosphere and then destroy ozone molecules, thinning the ozone layer, and increasing the radiation that reaches the earth's surface. Ozone near the surface of the earth forms a photochemical smog with hydrocarbons from vehicles and industries. Near the earth's surface it is a pollutant.

pollution, air, ultraviolet radiation, UV absorption, CFCs, chlorofluorocarbons, industrial pollution, halons, chlorine, bromine, clean air, ozone shield, photochemical smog, nitrogen oxides, ozone depletion, ozone thinning, plant damage, transportation, fuel combustion, air pollution standards, Montreal Protocol

ANIMALIA
Animals are many-celled and need to feed on other organisms. They are not photosynthesizers. Typically, they ingest their food to an internal cavity and are mobile. Over 1.5 million species have been described. More than 95 percent are invertebrates (without backbones), though humans call mammals "animals" and forget about the invertebrates or even the birds and reptiles and amphibians.

kingdom Animalia, fauna, endangered species, animal trafficking, fur industry, trapping, poaching, fauna, nondomesticated, ecology, habitat protection, conservation, population, wild animals, threatened species, endangered species, undomesticated, untamed, feral, plants, flora, wildlife

Amphibians (110): Amphibia is the class of carnivorous vertebrates that includes frogs, toads, and salamanders. Many amphibians spend the larval part of their lives under water and the adult part on land or in water. They must return to damp areas to reproduce. There are about 5,000 species. About 150 are threatened by climate change, UV radiation, a fungus pandemic, introduced species, animal trafficking, and loss of habitat.

class Amphibia, frogs, toads, salamanders, fauna, conservation, habitat, animals, newts, hellbenders, inland waters, ozone layer, pathogens, introduced species, climate change

Animal and Plant Trafficking (591): Animal and plant trafficking is the illegal trade in wild animals and plants. Illegal trade may be a threat to the conservation of the plant and animal species and their ecosystems. Trafficking includes illegal hunting for local herb, plant, and bushmeat markets.

illegal trade, black market trade, endangered species, fauna, smuggling, wildlife, animals, population threat, habitat threat, flora, conservation, poaching, aquarium fish, ornamental fish, primates, cacti, timber, bushmeat, turtle eggs, mammals, birds, medicinal plants, animal species protection, plant species protection

Animal Welfare and Rights (2,353): Animal welfare involves a respect for and a commitment to the humane treatment of all nonhuman animals, which includes proper housing, adequate nutrition, disease prevention and treatment, responsible care, and humane handling and can include humane euthanasia when necessary. Animal rights are those rights that some people consider to be inherent to animals, which should not be violated by humans, such as using animals for testing cosmetic products. Animal welfare organizations focus on laboratories, testing facilities, zoos, moviemaking, domestic animals (e.g. slaughterhouses, feedlots), and pets, with facilities for injured wildlife.

animal rights, fur, laboratories, trade in wildlife, fauna, animal species protection, wildlife rehabilitation, animal abuse, animal cruelty, animal rescue, wildlife treatment, animal species protection

Artiodactyls (213): The "cloven-hoofed" artiodactyls comprise 192 species of pigs, peccaries, hippopotami, llamas, camels, deer, giraffes, pronghorn, antelope, cattle, goats, and sheep. Many have horns or antlers and most are ruminants or cud-chewers that thrive on grass and leaves. They all are symbionts with bacteria that help digest the vegetation. Artiodactyls are crucial parts of human civilization, especially after the domestication of pigs, llamas, camels, cattle, goats, and sheep. Sustainable futures include preventing overgrazing and browsing by domesticated species, pollution from industrialized pig farms and feedlots, wildlife management of deer, antelope, and other hunted species, as well as conservation of rare and endangered species and domestic races.

order Artiodactyla, pigs, peccaries, hippopotami, llamas, camels, deer, giraffes, pronghorn, antelope, oryx, cattle, goats, domestic animals, sport hunting, endangered species, overgrazing, feedlots, industrialized agriculture, wildlife management, wildlife conservation, threatened breeds, bloodlines, fauna, mammals, family Bovidae, antelopes, buffalo, bison, goats, yak, cattle, sheep, conservation, habitat, animals, bushmeat, fiber, domesticated animals, bovids

Bats (94): Bats, order Chiroptera, are the only flying mammals with forelimbs developed as wings. Most are insectivorous or nectar/fruit eating. There are about 945 species, which are cosmopolitan except for remote oceanic islands and cold climates. Some species provide pollination, bat guano (fertilizer), and insect-control services. Some date-eating bats are considered pests. Some are long-distance migrants. Some species are threatened by habitat removal, flyway destruction, or destruction of roosting and nursery sites, especially on oceanic islands.

order Chiroptera, bats, habitat, conservation, animals, wildlife corridor, pollination, biocontrol, strip mining, deforestation, animal protection, wind turbines

Birds (1,729): Birds are members of the vertebrate class Aves that are bipedal, feathered, lay eggs, have forelimbs modified as wings, and are capable (most of them) of flight. There are about 9,750 living species. About 2,000 are threatened. They provide biocontrol of insects and rodents, pollination, seed dispersal, human food, decorative feathers, and aesthetic enjoyment. They can be transmitters of disease (avian flu) and crop pests. Hundreds of bird species migrate south to avoid the northern winter. Neotropical migrant wood warblers, waterfowl, shorebirds, as well as some raptors are the major migrants in the New World. For instance, over 20 million North American shorebirds head south in winter. Bird migration poses unique multinational conservation issues—protecting northern breeding areas, multiple-stop flyways, and overwintering grounds.

class Aves, raptors, birds of prey, fauna, conservation, habitat, animals, ornithology, migratory birds, neotropical migrants, endemic, deforestation, habitat conversion, bird migrants, migratory routes, bird migration monitoring, bird species monitoring, flyways, global conservation, deforestation, light pollution, shorebirds, waterfowl, ducks, warblers

Canids (98): Canids ("canines") are carnivorous and omnivorous members of the family Canidae and include wolves, dogs, foxes, jackals, and coyotes. They have a lithe, muscular, deep-chested body, usually long, slender legs, a bushy tail, slender muzzle, erect ears, and can walk tirelessly on their four front toes and five hind toes. There are thirty-six species and the domesticated dog. About one-third of the canids are vulnerable or threatened from hunting, trapping, and loss of habitat.

dogs, wolves, foxes, coyotes, jackals, family Canidae, fauna, conservation, habitat, animals, wolf, recovery, livestock, sport hunting

Cetaceans (209): Cetaceans are totally aquatic mammals in the order Cetacea. The eighty species includes whales, dolphins, and porpoises. Along with elephants, they are the only mammals with convoluted brains larger than humans'. All species have some level of protected status.

whales, dolphins, porpoises, order Cetacea, fauna, protection, conservation, habitat, mammals, animals, marine ecosystem, International Whaling Commission, Marine Mammal Protection Act, Japan, Iceland, Norway

Elephants (34): Elephants are members of the order Proboscidea and the family Elephantidae and have a long, muscular trunk that functions like a fifth limb. The males have a pair of huge tusks derived from upper incisors. The Asian and the African elephants are the remaining two species of elephants.

fauna, order Proboscidea, family Elephantidae, African elephants, Asian elephants, mammoths, protection, conservation, habitat, animals, mammals

Endangered Animal Species Protection (1,667): Endangered animal species are those animal species whose ability to survive and reproduce has been jeopardized by human activities. The size of the population is so low or the amount of remaining habitat so little that the species is at risk of extinction.

monitoring, biodiversity, extinction, conservation, habitat, habitat preservation, fauna, wildlife, animal species protection, endangered, legal protection, critical population, critical habitat, animals, threatened, vulnerable, viable population, habitat conversion, habitat loss, habitat destruction, habitat modification, alien species, introduced species, invasive species, exotic species, accidental introduction, animal trafficking, animal trade, climate change, habitat fragmentation, habitat corridors, conservation, Endangered Species Act, Convention on Biological Diversity, Convention on the Conservation of Migratory Species of Wild Animals, Convention on International Trade in Endangered Species (CITES), United Nations Convention on the Law of the Sea, animal species protection, wildlife corridors

Endemic Animal Species Protection (709): Endemic animal species are those animal species that evolved in a particular area and are restricted to it. The area may be a country or continent or more significantly a small area like a mountaintop or island or lake.

native animal species, habitat, conservation, ecosystem, ecology, genetic diversity, endemic species, species protection, viable population, invasive species, nonnative species, native species, biogeography, vulnerable species, animal species protection

Felids (130): Lions, tigers, cheetahs, cougars, and domestic cats are examples of the thirty-seven species of felids. Felids belong to the family Felidae and are carnivorous, with haired but not bushy tails, lithe, muscular, with five front leg digits and four behind.

They have retractile claws. Some are considered a threat to domestic animals and some hunted for the fur trade. Many are vulnerable or threatened (e.g. the Asiatic cheetah).

cats, leopard, cheetah, lion, tiger, family Felidae, mountain lion, cougar, puma, jaguar, fauna, lynx, conservation, habitat, mammals, animals

Fish (590): Fish are cold-blooded aquatic vertebrates with gills. The three major groups are eels and hagfishes, sharks and rays (cartilaginous fish), and finfish or bony fish such as sturgeons, carp, or Nile perch catfish. There are about 25,000 species of fish and at least 800 species are threatened. Inland waters and marine ecosystems are their major ecosystems. Many inland waters and their endemic fish species are threatened from habitat loss (dams), introduced species, and the aquarium trade.

class Chondrichthyes, sharks, rays, chimaeras, class Osteichthyes, bony fish, sturgeon, bowfins, catfish, Superclass Agnatha, lampreys, hagfish, fish, vertebrate, fauna, protection, conservation, habitat, animals, finfish, native, nonnative, endemic, biotechnology, migratory, fisheries

Insects (107): A class of animals with its body distinctly divided into head, thorax, and abdomen. Most have three pairs of legs and two pairs of wings. They are the only invertebrates that can fly. Insects provide ecosystem services (e.g. pollination) but are pests to crops and transmit specific diseases (e.g. malaria). There are over a million species including dragonflies, cockroaches, grasshoppers, crickets, termites, ants, true bugs, aphids, cicadas, beetles, moths and butterflies, bees, wasps, fleas, and true flies. Over six hundred species have been listed as threatened.

insects, pollination, pests, pathogens, agrochemicals, insecticide, conservation, habitat, animal protection, arthropods, animals, entomology, entomological, pollinator appreciation

Lagomorphs (10): The pikas, rabbits, and hares are small, terrestrial mammals that eat vegetation, mostly grasses and herbs. Humans hunt them for food and sport, and have domesticated them for pets, medical experiments and testing of products, and eating. Their soft fur skins have been favored for clothing. In large numbers, they are agricultural pests and competitors with other grazers. Their introduction to Australia was one of the most dramatic examples of problems caused by invasive species. A few island species are threatened.

order Lagomorpha, pikas, hares, rabbits, mammals, fauna, invasive species, sport hunting, threatened species, endemic species, habitat destruction, agricultural pests, animals

Lepidoptera (113): Butterflies, moths, and skippers are examples of lepidopterans. Lepidoptera is an order of insects characterized by a soft body, prominent eyes, mouthparts coiled into a proboscis for extracting nectar, and colorful wings made with scales. The caterpillar stage can provide services such as silk-making or can be a pest such as the gypsy moth. Beautiful lepidoptera are part of the illegal trade in species. Many "butterfly farms" hope to undercut the illegal trade.

butterfly, fauna, moths, skippers, order Lepidoptera, conservation, habitat, animals, animal protection, illegal trade, threatened species, butterfly farm, animals

Marsupials (47): Kangaroos, wombats, possums, and wallabies are examples of marsupials. Marsupials are metatherian mammals whose distinguishing features include the birth of their young at an early fetal stage of development inside an external pouch into which the incompletely developed young climb to nurse and finish developing. There are 250 species, mostly in Australia, and about 25 species are threatened. The Tasmanian wolf is extinct.

Metatheria, mammals, kangaroos, wallabies, wombats, opossums, numbats, Tasmanian wolf, Tasmanian devil, American marsupials, bettongs, cuscuses, bandicoots, billabies, protection, conservation, habitat, animals, koala

Mollusks and Crustaceans (19): There are over 100,000 living species of mollusks and crustaceans. Mollusks are highly valued as food, for their beauty, for making pearls, for medical research, and as collectibles, and were a form of currency. Well known are oysters, clams, scallops, squid, octopus, and snails. They live in every habitat: oceanic, rocky shore, freshwater, and terrestrial. Freshwater mussels have been threatened or are extinct from river manipulation in the United States. No marine mollusks are threatened with extinction but many are overexploited and overcollected, causing local populations to disappear. The most well known of the 40,000 crustaceans are lobsters, crayfish, crabs, and shrimps. As with mollusks, only a few isolated populations appear threatened with extinction but overexploitation is rampant. In addition, the introduction of crayfish has upset ecosystems and put other species in danger.

mollusks, crustaceans, oysters, shrimps, clams, scallops, squid, octopus, snails, lobsters, crayfish, crabs, isopods, habitat alteration, pollution, aquaculture, invasive species, endemism, fisheries, overexploitation, animals

Mustelids and Viverrids (55): Weasels, minks, ermines, ferrets, wolverines, badgers, skunks, and otters are examples of mustelids. Mustelids are members of the carnivorous mammal family Mustelidae (about sixty-five species). Civets, genets, mongooses, and fossas are members of the family Viverridae (about seventy species). They biocontrol rodents but may hunt poultry. Civets have been raised or hunted for their musk. Mongooses have become introduced pests. Many have been hunted or raised for their fur. Otters have suffered from habitat destruction, water pollution, pesticides, as well as hunting.

family Mustelidae, weasels, stoats, polecats, ferrets, mink, marten, fishers, tayras, wolverines, grisons, badgers, skunks, otters, sea otters, largest Carnivora family, plantigrade or digitigrade, protection, conservation, habitat, animals, mammals, family Viverridae, civets, genets, mongooses, fossas, musk

Perissodactyls (48): Rhinos, horses, donkeys, zebras, asses, and tapirs are examples of Perissodactyls. Perissodactyls are members of the order Perissodactyla and are odd-toed ungulates that are large browsing and grazing mammals with relatively simple stomachs and a large middle toe. There are seventeen species. Horses and donkeys have been introduced into the New World and Australia, sometimes domesticated, sometimes free-ranging. The provide meat, transport, and recreation. They are grazers and browsers with major impacts on grassland and savanna ecosystems. All rhinos, the Przewalski's horse, and a few asses are considered threatened.

order Perissodactyla, family Rhinocerotidae, rhinoceros, family Equidae, horses, asses, donkeys, zebras, family Tapiridae, tapirs, conservation, habitat, mammals, animals, endangered species, heritage farm animals

Pinnipeds (42): There are thirty-four species of seals, sea lions, and walruses. These pinnipeds are medium to large mammals that, as opposed to whales, must keep some connection to land for breeding. They are graceful and fully adapted to water, with the majority of pinnipeds living in colder waters. Humans have utilized pinnipeds for meat, oil, ivory, and fur and competed with them for fish resources. They are most easily hunted when they return to land to give birth. Young seals, hunted for their pelts, have been a major focus of animal rights and conservation sectors of the environmental movement.

walrus, seals, sea lions, hunting, conservation, animal rights, animal trade, CITES, fisheries, legal protection, extinction, animals

Primates (134): Primates, such as apes, monkeys, gibbons, and lemurs, are members of the mammalian order Primates. Humans are a member of the primate family Hominidae along with gorillas, chimpanzees, and orangutans.

order Primates, class Mammalia, apes, marmosets, tamarins, lemurs, bushbabies, gibbons, indris, sifakas, tarsiers, lorises, galagos, aye-ayes, monkeys, fauna, protection, conservation, habitat, mammals, animals

Raptors (249): Raptors are carnivorous birds of prey. Hawks, eagles, falcons, and owls are typical raptors, and are distinguished by strong hooked beaks and taloned feet. They are more endangered than any other type of bird. They are killed as varmints, illegally traded for skins or falconry, and impacted by pesticides and habitat destruction. Recent deaths have occurred at wind generators and transmission lines.

birds of prey, fauna, eagles, hawks, falcons, kites, order Falconiformes, owls, order Strigiformes, protection, conservation, habitat, animals, migratory, power poles, wind generation, agrochemicals, pesticides, species recovery, species trafficking, wind turbines

Reptiles (234): Turtles, snakes, lizards, alligators, and crocodiles are examples of reptiles. Reptiles are members of the vertebrate class Reptilia that are characterized by having scales, laying amniotic eggs, and engaging in predation. There are about 8,000 reptile species and about 300 are threatened. They have been killed for the pet trade, for their shells and meat, from habitat loss, and from fear. Specific reptiles are poisonous. Some species provide the service of rodent control. Marine turtles and the Galápagos iguana are typical of marine reptiles. Over 7,500 are terrestrial.

class Reptilia, Testudines, turtles, tortoises, Lepidosauria, snakes, lizards, cold-blooded, scales, Crocodilia, crocodiles, alligators, fauna, protection, conservation, habitat, animals, iguanas

Rodents (30): Rodents are in the largest order of mammals (order Rodentia). The teeth of rodents are specialized for gnawing with continuously growing incisors. There are about 1,700 species including mice, rats, squirrels, beaver, porcupine, lemmings, gerbils, and gophers. Rodents like beavers are eco-engineers; certain mice and rats have been domesticated and are crucial to human health testing; others carry disease and destroy crops; squirrels provide the crucial service of dispersing and planting tree seeds; prairie dogs play a controversial role in conservation ranching; and chinchillas are famous for their fur.

order Rodentia, suborder Sciurognathi, gophers, squirrels, marmots, scaly-tailed squirrels, mountain beaver, gundis, jerboas, jumping mice, pocket gophers, kangaroo rats, pocket mice, voles, dormice, springhare, squirrels, suborder Hystricognathi, hystricognath rodents, pacas, chinchillas, tuco-tucos, mole-rats, agoutis, capybaras, voles, lemmings, hamsters, gerbils, porcupines, pacarana, conservation, habitat, protection, animals, mammals, endangered species, public health

Sirenians (12): Dugongs, sea cows, and manatees, six species all told, are examples of sirenians. Sirenians are herbivorous aquatic mammals with massive, fusiform bodies that live in coastal waters and are members of the order Sirenia. They are scarce because humans have hunted them for food, hides, and oil, and have endangered their populations with powerboats. They are vulnerable to climate change.

order Sirenia, family Dugongidae, dugongs, sea cows, family Trichechidae, manatees, conservation, protection, habitat, mammals, animals, marine animals, coastal waters, hunting

Ursids (55): Ursids are omnivorous mammals in the family Ursidae, which includes eight species of bear including polar bears, black bears, brown bears, and pandas. Bears have been persecuted for killing humans (usually provoked), for their hide, meat, fat, and for ceremonies. All bears, except the American black bear, are threatened.

family Ursidae, subfamily Ailurinae, pandas, giant panda, subfamily Ursinae, bears, sun bear, sloth bear, spectacled bear, black bears, brown bear, polar bear, protection, conservation, habitat, animals, mammals, grizzly bear

Wildlife Ecology (320): Wildlife ecology is the scientific study of the human influences and ecological interactions of wildlife and the environment. Wildlife ecology studies, particularly at the level of wildlife population and communities rather than at the individual species level, inform wildlife management and conservation biology.

wildlife species, communities, populations, ecosystems, habitat management, research, conservation, fauna, undomesticated, interactions, environment, ecology, endangered species, protection, management, study, research, preservation, biodiversity

Wildlife Habitat Conservation (6,149): Wildlife habitat conservation is the practice of protecting and preserving the habitat of wildlife species, preventing habitat loss, and promoting land management that enhances wildlife habitat and prevents wildlife species extinction in the long term.

ecosystem, protection, restoration, population, undomesticated, animals, plants, land management, encroachment, parks, refuges, legislative protection, wildlife habitat easements, conservation agreements, habitat conservation planning, study, wildlife species niche, community, range, biodiversity protection, wildlife corridors, carrying capacity

Wildlife Law and Policy (320): Wildlife law and policy attempts both to produce a sustained yield of game animals and fish and to protect threatened nongame species. It must reconcile disputes between ranchers and those desiring the return of predators (wolves, grizzlies); between cattle raisers and buffalo that may carry brucellosis; and many upstream and downstream river basin users (e.g. salmon). Wildlife law and policy must reconcile mobile animal life that crosses privately owned, club-owned, state and federal lands as well as international boundaries. In the United States, all wildlife is owned by the state and laws pertain to common property management issues. Federal rules override state rules when it comes to endangered species. International agreements pertain to migratory birds, fish, cetaceans, trade in endangered species, and other border issues. Typical tools of wildlife managers include: compensatory payments for livestock killed by predators, safe harbor agreements with private landowners, refuges for migratory birds, and changing season and limits on game species.

wildlife management, game species, nongame species, refuges, seasonal limits, take limits, private lands, state lands, federal lands, permit buyouts, compensatory payments, safe harbor rules, conservation agreements, habitat conservation programs, wildlife corridors, Neotropical Migratory Bird Act, CITES, Environmental Quality Incentive Program (USDA), Grassland Reserve Program (USDA), Wildlife Habitat Initiative Program (USDA), Conservation

Reserve Program (USDA), National Resource and Conservation Service, Resource Conservation Districts, wildlife politics, wildlife conservation, wildlife policies

Wildlife Management (656): *Wildlife management* is an ambiguous label, usually applied to mammals, birds, and fish that are harvested by humans. Wildlife, in the United States, focuses on migratory birds, fur-bearing mammals, game fish, game animals, and, more recently, protection of nongame species. Wildlife managers work to ensure that land and water produce sustainable crops of these wild animals. Wildlife management tries to accomplish "sustainable yields" by monitoring predation and pathogens, studying the life histories of species, protecting and enhancing habitat, setting seasons and limits on fishing and hunting, providing supplemental feed and fish hatcheries, and patrolling illegal activities.

wildlife management, fish, fishermen, game fish, game, hunters, wildlife habitat, fur bearers, trapping, waterfowl, duck hunting, nongame species, farm wildlife, urban wildlife, wilderness wildlife, hunting regulations, fishing regulations, sustained yield, wildlife refuges, winter feeding grounds, fish hatcheries, wildlife species monitoring, wildlife corridors

ARTS
Arts refers to human creative activities and techniques for communication encompassing art, painting, sculpture, literature, criticism, theater, performing arts, and music.

art, sculpture, arts therapy, arts education, literature, performing arts, visual, written, oral, creative, criticism, music, opera, storytelling, communication tool, painting, human creativity

Art and Sculpture (123): *Art and sculpture* refers to the creative human activities of painting, drawing, illustrating, etching, and creating three dimensional sculptured objects using a variety of techniques and materials.

creative, visual arts, eco art, environmental, natural materials, communication tool, three-dimensional, drawing, etching, illustration, painting

Arts Activism (526): Arts activism is a form of direct action to bring about change, be it social, political, or environmental, using arts to educate, raise awareness, and promote the aims of the activists. Arts activism can include murals, storytelling, installation art, drama, and music to communicate the message of the activist organization.

education, communication, teaching, environment, social, aesthetic representation, solutions, creativity, environmental education, social justice education, community, using media for activism, music, radical art

Arts Education (304): Arts education is the activity of educating, teaching, training, and imparting knowledge, ideas, and skills to people about arts and cultural traditions, including music, oral history, storytelling, and sculpture.

art history, education, arts, criticism, appreciation, creative, environmental art, education through art, cultural traditions, oral history, storytelling

Arts Therapy (44): Arts therapy is the practice of using expressive and creative arts, such as painting, singing, or role-playing, to explore psychotherapy and to heal and enhance the life of people undergoing arts therapy.

healing, growth, art psychotherapy, music therapy, drama therapy, poetry therapy, photo therapy, dance therapy, rehabilitation, communication skills, expression

Literature (60): *Literature* refers to published written works of both nonfiction and fiction, prose and verse, and can refer to a body of written work about a specific subject, such as environmental literature, or to all written and oral work. Literature encompasses books, magazines, articles, journals, plays, short stories, and poetry.

writing, poetry, written work, oral, prose, verse, comment, criticism, arts, journals, articles, technical literature, fiction, nonfiction, written material, composition

Performing Arts (324): *Performing arts* refers to those activities involving live and public performance, such as theater, dance, music, and opera.

theater, public performance, dance, music, mime, oral history, concert, opera, drama, comedy, communication, storytelling, folk dance

BIODIVERSITY

Short for biological diversity, *biodiversity* refers to the number and kinds of organisms on the planet. It includes genetic diversity, taxonomic (species) diversity, and ecosystem diversity. Biodiversity can be viewed within marine, terrestrial, and inland water biomes, or by genetics such as plant, animal, and fungi diversity. Biodiversity is the source of all human food, domesticated breeds of animals and plants, as well as bio-based medicines. Only about 10 percent of all the species on earth are known.

biological diversity, conservation, biota, flora, fauna, plants, animals, fungi, prokaryotes, genetic diversity, species diversity, ecosystem diversity, species abundance, extinction, endangered species, threatened species, vulnerable species, sensitive species, rare species, marine ecosystem, terrestrial ecosystem, inland water ecosystems, aesthetic value, environmental ethics, genetic resource

Biodiversity Conservation (3,048): Biodiversity conservation is the practice of protecting and preserving the abundance of biological diversity of all species of plants and animals, of ecosystems, and of genetic diversity on the planet. It includes genetic diversity, taxonomic (species) diversity, and ecosystem diversity. Biodiversity can be viewed within marine, terrestrial, and inland water biomes, or by genetics such as plant, animal, and fungi diversity. Biodiversity is the source of all human food, domesticated breeds of animals and plants, as well as bio-based medicines. Conserving biodiversity is about preventing the extinction of any species of plants or animals.

biological diversity, conservation, biota, flora, fauna, plants, animals, fungi, prokaryotes, genetic diversity, species diversity, ecosystem diversity, species abundance, extinction, endangered species, threatened species, vulnerable species, sensitive species, rare species, marine ecosystem, terrestrial ecosystem, inland water ecosystems, aesthetic value, environmental ethics, genetic resources

Domesticated Animal Diversity (39): Farmer-breeders have selected for specific traits of many local varieties of domestic animals. The four major domesticated mammals—pigs, cattle, sheep, and goats—have diversified into four thousand recognized breeds. These varieties are rapidly being lost. Heirloom domestic livestock conservation protects these animals alive on farms and by freezing sperm and eggs.

animal biodiversity, biodiversity, sperm bank, heirloom genetic diversity, bloodlines, threatened breeds, genetic resources, food production, Food and Agriculture Organization, congeneric wild species, Convention on Biological Diversity, cattle, goats, sheep, horses, dogs, rabbits, chickens, pigs, camels, yak, water buffalo, llama, reindeer, ass, guinea pig

Domesticated Plant Conservation (79): Germplasm is any plant or plant part that might be used in breeding new varieties. It includes seeds, whole plants, and tissues. The domestication of plant species involves the selection of plants for specific traits to be cultivated for human use. Domesticated plant conservation is the practice of preserving the biological diversity of those plant species.

plant diversity, biodiversity, seed collection, seed bank, genetic diversity, cultural plants, nonhybrid seeds, open-pollinated plant varieties, plant genetic resources, agriculture, environmental sustainability, heirloom varieties, genetic reserves, genetic engineering, cryopreservation, germplasm, green revolution, cultivars, Convention on Biological Diversity, International Plant Genetics Resource Institute (Rome), U.S. National Plant Germplasm System (Colorado), National Gene Bank (Malawi), Food and Agriculture Organization (FAO), cereals, leaf vegetables, tubers and roots, beans, oil crops, sugar crops, spices, tree nuts

Seed Conservation (157): *Seed conservation* refers to the practice of protecting the abundance of biological diversity of plant genes and seeds from loss and extinction, and ensuring the preservation of those genetic resources and seeds for the use of future generations. There are one hundred seed banks on the planet with 3 million samples of seeds.

plant diversity, biodiversity, protection, preservation, seed collection, seed bank, genetic diversity, cultural plants, nonhybrid seeds, open-pollinated plant varieties, plant genetic resources, agriculture, environmental sustainability, tree nursery

BUSINESS AND ECONOMICS

Business refers to commercial and industrial enterprises and to the provision of goods and services for a profit. *Economics* is the social science that studies the production, distribution, and consumption of goods and services, and the allocation of scarce resources by individuals, companies, and states to satisfy wants.

goods, services, consumption, allocation, scarcity, scarce resources, production, competition, distribution, income, expenditure, money, commodities, costs, benefits, wealth, capital, commerce, enterprise, entrepreneur, industry, responsibility, profit making, sales, product, trade, venture, shares, shareholder, equity, investment, company, partnership, currency, corporate ethics and governance, corporate responsibility, finance, microcredit, natural capitalism, environmental economics, progressive taxation, responsible business practices, socially responsible investing, ecological economics

Business Firm and Organization Sustainability (51): A business firm has many components: shareholders, employees, trading partners, suppliers, insurers, regulators, customers, investors, and philanthropic activities. Any or all of these players can work toward sustainability and push the other players to act toward sustainability. Trading partners can demand "green products"; shareholders can demand social justice; and the philanthropic division can focus its attention on sustainability-oriented groups. It is through these subgroups of a business firm and government, consumer, or NGO pressure that businesses can change their way of operating.

business firm, green business, shareholders, employees, trading partners, insurers, regulators, business philanthropy, sustainable business, NGOs, investors, banks, insurance companies, regulations, boycotts, green purchasing

Corporate Ethics and Corporate Governance (300): *Corporate ethics* refers to the ethical and moral decisions that are faced in all business contexts. *Corporate governance* refers to the particular responsibilities and ethics related to running a business or corporation such as accountability to shareholders, transparency to improve credibility on environmental management, strategies to gain competitive advantage, environmental management guidelines, and financial management to avoid corruption and guide research and development. Many corporations sign respected, voluntary codes of conduct in order to set guiding ethics for operations.

standards, corporate governance, responsibility, good practice, accountability, transparency, monitoring, reporting, policy, board, management, ethical behavior, corporate responsibility, business, economics, social responsibility, capitalism, anti-capitalism, multinationals, codes of conduct, UNEP, ICC, CERES, ISO 1400, Natural Step, CEO statement, fair trade, money laundering, drug trafficking, tax evasion, oppressive regimes, tobacco, factory farms, animal rights, fur trade, blood sports, weapons, child labor, racial discrimination, gender discrimination, R&D spending, corporations redefined and limited

Ecosystem Services (279): Ecosystem services are those services provided by different ecosystems and natural capital that are critical to human welfare, such as the provision of clean water by wetlands, the absorption of carbon dioxide in forests, and the pollination of plants by insects. Existing economic models have not effectively valued ecosystem services. Internalizing the value of ecosystem services is a prerequisite for sustainability and for the health of the environment.

externalities, valuing environment/ecosystem functions, ecological services, clean air, erosion prevention, pollination, flood defense, wetlands filtration, seed dispersal, decomposition, nutrient cycling, oxygen production, soil formation, natural life-support structure, pollution mitigation, economic value, valuing services, carbon sequestration, ecological costs, ecological economics, ecocentric, water purification, shadow pricing, public good

Ecotourism (1,239): Ecotourism is ecological tourism that avoids harming the natural and ecological quality of a particular area. Ecotourism promotes enjoyment balanced with protection of the environment, where tourists seek out environmentally sensitive travel or tours which, in some way, improve or add to their knowledge of an environment.

ecotourism, low-impact, environmental education, natural resources, ecological impact, appreciation of nature, protection, sustainable travel, environmentally responsible, ecologically sensitive architecture, economic development, land preservation, nature conservation incentive, environmental conservation, socioeconomic benefits for local people, sustainable use of biodiversity, sustainable tourism, business, economics

Finance Policies and Institutions (909): *Finance* refers to the way in which businesses, individuals, and organizations allocate and use financial resources, and refers to the flow of money, assets, capital, banking, credit, investments, equity, and debt. Since the 1960s, the financial services industry has internationalized with the massive increase in world trade, the expansion of transnational companies, the proliferation of remittances from foreign workers to their home nations, and the vast institutionalization of savings in pension funds seeking maximum investment returns. In addition, the creation of offshore financial centers and the movement of illegal transactions from drugs and weapons have challenged government management of internal finances. Sustainability advocates have responded by developing green banking and other financial services, such as microfinance (which encourages capital to remain in communities), alternative currencies, shareholder advocacy, and critiques of major financial organizations such as the IMF.

money, debt, banking, lending, development banks, lending program impacts, international financial institutions, IFIs, capital, loan, transparency, microcredit, financial flows, business, economics, assets, capital, equity, debt relief, green enterprise, capitalism, credit union, international law, privatization, structural adjustments, equity and debt, offshore financial centers, furtive money, reserve requirements, tax shelters, electronic cash management services, remittances, transnational banks, commercial banks, investment bank/securities house, credit card company, accountancy firm, insurance company, outsourcing, IMF, World Bank, Eurodollar, International Banking Facilities (IBF), stock exchange

Fiscal Policies, Institutions, and Taxation (63): *Fiscal* refers to public finance and financial transactions. It includes a government's policy toward taxation, public debt, public appropriations and expenditures, and similar matters. Fiscal policies impact private businesses, NGOs, nonprofits, and the economy at large. Public appropriations and taxation are two powerful tools to encourage or discourage sustainability. In developing nations, public debt and monetary policy are crucial driving forces hampering or helping rid the nation of poverty. Taxation refers to the compulsory payment of money or assets from individuals, businesses, and organizations to the state or government. Taxation can be a tool used by governments to produce environmental and social benefits, such as the taxation of pollution production to create an incentive for businesses to reduce their output of specific pollutants.

public finance, national banks, national treasury, public appropriations, state budget, taxation, public debt, monetary policy, fiscal monopoly, money, credit, treasury bonds, business, economics, ecological taxes, green taxes, markets, fiscal tools, progress indicator, environmental taxes, externality taxation, tax policy reform, taxation equality, ecological tax reform, economics, business, social taxes, capitalism, progressive taxation, tax haven

Green Banking and Insurance (9): Banks are businesses formed to maintain savings and checking accounts, issue loans and credit, and deal in negotiable securities and other financial services. Insurance businesses collect premiums from customers in exchange for a promise to reimburse the customer in the event of loss. Both services can have great influence on sustainability by: financing "green" product investments or lowering insurance rates if "green" products are used; creating microinsurance packages for the poor; issuing "green" credit cards in which part of the profits are donated to sustainability NGOs; providing "green" mortgages at special rates for sustainable homes; preferential banking packages to organizations adopting "green" strategies (especially for energy and greenhouse gas reduction); or providing "green" investment funds at special rates for restoration projects such as brownfields.

green banking, green insurance, green lending, green project financing, green insurance, green investments, green products, green credit cards, green mortgages, green savings accounts, preferential banking packages, environmental liability/damage insurance, green insurance research, recycling write-offs, catastrophe bonds, green due diligence checks, microinsurance

Microcredit (17): Microcredit is the extension of very small loans (microloans) to the unemployed, to poor entrepreneurs, and to others living in poverty who are not bankable. These individuals lack collateral, steady employment, and a verifiable credit history and therefore cannot meet even the most minimum qualifications to gain access to traditional credit. Microcredit is a part of microfinance, which is the provision of financial services to the very poor; apart from loans, it includes savings, microinsurance, and other financial innovations. The intention of microcredit institutions is to help poor people and those denied access to credit to overcome poverty and to fund income-generating activities for self-employment in order to build social capital.

microcredit, microlending, microfinance, access to capital, collateral, credit as a human right, poverty reduction, financial sustainability, financial empowerment, economic independence, social capital, business, economics, credit societies, debt, credit union, Grameen Bank

Microfinance (1,323): *Microfinance* refers to the provision of financial services to clients who are excluded from the traditional financial system on account of their lower economic status. These financial services will most commonly take the form of loans (see *Microcredit*) and microsavings, though some microfinance institutions will offer other services, such as microinsurance and payment services.

microfinance, microlending, microcredit, access to capital, financial services to the poor, collateral, credit as a human right, poverty reduction, financial sustainability, financial empowerment, economic independence, social capital, business, economics, credit societies, debt, credit union, Grameen Bank

Natural Capitalism (201): Natural capitalism is the concept that natural capital—natural resources and the ecological systems that provide vital life-support services—is scarce and that business, industrial, and economic activity must value and use natural capital wisely to continue to operate efficiently. Natural capitalism encompasses increased resource productivity and material efficiency, redesigning industry by mimicking biological processes, providing services rather than products, and reinvesting in natural capital for a sustainable economy.

environmental economics, economy, change, sustainability, social responsibility, operational efficiency, alternate economies, valuing, externalities, scarcity, natural resources, green business, ecosystem services, sustainable economy, human well-being, biosphere well-being, resource productivity, material efficiency, biomimicry, reinvesting in natural capital, green enterprise, green economics, ecological economics, ecocentric, human capital, ecological cost, human development theory, material goods, bioregional economies, conservation economy

Responsible Business Practices (549): A responsible business strategy includes four steps toward sustainable development. First, the business can follow compliance-oriented regulations (a reactive stance). Then, it can adopt a precautionary or preventive strategy to avoid harmful environmental impacts. Third, it can use economic, social, and environmental strategies for competitive advantage (opportunity seeking). Finally, it can adopt a full range of responsible operating, product, financial, stakeholder, labor, environmental, and social strategies that make up sustainable development.

corporate social responsibility, CSR, ethics, eco-efficiency, Environmental Management Systems, EMS, green procurement, supply chain management, accountability, environmental auditing, environmental reporting, monitoring, waste reduction, environmental entrepreneurship, environmental consulting, green enterprise, capitalism, social enterprise, corporate codes of conduct, green accounting, health and safety reporting, eco-design, stakeholder involvement, life cycle assessment, LCA

Socially Responsible Investment (271): *Socially responsible investment* (SRI) refers to evaluating and choosing investment opportunities on the basis of social responsibility and environmental sustainability criteria in addition to financial criteria. The scope of SRI ranges from an investment fund manager choosing to buy shares or invest equity in socially and environmentally responsible companies, to activism by shareholders who campaign to make companies improve their social and environmental record in their business activities.

SRI, mutual funds, investment, shareholder activism, boycott, triple bottom line, equity, environmental sustainability, work conditions, responsibility, ethics, corporate governance, campaigns, attack, market, branding, organizing, advocacy, impact, name, business, economics, reputation, incentives for socially and environmentally responsible behavior, capitalism, multinationals, privatization

CHILDREN AND YOUTH
Children and youth refers to those rights and issues that concern children and young adults, such as children's rights, children's health, juvenile justice, and the empowerment of youth.

education, teenagers, programs, empowerment, counseling, skills, kids, young people, youth-led, rights of the child, children's health, child and youth protection, juvenile justice, youth education and empowerment, youth organization, girls, boys, training, leadership, capacity building

Child and Youth Protection (4,645): *Child and youth protection* refers to society's responsibility to protect children and youth from threats, be it physical abuse or lack of access to basic needs such as food, and to protect the rights of children to be free from harm.

abuse, orphanage, quality of life, care, legal protection, social services, monitoring, impact, support, access, childcare, health, family, housing, neglect, antisocial behavior, safe environment, shelter, exploitation, victimization, conflict, discrimination, outreach, kids, young people, preventing abandonment, girls, boys, infanticide, sexual exploitation of children, social exclusion

Child Labor (1,085): *Child labor* refers to children working in situations where they are exploited and are denied their human rights. The exploitation of child labor encompasses providing children with insufficient compensation, a lack of consideration for their health and safety, the loss of opportunities for children to receive education, as well as child prostitution and the use of children in the military.

underage, forced, systematic exploitation, oppression, loss of childhood, wages, working conditions, compromising child development, physical, psychological, mental, social, moral, deprived of access to education, poverty, child rights, youth, children

Children in Armed Conflict (366): *Children in armed conflict* refers to both the use of children as soldiers and the impact of armed conflict on the human rights, dislocation, health, education, social support system, freedom, safety, and security of children.

child soldier, child rights, child exploitation, family, refugees, conflict, war, children's disease, orphans, youth, children

Children's Health (2,632): *Children's health* refers to improving children's access to health care, educating families about children's health, and working on issues of health that are specific to children, such as the impact of environmental pollutants on child development, immunization, asthma, child obesity, and health education for children that will enable them to make healthier lifestyle choices for their health in the future.

health, children, toxics, development, disease, susceptibility, kids, young people, youth, educating about health, toxics, health advocacy, raising awareness, safe products, en-

vironmental hazards, child obesity, environmental health, chemical, health equity, health policy

Juvenile Justice (487): *Juvenile justice* refers to the treatment and rights of children, who are deemed too young to be held responsible for criminal acts, within the legal and criminal justice system.

children's rights, equity, abuse, social justice, youth rights, representation, young people, detention, custodial, youth development, girls, boys, legal system, criminal justice system, rehabilitation, social exclusion

Rights of the Child (4,631): Rights of the child are those basic human rights that apply to children such as the right to survive and to have basic needs met, the right to develop and to reach their fullest potential, the right to be protected from harm, exploitation, and abuse, and the right to participate.

children's rights, child labor, justice, educating about rights, abuse, discrimination, violence, protection, kids, youth, children, young people, social justice, access to education, access to health care, fundamental rights, human dignity, nondiscrimination, best interests of the child, survival, development, protection, participation, freedom of expression, freedom of speech, freedom of thought, girls, boys, sexual exploitation of children, race, gender, sexuality, disability, political rights, identity rights, social exclusion, leisure

Youth Capacity Building (2,352): *Youth capacity building* refers to the practice of developing the skills and potential of youth to take the lead in their own development and the development of their communities by enhancing their expertise and awareness about relevant issues and their community, such as youth training on cultural diversity.

training, skills, teaching, knowledge, information, vocational training, mentoring, empowerment, partnership, cultural understanding, development, inclusion, identity, acceptance, tolerance, communication, youth, child, children, developing potential, confidence, leisure, play

Youth Education and Empowerment (5,862): *Youth education and empowerment* refers to the activity of educating, teaching, training, and imparting knowledge, ideas, and skills to children and youth, both within the formal education system and informally to give youth the skills to be empowered to fully participate in society.

leadership, mentoring, training, civic, engagement, recreation, responsibility, organizing, rights, participation, skills, inclusion, equality, access, services, capacity building, identity, culture, diversity, integration, family, counseling, voice, advocacy, gender, personal development, teambuilding, challenge, adventure, outreach, kids, young people, children, activism, grassroots, action, lifelong learning, social exclusion, leisure

Youth Leadership (1,337): *Youth leadership* refers to fostering the ability to lead youth to empower them to fully participate in society. Youth leadership training is the practice of teaching, imparting knowledge, and providing instruction in leadership skills to youth in the community to empower them to organize and lead community activities.

leadership, skills, training, knowledge, development, youth, child, participation, engagement, understanding, fostering partnership, organizing, children

Youth Participation (1,997): Youth participation addresses the ability of children and youth to actively participate in community, local, and national institutions, and to encourage youth to tackle social, cultural, and economic barriers to civic and demo-cratic participation. Effective civic participation of youth strengthens institutions and empowers youth to be leaders within their communities.

civic, participation, promoting, engaging, community organizing, involvement, ac-tivism, partnership, promote understanding, responsibility, engagement, democracy, fostering participation, civic understanding, child, development, community values, empowerment, civil disobedience, direct action

Youth-led Organizations (882): Youth-led organizations are those groups that are founded, organized, operated, or led by children and youth, and are concerned with youth issues.

youth-led, youth organization, children, youth empowerment, participation, civic, grass-roots, leaders, action, youth leadership

CIVIL SOCIETY ORGANIZATIONS—NONPROFITS, PHILANTHROPY, AND SOCIAL ENTREPRENEURS

Civil society organizations refers to a variety of institutions, groups, foundations, trusts, and associations formed by citizens to address social and environmental issues, prob-lems, and needs that are either caused by business or government or are insufficiently addressed by them. These organizations are primarily known by terms that indicate what they are not: In Europe, the term "nongovernmental organization (NGO)" arose; in the United States the term "nonprofit organization (NPO)" is used to designate citi-zen initiatives. Although both describe what organizations are not, they do not describe what they do. *Civil society* or *citizen sector* are preferred terms used by many organiza-tions and institutions. They are active ways to describe the NGO and NPO world. Civil society is part of the "sustainability triangle" of government and business, which to-gether are the three most powerful driving forces in fostering change on the planet. Civil society includes community-based organizations (CBO), educational institutions, faith-based organizations, foundations, networks, coalitions, alliances, research insti-tutes, UN organizations, village-based organizations (VBO), activist organizations, and virtual organizations that function on the Internet but have no geographical location. Some are tightly linked to the nation-state, while some are fiercely independent. They provide an array of services to marginal and neglected populations and tackle some of the most pressing environmental issues on the planet. Some scholars include local groups that perform specific tasks, like irrigation groups in India or New Mexico or religious groups in Indonesia or California. (The term *civil society* excludes and is con-trasted to ideological organizations that advocate violence, further ecosystem destruc-tion, or promote prejudice.)

civil society organization (CSO), global civil society, civilis societas, citizen sector, global networks, social movements, environmental movement, human rights, crisis relief, sustain-ability, nongovernmental organization, NGO, ONG

Communications Training (1,344): *Communications training* refers to the activity of supporting and instructing civil society organizations and nonprofits to improve their methods of communication externally, such as through marketing or promotion, and internally, to raise their profile and to enhance their effectiveness as an organization.

nonprofit marketing, raising awareness about nonprofits, nonprofit organization marketing, communications training, communications quality, information sharing, skills, effective communications, enhancing civil society skills, NGO services, The Chronicle of

Philanthropy, American Philanthropy Review, Philanthropy, *NGO, ONG, citizen sector, civil society*

Nonprofit Law (6): The nonprofit sector has grown and continues to grow exponentially, but the sector's growth worldwide has been uneven. There are still many countries where the right to associate is not protected in a country's basic legal system; the provisions governing the basic establishment of nonprofit organizations is arduous and onerous; there are few or no fiscal and related advantages accorded to such organizations; and the nonprofits operate under strict guidelines and control. Legal systems influence the financial viability of nonprofit organizations. In an enabling legal environment, nonprofit organizations not only receive tax exemption on their income, but there are also special tax treatments for contributions by others to the organizations. Enabling legal systems also support the nonprofit sector by establishing accountability standards and ensuring that nonprofits operate in ways that are legitimate and credible.

associations, corporation, tax benefits, public benefit organization (PBO), memberships, registration requirement, NGO law, nonprofit law, foreign funding, domestic funding, tax-exempt organization, nonprofit organization (NPO), community-based organization (CBO), civil society of laws, legal framework for civil society, public charitable trust, freedom of association, freedom of information, freedom of expression

Organizational Funding (1,907): *Organizational funding* refers to the practice of providing financial support and nonfinancial assets to civil society organizations to enable them to achieve their mission. Funding may include memberships, gifts, personal giving, payroll deductions, volunteer time, tithes, bequests, stock donations, conservation easements; foundation, government, and corporation grants; loans; and in-kind funding such as equipment and other nonmonetary resources. In the United States, 80 percent of the contributions to nonprofits, NGOs, and civil society organizations come from individuals.

funding, financial support, in-kind funding, match funding, leverage funding, civil society effectiveness, philanthropy to nonprofits, giving, nonprofit services, nonprofit fundraising, memberships, gifts, personal giving, payroll deductions, volunteer time, tithes, bequests, stock donations, family foundations, community foundations, charitable trusts, federated campaigns, government grants, corporations, grants, loans, in-kind funding, religious tithes, religious personal giving, nonprofit tax breaks, dedicated sales tax, philanthropic income tax write-offs, donated financial securities, partnerships, endowments, reserve funds, development assistance contributions, NGO, ONG

Organizational Governance (206): *Organizational governance* refers to the rules and regulations set up by a particular nation that allow civil society organizations to exist. In the United States, some function as nonprofits and avoid taxation of donations. Others can lobby politically and promote candidates (raising election funds) as they have different reporting requirements. In many developing nations and in the former Soviet Union, nonprofit organizations led by citizens need permits to operate. Internal governance structures for nonprofits are sometimes dictated by national government requirements and sometimes invented by the organization itself. Internal governance standards vary widely depending on if the citizen organization is multinational or local; supports a religious group or is secular; or receives funding from private, business, and/or government donors.

nongovernmental organization, nonprofit governance, nonprofit board, nonprofit

*management, nonprofit accountability, nonprofit transparency, nonprofit rules of incorpora-
tion, 501(c)3 (United States), political NGOs, for-profit NGOs, religious NGO, ONG*

Organizational Support and Management (4,076): *Organizational support and management* refers to the practice of supplying nonfinancial support to nonprofits and civil society as a sector to enable individual organizations and the broader movement to achieve their goals more effectively. Nonprofit support can include consulting services to individual organizations as well as advocating for nonprofits and civil society to national governments and international institutions.

civil society building, supporting nonprofit activity, promotion, advertising nonprofit work, representing nonprofits, nonprofit promotion, nonprofit work, capacity building, technical assistance, consulting services, advice, enhancing nonprofit effectiveness, networking, advocacy, funding, nonprofit services, building the third sector, movement building, NGO, ONG, strategic grantmaking, effective grantmaking, social change grantmaking, grantmaking, philanthropy, effective fundraising, foundation training, organizational effectiveness

Philanthropy (907): Philanthropy is the act of giving money, goods, time, or effort to support a charitable cause, usually over an extended period of time and in regard to a defined objective. In a more fundamental sense, philanthropy may encompass any altruistic activity that is intended to promote good or improve human quality of life. Although philanthropy is associated with people and institutions of wealth, people may perform philanthropic acts without possessing great wealth. And altruistic giving toward any kind of social need that is not served, undeserved, or perceived as such by the market can be considered a philanthropic act. However, philanthropy is most associated with foundations and trusts that act on behalf of wealthy persons, living or deceased. Corporations such as Google are creating a new, non-tax-exempt branch of philanthropy called for-profit philanthropy. Along with individuals such as Jeff Skoll and Pierre Omidyar, it makes for-profit investments that are considered philanthropic because their purpose is social benefit instead of monetary profits (see *social entrepreneurship*).

altruism, foundations, charity, social philanthropy, nonprofit organization, philanthropist, volunteerism, trusts, grantmaking, endowment, humanitarian cause, Clinton Global Initiative, for-profit philanthropy, social capitalism, social commons

Social Entrepreneurship (96): *Social entrepreneurship* refers to the activity of activists who use entrepreneurial methods to address systemic social problems. The scope of social entrepreneurship encompasses the entire listing under Area of Focus. The term has been associated with risk takers who use ideas, resources, and opportunities in novel ways in order to produce outcomes that benefit society. Although social entrepreneurs can work in both the for-profit and nonprofit realms, their success is measured by social profit not monetary gains. The term is relatively new, dating back to the 1950s and the work of Michael Young, who created sixty different social-benefit organizations in the world, including the School of Social Entrepreneurs in 1997. The practice of social entrepreneurship extends back to the public health movement during the Industrial Revolution and would include such notables as Florence Nightingale, Susan B. Anthony, and M. K. Gandhi. The greatest single proponent of social entrepreneurship has been Bill Drayton of Ashoka. The best-known practitioner of social entrepreneurship is Muhammad Yunus, the creator of microfinance and microcredit, the founder of Grameen Bank in Bangladesh, and the winner of the 2006 Nobel Peace Prize. Despite the recent publicity, the work of most social entrepreneurs remains largely unnoticed.

social change, social profit, social benefit, social capital, social entrepreneur, nonprofits, social enterprise, public benefit, citizen sector, Ashoka, Skoll Foundation, Omidyar Network, Schwab Foundation, Michael Young, citizen sector, civil society, nonprofits, foundations, nongovernmental organizations, NGOs, change-makers, social innovation, social development, poverty, social capitalism, community development

Training for Nonprofits (909): Nonprofit training occurs on many levels: the board, the staff, the fieldworkers, the clients, even the donors. Training ranges from improving leadership skills, improving grant writing, capacity building, learning a host nation's language and culture, acquiring technical and mechanical skills for work in the field, practicing basic survival skills, learning communications and media skills, and the style of diplomacy needed in conflict resolution. A subculture of nonprofits offers the various training skills in workshops and consultancies.

civil society building, strengthening, growing, supporting nonprofit activity, promoting, advertising nonprofit work, representing nonprofits, nonprofit promotion, nonprofit work, capacity building, technical assistance, consulting services, advice, enhancing nonprofit effectiveness, networking, advocacy, funding, NGO services, building the third sector, communications, media, marketing, ONG, NGO, strategic grantmaking, effective grantmaking, social change grantmaking, grantmaking, philanthropy, effective fundraising, foundation training, organizational effectiveness

COASTAL AND MARINE ECOSYSTEMS

Coastal and marine ecosystems refers to the interdependent and dynamic relationships of all living organisms in ocean environments, from tidal pools to salt marshes to continental shelves and marine current systems.

wildlife, protection, preservation, threats, pollution, fishing, coastlines, ocean science, technology, ecosystems, coral reefs, species, population, ecological health, research, ecology, biodiversity, protecting fisheries, fish populations, marine products, sustainable fishing, tropical fish trade, mangroves, spawning, habitats, shorelines

Coastal Ecology (466): Coastal ecology is the scientific study of the relationships between organisms and their environment along coasts and in coastal waters. Coastal ecology is concerned with the distribution and behavior of individual species as well as with the structure and function of coastal systems at the level of populations, communities, and ecosystems.

ocean, regeneration, change, beach, dunes, conservation, environment, erode, land, wind, wave, tide, pollution, nutrients, salinity, salt marshes, species, ecosystems, ecology, biodiversity, coast, siltation, human, nature, habitats, interactions, sea defenses, delta, ocean, sea, marine, fish, marine conservation

Coastal and Marine Human Impacts (39): The coasts of the world are increasingly impacted by migration of humans to the coast, the growing importance of coastal waters for recreation and aesthetic enjoyment, and the growing use of coastal and marine resources for transport, commercial fisheries, aquaculture, nearshore energy development, and mineral/salt extraction. Many coastal ecosystems have been destroyed by landfills, pollution, invasive species, watershed alterations, and industrial development. Marine environments have been increasingly impacted by overfishing, waste dumping, sonic boom testing, and climate change.

human migration, coastal ecosystems, mangroves, estuaries, deltas, coral reefs, gulfs, recreation and conservation, shipping, oil and gas development, marine fisheries, aquaculture, wind energy, beaches, water quality, navigation, boating, coastal protected areas, marine endangered species, fish stocks, shellfish stocks, invasive species, sea level, wildlife viewing, dredging, landfill, waste discharge, toxic spills, endocrine disruptors, sewage treatment plants, power plants, agricultural runoff, coastal wetlands, dams, sediment, eutrophication, algae blooms, sediment, point and nonpoint pollution, maritime industry, Navy, jet ski operations, harbors

Coastal and Marine Invasive Species (18): The deliberate or accidental release of alien species to the ecosystem is having major impacts on native biota. Over five hundred marine species, for instance, have moved into the Mediterranean through the Suez Canal. Other sources include: deliberate introduction of harvestable species, accidental escapes from aquaria and aquaculture, escape from ship ballast, and release of fouling species on the hulls of ships.

invasives, alien species, exotic species, Lessepsian migrants, Suez Canal, ballast, aquaculture, aquarium fish, fish releases, fouling species, transport impacts, ballast, marine biodiversity, food chains, pathogens

Coastal and Marine Law and Policy (48): Coastal and marine environments have not been managed on an area-wide, multipurpose, or ecological basis. There are few agreed-upon processes to resolve conflicts between interests (e.g. private property rights/public interests, protection/development). There are problems of governance (ultimate authority), and implementation (enforcement on the open seas). In the United States, the coast out to about three miles is governed by local authorities; from three miles to two hundred miles it is governed by the nation; and beyond that are poorly governed "open" seas.

Exclusive Economic Zone, Coastal Zone Management Act, Outer Continental Shelf Act, Fisheries and Conservation Management Act, UN Convention on the Law of the Sea, marine sanctuaries, National Oceanic and Atmospheric Administration, U.S. Environmental Protection Agency (EPA), National Park Service, U.S. Coast Guard, U.S. Navy, marine boundaries, transboundary issues, migratory resources, weapons and missile testing areas, National Marine Sanctuaries, National Estuaries Program, coastal zone management, oil pollution, fisheries management, outer continental shelf management, marine sanctuary management, Clean Water Act, Clean Air Act

Coastal and Marine Pollution (526): Approximately 85 percent of commercially harvested fish depend on estuaries and coastal waters. Coastal and marine pollution is a change in the physical, chemical, or biological characteristics of the ocean or sediments, which results from contamination by human wastes and can affect the health and survival of all forms of life, degrading the natural quality of the ocean and coastal environments. Physical pollution includes plastic bags and discarded fishing nets that can strangle sea mammals and cause smothering from oil. Physical pollution can also be the lack of nutritious sediments held back by dams. Severe water pollution has come from urban and agricultural runoff that lowers oxygen levels in estuaries, sounds, and gulfs, harming fish and shellfish stocks, as well as from offshore waste dumping and oil spills.

oil development, shipping, cruise ships, aquatic toxicology, pollutant, estuaries, coastal, health impacts, dumping, oil spills, toxic waste, cleanup, industrial wastes, ocean, sea, fish, birds, nonpoint source pollution, coastal wetlands, sewage outfall, industrial outfall, storm sewers, beaches, offshore gas and oil, shellfish, swimming, boating, tourism, nutrients, persis-

tent organic pollutants, heavy metals, food webs, eutrophication, fertilizer runoff, sewage disposal, algal blooms

Coral Reef Conservation (147): Coral reefs are extensive oceanic limestone structures that are formed by corals and calcareous algae in shallow tropical waters, providing habitat for a large variety of marine organisms. Coral reef conservation is the practice of preserving, protecting, and sustainably managing coral reefs, which are under threat from coastal and marine pollution, destructive fishing practices, and damage from tourism.

coral reef, protection, preservation, conservation, sustainable livelihoods, ecotourism, biodiversity, fish, ecosystem, ecology, fishing, sustainable fishing practices, pollution, habitat, nutrients, species, marine protected areas, marine parks, sustainable island economies, ocean, sea, marine, fisherfolk, marine conservation

Mangrove Conservation (73): Mangrove habitat occurs in tidal estuaries and along coastlines in tropical regions where the soil and sediments are saline and waterlogged. Mangrove species, also known as "mangal," include trees and shrubs that are able to grow in tropical and tidal saltwater, trapping sediment between their roots and providing defenses against coastal erosion. Conserving and preserving mangrove ecosystems involves protecting species habitat, providing buffers against waves, wind, tsunamis, and coastal erosion, and maintaining the spawning grounds for marine species that sustain local livelihoods.

tidal, tropical, subtropical, coastal, protection, preservation, threats, ecosystem, mangrove ecology, salt-tolerant tree species, tidal marine forests, prop roots, breeding habitat, biome type, mangrove forest, estuarine, pollution, habitat, storm surge protection, spawning ground, mangrove destruction, restoration, reforestation, sustainable aquaculture, management, ocean, sea, fish, fishing, fisherfolk, marine conservation

Marine Ecology and Conservation (762): Marine ecology is the scientific study of ocean ecosystems and the interactions of living organisms with each other and with the marine environment.

protection, preservation, coral, reef, restoration, disease, destruction, nutrients, environment, habitat, conservation, biodiversity, protection, fish, fisheries, marine mammals, fishing nets, rescue, pollution, species interaction, ocean, sea

COMMUNITY DEVELOPMENT
Community development is the process of building the participation and capacity of local people to identify and work toward their own solutions to enhance the long-term social, economic, and environmental conditions of their community.

empowerment, skills, consultation, leadership, stakeholder, capacity building, networking, development, consensus, cooperation, conflict resolution, engagement, participation, problem solving, tools, resources, strategic tools, information, knowledge, collaboration, accountability, partnerships, accreditation, standards, sustainable, housing, organizing, quality of life, communication, citizen-led, social networks, campaign, grassroots, community building, community enterprise, training, resources, infrastructure, fundraising, housing, dispute resolution and consensus-building, community service and volunteerism, social exclusion, community cohesion

Community Enterprise (2,127): *Community enterprise refers to the activity of starting an organization or business venture within a community that is socially and economically* important for that community, such as setting up a small business to produce traditional crafts and employing community members, bringing employment, income, and self-sufficiency to the community.

local enterprise, job creation, community benefit, entrepreneurial, community empowerment, economic development, cooperatives, self-reliance, capacity building, community business ecosystem, small business, venture, social entrepreneurship, green enterprise, social enterprise, co-operative, income generation

Community Participation (10,053): *Community participation* refers to the ability and opportunities for community members to participate directly in the decision-making activities that steer the development of the community. Participating in community decisions gives ownership to community members and empowers people to develop their own community.

promoting civic participation, consensus-building, stakeholder, activism, project identification, input, decision-making, community welfare, fostering, community empowerment, capacity building, self-reliance, community independence, sustainable community empowerment, community problem-solving, community organizing, participatory, community led, popular base, community power, grassroots organizing, coalition building, civil disobedience, direct action, participatory budgeting, PB, people power, common vision, local action

Community Resources (7,804): *Community resources* refers to the provision, availability, and accessibility of resources for communities that enable the development of a thriving and participatory community, such as access to community development consulting services, identifying intellectual and social capital, networking, strategies to identify and solve community issues, and funding for community projects.

funds, equipment, capacity building, enabling community development, provision of resources, community services, community reinvestment, advice, communication structures, communication strategies, technology, community empowering, strategies, stakeholders, community empowerment, co-operative, participatory budgeting, community gardens

Community Service/Volunteerism (4,663): *Community service* and *volunteerism* refer to the practice of encouraging and facilitating community members to give their time to impart their knowledge and offer their skills, talents, and effort to contribute to a more cohesive and strong community.

participation, community, contribution, benefit, time, volunteering, giving back, community benefits, service, skills

Community Training (3,615): Community training is the practice of teaching, imparting knowledge, and providing vocational skills instruction to members of the community to empower them to more effectively participate in or to run community activities. Community training can include instruction on how to form a community group, how to manage community group meetings, and how to organize community members to work collectively to achieve benefits for the community.

workshops, classes, forums, conferences, capacity building, leadership training, human capital, mentoring, facilitating community action, skills building, information sharing, empowerment, indigenous teaching, traditional teaching, IT, computers, information and computer technology, ICT, lifelong learning

Dialogue, Deliberation, and Consensus Building (461): Dialogue and deliberation describe a set of tools used to increase citizen participation by fostering dialogue to generate collective intelligence and community understanding. The intention of dialogue and deliberation is to inquire, to explore, and to discover shared solutions to community concerns based on understanding and learning from other peoples' perspectives. Consensus-building is the process of finding common ground and a permanent outcome that is satisfactory to all parties involved in a dispute.

dialogue and deliberation, negotiation, consultation, agreement, stakeholder, collaboration, peace, violence prevention, nonviolent, negotiation, stakeholder, community building, mutual respect, affirmation, alternatives to violence, anger management, tolerance, mentoring, equal representation, active listening, reconciliation, participatory budgeting, conflict resolution, common vision, process arts, whole systems change, talk across differences, meaningful action, living network of conversation, collaborative dialogue, common ground, mind map, responsible citizen, active participation, respectful relationships, facilitated dialogue, transformative approaches to conflict, transformative mediation, nonviolent communication

Fundraising (2,153): Fundraising is the process of seeking financial support for a cause, community organization, or enterprise through identifying potential sources of money or in-kind contributions, networking, and accessing funds such as by grant writing and applying to foundations.

grants, gifts, foundations, philanthropy, fundraising training, raising funds, community fund-raising, resources, community action, scholarships, sponsorship, grant writing, financial support, donations, in-kind contributions

Leadership Training (1,228): Leadership training is the practice of teaching, imparting knowledge, and providing instruction in leadership skills to community members to empower them to more effectively participate in and to organize and lead community activities.

training, development, leaders, skills, knowledge, learning, social change, enabling, empowering, community, organizing, effectiveness, working with others, personal development, community training

Sustainable Communities (8,999): Sustainable communities seek community development that enhances the local environment and quality of life as well as developing a local economy that supports both thriving human and ecological systems. Sustainable communities use natural resources to meet current needs while ensuring that adequate natural resources are available for future generations.

land use, transportation, urban, planning, development, resource recycling, sustainable, integrated, ecological footprint, consumption reduction, cooperation, environmental, equitable, sustainable community development, sustainable livelihoods, thriving, participation, public health, design for people, environmental well-being, cities, economic well-being, community well-being, resource efficiency, locally based suppliers, local food systems, sustainable infrastructure, quality of life, restorative economics, natural capitalism, community business ecosystem, bioregionalism, intentional community, ecoplanning, cooperative, participatory budgeting, PB, neighborhood renewal, community cohesion, participatory deficit, bioregional economies, ecoliteracy, common vision, green cities, sustainable design, economic relocalization, environmental literacy, ethical lifestyle, sense of place, traditional economy, physical environment, community, environment, green city movement, social accountability, local self-reliance, reducing food miles

CONSERVATION

Conservation is a very general concept covering the careful management of natural resources. It is commonly used to describe the careful management of ecosystems, watersheds, parcels of land, specific water bodies, species, even genomes. At times, it is also used to describe the careful management of museum specimens, art, historic buildings, and nonrenewable or scarce resources (e.g. water conservation).

biome, ecosystem, tundra, taiga, temperate rain forest, temperate deciduous forest, grassland, savanna, chaparral, desert, thorn forest, tropical rain forest, arctic, Antarctic, marine ecosystems, terrestrial ecosystems, inland aquatic ecosystems, watershed, species, gene banks, museum conservation, art conservation, historic building conservation, conservation biology, land protection, sustainable conservation

Conservation Area Creation (1,033): Conservation area creation is the practice of identifying and protecting terrestrial, wetland, and marine areas that have biological and cultural value by securing government or private land tenure and management rules that specifically address the ecological integrity of the area. The rules include protecting an area from loss, damage, or depletion, and ensuring its preservation for future generations. Conservation areas include habitats and ecosystems, geological and physiographical features, and areas that have scientific, aesthetic, and cultural value. There are now over three thousand national parks, sanctuaries, protected forests, and similar set-asides in the world.

reserve, refuge, park, national park, new reserve, open space, environmental protection, green corridors, designated conservation areas, conservation protection policy, land acquisition, biosphere reserves, United Nations Convention on the Law of the Sea, Convention Concerning the Protection of the World Cultural and Natural Heritage, Convention on Wetlands of International Importance (RAMSAR), UNESCO Man and the Biosphere Program, conservation area establishment, new park formation, greenway, green corridors

Conservation Area Protection (2,931): Conservation area protection is the practice of protecting existing terrestrial, wetland, and marine conservation areas through real estate easements, inheritance conditions, trusts, zoning, public/private agreements between landowners and governments, lobbying, policy change advocacy, new laws and regulations, and lawsuits from development and uses that degrade the conservation value and long-term viability of the conservation areas.

park land protection, wilderness protection, permanent protection, open space, functional conservation area, park, refuge, reserve, national park, habitat conservation, existing conservation areas, protecting conserved land, development, encroachment, threats to protected land, threats to protected parks, environmental protection, legal defense, conservation easements, zoning laws, safe harbor agreements, land trusts, land management, natural heritage, cultural heritage, marine protection, conservation values, restoration ecology, Endangered Species Act, Habitat Conservation Plans, World Conservation Strategy, wildlife management, flyways, ecosystem diversity, state park, private reserves

Conservation Biology (1,077): Conservation biology is a multidisciplinary science that studies how humans impact and threaten habitat and species diversity, in order to conserve and preserve animal and plant biodiversity.

biological diversity, biodiversity, habitat preservation, genetic diversity, flora, fauna, threats to biodiversity, conservation for biodiversity protection, human impacts, field of study, extinction prevention, stewardship of biological communities, environmental protection, in situ conservation, ex situ conservation, restoration ecology, zoos, aquaria, botanical gardens, seed banks, wildlife management, flyways, biogeography, game farming, wildlife ranching, captive breeding, patch dynamics, endemism, fragmentation, edge effect, keystone species, intrinsic value, extinction

Conservation and the Commons (10): The commons is a shared resource (common pool resource) and a conscious collective choice to conserve and maintain the resource (common property management). There are global commons (e.g. air, oceans) and place-based commons (e.g. smoke in an office, watersheds). Many commons are contested: Where is the boundary? Who has access? How much can a person or group harvest or take? How is the resource allocated? Commons are connected to democracy: Who is eligible to make the rules? What are the payoffs or punishments for following or defying the rules? Indigenous peoples, in particular, must confront issues of the commons in areas without clear property rights.

commons, global commons, local commons, air, oceans, biodiversity, common pool resources (CPR), common property management, natural resource management, groundwater, watershed management, universal norms, working rules, Kyoto Protocol, United Nations Convention on the Law of the Sea, Convention Concerning the Protection of the World Cultural and Natural Heritage, Convention on Wetlands of International Importance (RAMSAR), UNESCO Man and the Biosphere Program, new enclosures, corporate concentration

Conservation Policy (1,167): *Conservation policy* refers to the practice of creating new regulations and procedures, and of changing existing ones through campaigning, lobbying, advocacy, and direct action, that have been adopted by a government or an organization, and that govern activity related to conservation of land, water, and other natural resources.

conservation, new policy, reform, planning, integrated, land protection, environmental protection, land management, biodiversity protection, landscape diversity, legislation, land use, natural heritage, cultural heritage, marine protection, conservation values, wildlife corridors, greenbelt

Conservation and Recreation (2,632): Recreation is a major driving force protecting the environment, but can also cause environmental impacts and significantly alter local economies. Recreation can be largely passive (e.g. birding, vista points, or photography), low impact (e.g. designated hiking trails with limited permits), or high impact (e.g. horse trails/off-road vehicles or motorboats). The level of access and kind of access (e.g. walk-in, bus, individual car, canoe, motorized boats) alter the kinds of impacts and conservation potential. Hunters, sports fishers, rock hounds, and mushroom collectors are extractive forms of recreation. In developing nations, the taking of lands for national parks can have major impacts on local peoples and their economies.

recreation, protected areas, public lands, public use, road closure, land management, multiple use, sacred lands, ecotourism, national parks, swimming, hiking, biking, walking, skiing, snowmaking, equestrian trails, motorized boating, fishing, hunting, rock collecting, gem collecting, mushroom collecting, off-road vehicles, all-terrain vehicles, disabled access, rock climbing, jogging, leisure

Land Restoration (690): Land restoration is the practice of restoring the ecosystem functions, such as nutrient cycling, the hydrological balance, and ecosystem resilience, to land that has been degraded, contaminated, or disturbed by human development and industrial activity. Restoring the original flora of the site may be a goal of land restoration.

remediation, reclamation, brownfield, conservation, preservation, land improvement, pollution mitigation, habitat restoration, open space creation, planting, revegetation, indicator species, habitat quality, habitat management, invasive species, endangered species, biogeochemical cycles, erosion, community development, ecosystem restoration, ecological restoration, environmental restoration, bioremediation, Superfund

Land Stewardship (2,062): Land stewardship is the practice of using and managing land to ensure that natural systems are maintained or enhanced for future generations. Land stewardship incorporates the principles of caring for natural systems as a whole, conserving resources, enhancing the stability of ecosystems, and instilling the cultural ethic of taking care of the land for long-term use and ecosystem stability.

steward, land ethic, ecological responsibility, sustainable land management, caretaker of the land, ecosystems, resource management, landscape, restoration, conservation, easement, landowners, ranchers, farmers, foresters, infrastructure, historic, rural, traditional, natural heritage, land trust, buffer zone, nature reserve, low-impact recreation, sustainable agriculture, managed access, viable farms, rural economy, earth respect

Natural Heritage Conservation (5,164): Natural heritage conservation is the practice of protecting natural resources from loss, damage, or depletion, and of ensuring the preservation of those natural places for future generations. Natural heritage includes indigenous species, habitats, and ecosystems, and geological and physiographical elements, features, and systems that have scientific, aesthetic, and conservation value.

environmental protection, conservation, preservation, habitats, ecosystems, geological, physiographical elements, physical landscape features, biological formations, outstanding aesthetic value, scientific value, habitat of threatened species, animals, plants, natural beauty

Natural Resource Conservation (11,393): Natural resource conservation is the practice of protecting natural resources, such as water, air, land, plants, forests, topsoil, wildlife, fish, minerals, and sources of energy, on which humans depend for their needs, from loss, damage, or depletion, and ensuring the preservation of those resources for future generations.

natural resources, water, air, land, soil, minerals, coal, petroleum, iron, plants, forests, ecosystems, protection, preservation, trees, scarcity, renewability, consumption, monitoring, supply, natural systems, environmental protection, natural resource management

Practical Conservation (2,221): Conservation requires the application of every aspect of human activity from theoretical science (conservation biology, landscape ecology) to technology (gabions, fire equipment) to labor-intensive restoration (planting riparian zones, pulling invasives). It can include changing rules (in-stream flow rights, hunting regulations, international migratory bird protection). Funding can be governmental, private (from NGOs or business), and from volunteer labor. Practical conservation is the practice of physically carrying out work to conserve natural resources, protect natural and cultural heritage from loss, damage, or depletion, and ensure the preservation of those resources for future generations.

volunteering, practical conservation skills, activist conservation, tree planting, vegetation management, habitat restoration, native species, practical land management, community conservation, enhancing local environments, environmental protection, conservation tillage, watershed management, aquifer protection, easements, instream flows, prescribed fire, reclamation, rehabilitation, forestry, invasives, biodiversity, land trusts, land easements, debt-for-nature swap, stream restoration, ecosystem restoration

Wilderness (864): *Wilderness* refers to those areas of land that have not been significantly changed directly or indirectly by human activity; usually, areas that humans may visit but not permanently inhabit. The ecosystem characteristics and natural habitats of wilderness have remained relatively intact. *Wilderness* is also a designation for protected areas in North America. The Asian and African continents have been inhabited for so long that the term is rarely employed there.

wilderness area, wilderness designation, conservation area, refuge, access, management, permanent protection, wilderness ethic, minimal direct human activity, minimal human interference, uninhabited, undeveloped land, conservation, environmental protection, Wilderness Act, National Wilderness Preservation System, wilderness impacts, trails, off-road vehicles, invasive plants, Wild and Scenic Rivers Act, Land and Water Conservation Fund Act, natural regulation, let-burn policy, Bureau of Land Management, U.S. Forest Service, U.S. Fish and Wildlife Service, National Parks Service

CULTURAL HERITAGE
Cultural heritage encompasses the qualities, attributes, and significance of places and cultures that have aesthetic, historic, scientific, or social value for past, present, or future generations, including conserving historic monuments, cultural sites, language, customs, and traditions.

cultural diversity, differences, race, ethnicity, language, nationality, religion, gender, orientation, physical ability, intracommunity differences, multicultural communities, variety, human cultures, multiculturalism, enhancing community, thriving, community fabric, cultural heritage, endangered languages, language revitalization, customs, cultural significance, cultural appreciation

Cultural Diversity (4,531): *Cultural diversity* describes the differences in race, ethnicity, language, nationality, beliefs, values, religion, and the variety of human cultures within a community, organization, nation, or region. It refers to the practice of enhancing the acceptance, tolerance, and understanding of cultural differences within a community, organization, country, or region.

cultural exchange, understanding, community, preservation, multicultural, conflict resolution, inclusive, ethnicity, minority, racism, tolerance, cultural heritage, identity, cross-cultural understanding, dialogue, youth, children, developing understanding, tolerance, equality, cultural appreciation, interfaith tolerance, hate crimes, racism, racial healing

Cultural Heritage Conservation (2,427): Cultural heritage conservation is the practice of protecting cultural heritage from loss or damage, and ensuring the preservation of that cultural heritage, including structures, places, traditions, and objects with cultural meaning, for future generations.

industrial heritage, aesthetic value, historical value, scientific value, social value, site conservation, protection, human activity, significant historical value, sense of place, ethnographic heritage, indigenous site protection, site restoration, rehabilitation, monuments, architectural works, archaeological structures, ethnological value, anthropological value, cultural plants, cultural plant diversity, built environment, cultural appreciation, cultural keystone species, ethnosphere, cultural systems, cultural identity, cultural foundation species, cultural landscapes, culture

Culture and Sustainability (495): At times, cultural diversity has been a major barrier to the achieving of sustainable communities, national policies, or international agreements.

Bridging cultural attitudes, even toward the meaning of sustainability or nature, can be challenging. In addition, many "cultures" are actually a mix of "subcultures" that may find it hard to find common ground on many issues of concern (e.g. family planning, male/female employment, diet, clitoridectomy, gun control, hunting, material wealth). Finally, there are elite subcultures, especially in science, technology, and business, that may have little experience with sharing power and decision-making tasks that involve more than one kind of social group. Education and conflict resolution have been two tools to try to overcome these barriers.

culture, race, social class, decision-making, sustainability, race relations, tolerance, conflict resolution, cultural elite, cultural diversity, long-term survival, racial healing

Language Revitalization (44): Language revitalization is the revival of a language that is no longer learned or is endangered as fewer people use and pass on the language to younger generations.

cultural systems, language documentation, language revitalization, language revival, language policy, native speakers, linguicide, death of a language, intangible cultural heritage, language diversity, language promotion, language as an instrument of education and culture, language endangerment, indigenous languages, dialect

Traditional Culture (808): *Traditional culture* refers to the practice of preserving and promoting cultural practices, traditions, and folklore, including traditional dance, handicrafts, art, storytelling, myths, oral history, customs, ceremonies, beliefs, and rites.

folklore education, folklore training, expressive behavior, cultural memory, aesthetic appreciation, cultural diversity, folk painting, tribal, ritual, mythical, folk dance, crafts, traditional music, noninstitutional, collected wisdom, oral traditions, storytelling, ritualistic, rites, legends, myths, beliefs, customs, ethnographic concept, ethnic, tribal, cures, medicine, social fabric, validate social identities, fostering intergenerational communication, native people, handicraft, cultural appreciation, traditional knowledge

DEMOCRACY AND VOTING

Democracy and voting refers to the opportunity and ability of people to participate in and to steer the governance of their region and country, either directly or indirectly, through voting and electing a representative.

fair election, process, rights, reform, civic participation, democratic, representation, citizen power, political control exercised by the people, transparent, vote-rigging, democratic participation, democracy education, democratic reform, fair electoral process, democratic values, rule of the people, participatory democracy, democratization, autonomy, direct democracy, democratic deficit

Democracy and Civil Society (889): Democracy, in practice, takes many forms. Illiberal democracies elect leaders who are not bound by rules or a constitution and can violate basic human rights. Liberal democracies elect representatives who speak for their constituencies within the context of established rules and a constitution. In representative democracies, individual voters do not vote on the individual decisions themselves. In direct democracies citizens vote directly on an issue. Deliberative democracy encourages meetings, discussions, and research before voting. All democracies must confront the "tyranny of the majority" issue and find ways for minorities to have a significant voice in decision-making. Lebanon, for instance, reserves specific (elected) roles in the government for specific religious orders in order to ensure that even minority religious leaders will be heard at top levels. All democracies must also confront voting fraud and the power of financial influence to modify the will of citizens.

illiberal democracy, liberal democracy, representative democracy, direct democracy, deliberative democracy, tyranny of the majority, voting fraud, voter eligibility, participatory democracy, green democracy, civil society, voter participation

Democracy Education (1,448): Democracy education is the activity of educating, teaching, training, and imparting knowledge and ideas to people about the concepts of democracy, including ideas about civic participation and people's rights within a democratic state.

democratic principles, democratic values, civic education, democracy movement, citizen representation, local democracy, civic responsibility, participatory democracy, democratization, direct democracy, representative democracy, civil disobedience, direct action, democratic deficit, electoral democracy

Democratic Participation (3,448): Democratic participation requires answers to a series of questions: Who is eligible to participate? Age, sex, race, property ownership, a criminal record, literacy, and citizenship have all been used to separate voters from nonvoters. What actions can voters vote on? What actions are left to others to decide (e.g. by their representatives, bureaucrats, appointed officials)? What is beyond the reach of citizen influence? What rules and punishments exist to manage the constituency's representatives (e.g. moral codes, right to recall, impeachment, ability to sue for misconduct)? What rules exist to change the rules of representation (e.g. from an electoral college to direct election by the majority in the American presidency)? What knowledge (information, analysis) must be provided to the voter? Is the knowledge provided by public media and political advocacy groups slanderous or deceitful? Who are the monitor and enforcer in times of dispute or violation of democratic rules? Do they have the independence and resources to monitor and enforce fairly? Every democracy creates rights and responsibilities in order to encourage participation and to avoid a tyranny of the majority that might set rules that make voting a sham. Sometimes minorities cannot change the rules without civil disobedience or direct action against the government.

voting, representation, civic participation, right to vote, access to voting, fair election, democracy, democratic process, civic engagement, civil society, enabling participation, freedom from voting intimidation, participatory democracy, democratization, autonomy, democratic deficit, direct democracy, representative democracy

Democratic Reform (3,086): *Democratic reform* refers to the practice and process of lobbying, advocating changing policy and political systems to reflect democratic values and to empower citizens to guide the governance of their region through participation. Democratic reform includes making a state hold regular elections, upholding the rule of law, and having an independent judicial system.

electoral reform, transition to democracy, policy reform, independent judiciary, universal suffrage, civil society, democratic system reform, political funding, politics, fair representation, electoral corruption, campaign finance, pluralism, political accountability, authoritarian regimes, democratization, participatory democracy, autonomy, direct democracy, civil disobedience, direct action, democratic deficit

Fair Electoral Process (619): *Fair electoral process* refers to that part of a democratic system where citizens are able to elect their chosen representative free from intimidation and barriers to voting, including discriminatory voter registration processes, corrupt voting practices, and limits to the freedom of citizens to stand as democratic representatives.

electoral transparency, electoral accountability, voter registration, vote protection, access to voting, free and fair elections, gerrymandering, universal suffrage, citizen rights to vote, independent election monitoring, vote counting, voter fraud, voting secrecy, democratic elections, choice of candidates, participatory democracy, democratization, direct democracy, democratic deficit

ECOLOGY

Ecology is the scientific study of relationships between organisms and their environment. Ecology is concerned with the distribution and behavior of individual species as well as with the structure and function of natural systems at the level of populations, communities, and ecosystems.

relationships, organisms, environment, science, field of study, earth, ecological, ecosystems, microbial ecology, restoration ecology, pollination ecology, evolutionary ecology, fire ecology, molecular ecology, urban ecology, living organisms, interactions, organism populations, biotic, abiotic, population distribution, biotic communities

Evolutionary Ecology (71): Evolutionary ecology approaches ecology—the interrelationships between organisms and their relation to the environment—in a way that explicitly considers the evolutionary history of species and the interactions between the evolving species (coevolution). The distinction between ecology and evolution is largely artificial.

evolution, plants, animals, natural environment, organisms, species, evolutionary biology, relationships, interactions, ecosystems, coevolution, natural selection

Fire Ecology (54): Fire ecology is the relationship between fire, the physical environment, and living organisms. Fire can be a natural and human-caused process. Depending on the intensity, duration, extent, season, and frequency, fire can be beneficial or destructive. Humans have employed fire to maintain grasslands, fertilize farms, and as a weapon of war. Fire prevention in certain forests has caused ecosystem, safety, and financial problems.

grasslands, rangelands, forests, regeneration, fire regime, fuel, prescribed burning, fire management, conservation objectives, fire dependence, wildland fire, fire in a natural environment, forest fire, controlled burn, climate change and fire, agricultural fire

Landscape Ecology (250): Landscape ecology has two approaches: 1. The multidisciplined study of landscapes applying concepts from ecology, wildlife biology, cultural anthropology, landscape planning, and economics; 2. The study of landscape mosaics through time and the influences that govern the biological and nonbiological components and configurations of components. Landscape ecology can inform environmental sustainability, city growth, wildlife refuge management, etc.

ecosystems, biome, ecology, landscape, holistic, populations, communities, ecological processes, landscape structure, interactions between ecosystems, spatial arrangement, landscape heterogeneity, landscape scales, landscape mosaic

Microbial Ecology (17): Microbial ecology is the branch of ecology that deals with the study of the interactions of living microorganisms with one another and with their non-living environment of chemical compounds, water, and energy. Microorganisms are usually organisms requiring a stereo or light microscope; for viruses, a scanning electron microscope or transmission electron microscope. Microorganisms make significant contributions to climate change, carbon cycling, soil fertility, plant health, the degradation of toxic pollutants, as well as providing humans with antibiotics, free erosion-control, fertility, and air-cleaning and water-cleaning services. They also cause many plant and animal diseases that challenge sustainable practices.

ecosystem, microbe, environment, microorganisms, symbiosis, symbiotic relationships, bacteria, viruses, microscopic organisms, nutrient cycling, protozoa, fungi, algae, prokaryotic cells, nonnucleated cells, eukaryotic cells, nucleated cells, ecosystem services, carbon cycling, soil fertility, pollution control, greenhouse gases, climate change, Gaia hypothesis

Molecular Ecology (13): Molecular ecology is the study of the major cycles and physiology of life on the molecular level; and the use of molecular and genetic tools to address ecological, evolutionary, behavioral, and conservation questions.

biochemical indicators, contaminant, exposure, genetics, population, species, gene flow, genetically modified organisms, GMOs, genetic markers, molecular adaptation, speciation genetics, conservation genetics, carbon cycle, phosphorus cycle, water cycle, sulfur cycle, nitrogen cycle, biodegradation, acid rain, polymerase chain reaction (PCR), DNA fingerprinting, mitochondrial DNA, ribosomal RNA, DNA sequencing

Mycology (62): Mycology is the study of fungi, their genetic and biochemical properties, their taxonomy, and their use to humans as a source of medicinals and food, as well as their dangers, such as poisoning or infection.

fungi, lichens, algae, phytopathology, mycorrhiza, symbiosis, symbionts, antibiotics, penicillin, ethnomycology, yeasts, eukaryotic model organisms, Kingdom Fungi, medical mycology, fungal, parasitic fungi, symbiotic fungi, mutualism, mycorrhizal association

Pollination Ecology (24): Pollination ecology is the scientific study of relationships between pollinators and their flower hosts. Pollination ecology is concerned with the life histories, distribution, and behavior of individual species as well as the structure and function of natural systems at the level of populations, communities, and ecosystems. It is a crucial aspect of sustainable food supply for humans.

pollinators, environment, interactions, ecology, plant-pollinator interactions, hoverflies, solitary bees, bumblebees, pollinator communities, tropical pollination ecology, nonnative plants, alien pollinators, butterflies, pollinator bees, pollinator birds, pollinator mammals, pollinator appreciation

Restoration Ecology (561): Restoration ecology is perhaps the largest task of sustainability. It is the study and practice of how to repair human damage to ecological systems. Three levels include "reconstruction" of destroyed landscapes (e.g. brownfields, strip mines, levied estuaries, cemented rivers); "rehabilitation" of modified landscapes (e.g. grazed rangelands, damaged streams, widespread invasives); and the fine-tuning of "restoration" (e.g. remove invasives, restore natives, stop leakage of soil and nutrients, restore natural fire regimes, etc.).

conservation, native species, ecosystems, restoration goals, human site damage, anthropogenic damage, degradation, reconstruct, rehabilitate, brownfields, strip mining, levied estuaries, invasives, stream restoration, conservation ranching, fire regime, soil restoration, watershed management, toxics, conservation, ecosystem restoration, environmental restoration, ecological restoration

Soil Ecology (70): Soil ecology is the study of the interactions of living organisms with each other within the complex soil environment, from nutrient cycling, to soil formation and structure, to plant productivity. Bacteria, algae, fungi, and micro/mesofauna like worms are the major inhabitants that require tending. Soil ecology is crucial to sustainable agriculture, grassland forage, and toxic removals from brownfields. From a soil ecologist's perspective, soil as the nexus of water, air, geology and the biogeochemical cycles of the planet. There are more chemoelectric reactions in a teaspoon of moist, healthy soil than in the human brain. Soils can nurture greater biodiversity per volume of soil than any other microhabitat. Soils act as a biological valve connecting the microbial microcosmos to the atmosphere of the planet. Soil disturbances influence climate change and climate buffering, nutrient cycling and pollution, toxic cleanup, as well as soil stability and fertility (humus). In restoration, soil microbes can alter groundwater flow (ultrabacteria) or clean up toxics (phytoremediation).

minerals, agriculture, pesticides, erosion, nitrogen cycle, nutrient cycling, decomposition, biota, ecosystems, aeration, hydration, weathering, bedrock, carbon, soil temperature, viruses, rhizosphere, decomposition, compost, legumes, mychorrhiza, phosphorus, sulfur, salts, saline soils, soil profile, soil structure, soil as a medium for life, soil ecosystems, microbes, microfauna, microflora, micronutrient, biogeochemical cycling, soil moisture, climate change, phytoremediation, ultrabacteria, brownfields, toxics, farming, water pollution, biodiversity, groundwater, soil conservation, top soils, soil profile, soil symbiosis

EDUCATION
Education is the activity of educating, teaching, training, and imparting knowledge, ideas, and skills to those who are being educated.

teaching, training, skills, development, instruction, access, equity, academic, community, assessment, standards, information, learning, study, resources, experience, tools, science, environmental, ecotourism, literacy, green schools, vocational training, imparting knowledge, lifelong learning

Access to Education (3,837): Creating or enhancing access to education enables both children and adults to overcome physical, financial, and cultural barriers that prevent them from being educated or receiving an education. The right to education is considered a human right and facilitating access to education enables the fulfillment of that right.

grant, school, equity, capacity building, scholarship, public, private, higher education, literacy, basic education, school supplies, reading, writing, education for all, school infrastructure, economic access, correspondence, teacher supply, time to attend school, transportation, education as a human right, lifelong learning, education policy

Education, Government, and Sustainability (113): Education and information issued by government agencies is intended to create a responsible citizenship. Education helps close the gap between what people know or are able to know on their own and what they need to know to make well-informed voluntary decisions, especially how to vote, consume, and produce goods and services. Education has the following goals: ensure information quality and reliability; correct erroneous perceptions; and expand audiences for education. Government education has been crucial in projects such as toxic releases, harmful drug consequences, large infrastructure projects financed by taxes (highways, dams, levees), water and air pollution standards, environmental justice, and security decisions. Government education must be monitored by private, nongovernment groups as it can be biased. Information is just one ingredient in contested arenas of sustainability and may not carry sufficient political power to cause social change.

government documents, talking heads, dueling experts, national research councils, environmental education, public education, eco-labels, right-to-know, transparency, citizen awareness, ministries of education, voluntary decisions, National Research Council, National Academy of Sciences, National Academy of Engineering, Institute of Medicine, Environmental Protection Agency (U.S.), General Accountability Office (GAO), Office of Economic Cooperation and Development (OECD), UNCTAD, U.S. Superintendent of Documents (U.S. Government Printing Office)

Environmental Education (11,789): The goal of environmental education is to create a world population that is aware of, and concerned about, the environment and human impacts on the environment. An environmentally educated population will have the knowledge, skills, attitudes, motivations, and commitment to work individually and collectively toward solving current problems and preventing new ones. The objective of environmental education is to generate knowledge to enable the sustainable management of global natural and physical resources.

EE, demonstration, public, knowledge, issues, skills, responsibility, action, schools, youth, children, adults, ecotourism, conservation ethic, respect for nature, sustainable development, raising awareness, experiential learning, outdoor classroom, green purchasing, sustainability, environmental consumer, ethical consumption, publication, information, informed, advocacy, teaching, environmental literacy, raising awareness, ecoliteracy, experience nature, contact with nature, environmental problem solving, green maps, international covenants, Stockholm, Rio, Kyoto, direct action, international law, ecological cost, ecocentric, renewable energy, green values, green festivals, environmental curricula, curriculum

Environmental Resource Center (740): An environmental resource center is a building or facility, often with outdoor space, where environmental education takes place.

ecology center, education center, information, knowledge, infrastructure, programs, environmental education, awareness raising, protection, preservation, community resource, outdoor classroom, nature garden, reserve, interpretation, interactive displays, conservation ethic, outdoor education

Green Schools (137): Green schools are schools that teach and foster environmental awareness and sustainability by creating a small-scale model of an ecologically sustainable society in the school, encouraging students to take environmentally responsible action, and by physically greening the school environment. Environmental responsibility is part of the green school ethos and may be demonstrated by activities such as having an environmental curriculum, building green school buildings, and implementing recycling programs in the school and local community.

ecoliteracy, nature, restoring, community involvement, children, environment, outdoor learning, school grounds enhancement, environmental education, community-based education, school sustainability, water, energy, food, solid waste reduction, recycling, toxics, indoor air quality, transportation, interdisciplinary, habitat improvement, student environmental responsibility, youth, school gardens, outdoor education, environmental curricula, environmental curriculum

Literacy (713): Literacy is defined as the basic ability to read and to write. Each nation has various standards for its desired level of literacy.

basic skills, reading, writing, teaching, human right, understanding, communication, comprehension, information, youth, children, adults, ability to communicate, life skills, education

Natural Resource Education (3,457): Natural resource education is the activity of educating, teaching, training, communicating, and imparting knowledge and ideas about natural resources, such as water, air, land, plants, forests, topsoil, and minerals, that are used and valued by humans, other animals, and plants. The goal of natural resource education is to educate people about natural resources, their consumption, the impact of humans on the availability of natural resources, and the sustainable whole system management of natural resources to ensure healthy ecosystems and the availability of natural resources for future generations.

scarcity, environment, protection, renewability, consumption, education, information, study, scientific, research, monitoring, knowledge, supply, experiential learning, outdoor classroom, natural systems, youth, adults, conservation, environmental protection, management techniques, environmental awareness, environmental education

Public and Government Education (108): Public education on projects funded by taxpayers have specific requirements: requiring public notification; determining what information must be made available by whom, when, and how; removing constraints to accessing information; and requiring answers by the lead agency to all reasonable questions. Depending on the government, information may be easily accessible, difficult but accessible, or inaccessible. In the United States, the National Environmental Policy Act has been the most important tool for sustainability by requiring full disclosure, a dialogue between citizens and sponsors of projects funded by taxes, and a consideration of alternatives and mitigating measures to protect the environment and ensure human rights. The ideal government intention for education is to empower, not coerce.

public notification, full disclosure, transparency, environmental impact statements, right-to-know, environmental education, public education, citizen awareness, ministries of education

Sustainability Education (2,045): Sustainability education is the activity of educating, teaching, training, and imparting knowledge, ideas, and skills about concepts of sustainability to enable the management of global natural, physical, and social resources to meet the needs of present generations without compromising the ability of future generations to meet their own needs. Achieving a sustainable society relies on an environment, economy, and social systems that can be maintained in a healthy state indefinitely, which involves addressing human management of the environment, social justice, and economic equity.

environmental sustainability, environment, social justice, education, promote behavior supporting a sustainable environment, just, healthy world, information, knowledge, sustainable development, basic needs, fair, equitable, access to resources, quality of life, human rights, conservation ethic, environmental education, green values, conservation economy, long-term survival, progress as if survival matters

ENERGY
Energy refers to the conversion of energy for power production, power generation sources, and the use of energy by consumers, as well as to policy relating to power production and use.

work, efficiency, power, capacity, activity, force, policy, potential, kinetic, mechanical, electrical, chemical, thermal, geothermal, biomass, conservation, fossil fuels, alternative fuels, renewable, climate change, global warming, consumption, infrastructure, energy conservation, nuclear energy, power production

Alternative Fuels (247): Alternative fuels are substitutes for traditional liquid, oil-derived motor vehicle fuels like gasoline and diesel. Examples of alternative fuels are hydrogen, natural gas, hithane, and ethanol, which produce lower emissions and are less polluting than traditional fuels.

climate change, mobility, ecoefficiency, biodiesel, biofuel, biomass, residues, organic, renewable, biodegradable, emissions, policy, hydrogen, transportation, fuel cell, gasoline replacement, ethanol, alcohols, LPG, compressed natural gas, methanol, propane, hithane, energy, appropriate technology, energy plantations, clean air, oxygenated fuels, benzene, methyl butyl tertiary ether (MBTE), greenhouse gases, photochemical smog, syngas fuels

Electric Power (8): Electricity is one of the defining markers in the transition from developing to developed nations. Electric power supply for burgeoning megacities is a central concern. Sustainable infrastructure must choose the right components (e.g. power plants, power grid, transmission lines, cables, etc.) in the right configurations to prevent brownouts and blackouts and to maximize efficiency. Decentralized generators such as solar cells on roofs may be able to leapfrog power grids or return power to the grid.

electricity, power grid, electric grid, power stations, transmission lines, transformers, metering, energy conservation, energy growth, decentralized electricity, blackouts, voltage, watts, electric current, peak demand, electric reserve margins, power consumption, distributed generation

Energy Efficiency and Conservation (1,000): *Energy efficiency and conservation refers to promoting both more efficient technology that uses less energy to perform the same task and reducing unnecessary energy use, with the overall aim of reducing energy demand without detracting from end-use benefits.*

education about energy efficiency and conservation, green building, energy consumption reduction, technology substitution, energy star rating, demand side management, energy demand reduction, climate change, energy standards, industry efficiency, residential efficiency, commercial/public efficiency, agricultural efficiency, transportation efficiency, insulation, green chemistry, energy prices, grid inefficiencies, grid loss, subsidies, Corporate Average Fuel Economy (CAFE), tax credits, low-interest loans, ecolabeling, theoretical potential, technological potential, market trend potential, boilers, heat exchangers, electrical motors, electric appliances, pulp and paper, chemicals, cement, iron and steel

Energy Flow in Ecosystems (12): The crucial energy for life is the radiant energy of the sun. Its energy is captured by photosynthesis and is converted into sugar, which enables all work and action. The sun provides the ultimate ecosystem system. Energy cascades from the producers to the consumers to the decomposers. To survive in large populations, humans have switched solar energy from natural communities to agricultural and rangeland uses.

chemical energy, radiant or solar energy, heat energy, mechanical energy, nuclear energy, electrical energy, energy flow, photosynthesis, consumers, producers, food webs, primary productivity, decomposers, renewable energy

Energy Policy (364): *Energy policy* refers to the practice of creating regulations and practices, and of changing existing policies, which have been adopted by a government or an organization, and which govern activity related to energy production and energy use. Energy policy reform may include changing policies through campaigning, lobbying, advocacy, and direct action, which determine power plant emissions, energy conservation strategies, preferred types of power production, energy distribution and infrastructure, and sustainable and equitable energy production.

demand side management, energy supply, distributed generation, fossil fuel generation, alternative fuel policy, renewable energy, legislation, energy conservation, energy efficiency, advocacy, climate change, lobbying, campaigning, clean energy, Corporate Average Fuel Economy (CAFE), low-interest loans, ecolabeling, industrial licensing, energy pricing, energy audit, energy standards, energy trading, energy strategy, tax credits, energy subsidies

Energy Security and Sustainability (69): A nation's long-term energy security and minimization of adverse consequences depends on five objectives: 1. increasing energy efficiency and conservation; 2. developing alternatives to fossil fuels (e.g. renewables, maybe some nuclear generation); 3. assuring a petroleum reserve large enough to prevent a fuel shortage crisis; 4. advancing technologies with low emissions of greenhouse gases; and 5. assuring that future energy development does not harm the environment. National strategies have considered energy taxes, shifting subsidies, emission standards, and fuel rationing.

Energy Policy Act (U.S.), carbon taxes, energy taxes, energy subsidies, emissions standards, carbon sinks, market price, domestic price, renewable energy, nonrenewable energy, strategic reserve, energy independence, distributed generation, biofuels, transportation energy security

Nuclear Power (155): Nuclear power is the production of electrical energy using nuclear fission within a nuclear reactor to generate heat to power a steam turbine for electricity generation. Radioactive waste is produced as a by-product of nuclear power production. It has challenged sustainable energy thinkers because it produces few greenhouse gases in its life cycle but is a long-term waste storage hazard.

radioactive waste, energy, power plant life cycle, persistent hazardous substance, low emissions, water use, enriched uranium, fuel rods, reprocessing, storage, safety, nuclear waste disposal, nuclear power plant phase-out, nuclear hazards, climate change, greenhouse gases, energy subsidies, uranium reserves, International Atomic Energy Agency, World Energy Council, plutonium, fast-breeder reactors, electricity, risk assessment, fuels

Renewable Energy (963): Renewable energy is energy that is converted from renewable power sources predominantly for electricity generation. Renewable power sources are naturally replenished and are not depleted by use, unlike fossil fuel resources. Geothermal, wave, tidal, ocean current, hydro, wind, solar, and biomass are the most common renewable resources used to generate renewable energy.

power generation, geothermal, heating, solar thermal, photovoltaic, PV, hydrogen, tidal, distributed generation, fuel cell, renewable, replenishable, undepleted, biomass, hydroelectric, waste-to-energy, regenerative, emissions, climate change, appropriate technology, clean energy, wind energy, solar energy, hydropower, marine energy, environmental impacts of energy, electricity, heat energy

Sustainable Energy Development (155): Energy production and transformations significantly influence poverty, prosperity, women's work, and lifestyles, and are affected by demand from urbanization and the population. Sustainable energy development attempts to provide energy services to satisfy basic human needs without compromising future generations or the environment. In Africa, per capita energy use is one-tenth of North America's and has not grown significantly in twenty-five years. Sustainable energy development also overlaps with energy security—the need for a reliable and adequate supply of energy.

sustainable energy development, energy security, electricity, fuelwood, social equity, politically unstable regions, pollutants, acid rain, air pollution, climate change, biomass, developing nations, land use, affordable energy, energy technology, energy infrastructure, natural resources, coal, oil, solar energy, energy trade, globalization, energy investments, transportation energy security

FISHERIES

Fisheries are the areas of the ocean with populations of fish and other aquatic species that are cultivated and caught for both commercial use and for recreation. A sustainable fishery is one that is not overfished, where the fish population is sustained at a healthy level over the long term while supporting fishing activity.

fishery, conservation, sustainable fish populations, aquatic, ecosystem, science, research, education, overfishing, sustainable aquaculture, sustainable fishing, bycatch, fishing practices, commercial fishing, industrial fishing, recreational fishing

Aquaculture (368): Sustainable aquaculture is the farming of aquatic organisms including fish, mollusks, crustaceans, and aquatic plants, in a way that does not impair the long-term ability of the environment to support these organisms and does not significantly harm the ecosystem. Aquaculture involves intervention such as regular stocking, feeding, and protection from predators to enhance production of these species for harvesting. It may require ecosystem conversion (e.g. mangroves to shrimp farms) and has led to invasive species, gene mixing, pollution from feeds, and the hunting of predators.

farming, fish, shellfish, algae, breeding, hatchery, fishing, stocking, water quality, food safety, aquatic animal health, organic loading, fisheries, fisherfolk, shrimp, mangroves, freshwater aquaculture, marine aquaculture, pond aquaculture, commercial aquaculture, fish food

Aquarium Trade (1): About 2 million people keep aquariums, mostly stocked with wild-caught species. About two-thirds of the fish (as well as live coral) come from coral reefs. Fish are caught live often with the use of nonselective stupefactants. Shipping leads to high mortalities. The aquarium fish trade has a high potential for sustainable production and income generation in local villages. Certification and value-chain reorganization is needed to protect fish species.

aquarium trade certification, aquarium fish, sustainable development, coral reefs, animal welfare

Sustainable Fishing (770): Sustainable fishing can take place in the open seas (marine), coastal waters, and inland freshwater bodies. It includes crustaceans and mollusks (lobster, crabs, clams, shrimps) as well as bony fish, sharks, eels, whales and porpoises, and sea turtles. The "fish" may be wild, domesticated as in "aquaculture," or partially domesticated such as released stock from hatcheries. Sustainable "fisheries" attempt to provide humans with food, fish oil, animal feed, fish meal, aquarium fish, coral, recreation, and other products in a manner that is ecologically friendly, economically viable and equitable, and socially responsible to future generations.

fish populations, stocks, conservation, harvesting, environmental protection, habitat protection, responsible fishing practices, fisherman, overfishing, marine biodiversity, sustainable fisheries, coral reef protection, fisherfolk, inland aquatic biodiversity, marine fisheries, coastal fisheries, Magnuson Fishery and Conservation Act, U.N. Fish Stocks Agreement, subsistence fisheries

World Marine Fisheries (31): Marine fish provide humans with food, oils, livestock fish meal, and recreation. Fish also supply the aquarium trade. In some nations, 50 percent of animal protein comes from fish. Over 25 percent of the 441 recorded marine fish stocks are overexploited. Open access regimes, technology, wasteful bycatches, and inappropriate investment in fishing fleets continue to deplete fishing stocks.

herring, anchovies, cod, hake, haddock, jacks, mullets, mackerel, pollock, tuna, salmon, fish stocks, sardine, aquaculture, overexploitation, capture fisheries, climate change, post-harvest loss, threatened species, sea turtles, seabirds, cetaceans, bycatch, inland water fisheries, marine fisheries, sharks, eels, sustainable fisheries

FOOD AND NOURISHMENT
Food and nourishment refers to those issues and rights that concern food quality, access to food, food supply, food production, education about food, and the right of humans to be free from hunger.

sustenance, human right, equity, access, health, nutrition, nourishment, diet, food, hunger, malnutrition, food supply, food security, diet and health education, food literacy, biotechnology, GMO, genetically modified organism

Food Aid (239): Foreign aid programs are created during food crises and for longer-term development of food security. During food crises, aid programs replace protein losses (usually with dried milk), carbohydrate losses (usually with grains), try to encourage destocking, and find alternative sources of incomes for victims of famine. In some cases, the lack of fuelwood limits the amount of food that can be cooked. Longer-term food security starts with an understanding of malnutrition and its connection to prevalent diseases.

food deficits, agricultural support, agricultural extension services, World Food Program, famine, food distribution, food-for-work programs, cash-for-work programs, structural adjustment

Food Literacy (167): *Food literacy* refers to the degree to which people are able to obtain, process, and understand basic information about food in order to make appropriate health decisions. Food literacy encompasses understanding labeling on food and knowledge of nutrition.

education, youth, food, nutrition, schools, food labeling, knowledge, healthy food choices, information, understanding food information, terminology, balanced food intake, food contents disclosure, additives, vitamins, carbohydrates, fat, protein, dairy, fruit, vegetables, food autonomy, local agriculture, slow food

Food Supply (1,671): *Food supply* refers to the origin, method of production, distribution, and delivery of food to consumers, encompassing food safety, food hygiene, food production methods, security of the food supply, and the geographical scale and environmental impact of food production and delivery systems. Sustainable food supplies are priced fairly so that producers receive equitable profits and do no harm to the environment or human health.

nutrition, standard, eat, additive, storage, bacteria, spoilage, foodborne, salmonella, agriculture, small farmer, organic, low input, food bank, malnutrition, quality, food origins, suppliers, food production, hunger, farmer's markets, food safety, food systems, community agriculture, sustainable food supply, local food systems, food security, environmental impacts, consistency of supply, cooperative, food autonomy, local agriculture, slow food, community gardens, food miles

Global Food Supply and Sustainability (68): The crucial long-term questions for sustainability are: Can food production keep pace with population growth? And, given the continuing depletion of soils and scarcity of irrigation water, will the future see a lower nutrient carrying capacity for humans? Can technological breakthroughs keep up with food demand? The answers are complex. The Green Revolution (fertilizer, irrigation, pesticide) provided more time. Today, reduction of post-harvest losses, aquaculture, changes in the food commodities trade system, research on new food cultivars and alternatives to slash-and-burn agriculture, and incentives and loans through foreign aid programs will hopefully converge to meet the population's needs. Whether global food security can be accomplished without serious environmental damage cannot be known. In addition, climate change has made all global food security planning less predictable.

green revolution, fertilizer, irrigation, pesticide, genetic engineering, post-harvest losses, diverting feed to food, global warming, maldistribution, seafood, foreign aid programs, food research, trade system, food technology, aquaculture, climate change, genetically modified organisms (GMO), biotechnology, sustainability

Hunger and Food Security (1,992): Food security is a state in which all people have reliable access, both physically and economically, to a nutritionally adequate and culturally acceptable diet at all times through local nonemergency sources and are free from hunger. Food security can become a complex balance between national self-sufficiency (to ensure food supplies in times of shortages and high prices) and land set aside for export farm commodities. Some government subsidies pertain to self-sufficiency, some to trade advantages, and some to both.

food justice, adequate nutrition, culturally acceptable diet, food availability, environmentally sound food production, socially just food production, adequate diet, hunger, community agriculture, food security as a human right, food self-sufficiency, food quality, nutritional quality, food commodities, food exports, food imports, food distribution system, food prices, food price supports, food deficits, world food security

Local Food Systems (446): Local food systems are the local and regional interconnected systems of food production, distribution, and consumption. Local food systems foster direct links between farmers and consumers, allowing consumers to know more about where their food comes from and how it is produced, while keeping income within the local economy and reducing the environmental damage caused by the long distance transportation of food. Local food systems rarely provide all food. Usually grains come from more distant commercial markets. Their main goals in developed nations are improved quality of regional food and support for local rural farming communities.

food supply, ecoregional, transport impact, life cycle assessment, nutrition, food production, community agriculture, food diversity, local purchasing, sustainable agriculture, rural economy, farming communities, food self-reliance, sustainably grown food, supporting rural farming communities, local economies, cooperatives, food autonomy, local agriculture, community service agriculture (CSA), organic food, local/regional food, bioregional economies, slow food, farmer's markets, community gardens, small organic farming, reducing food miles

Malnutrition, Diet, Disease, and Education (853): Life loss from malnutrition has been calculated in disability-adjusted life years (DALYs). The World Bank conservatively estimates about 14.5 million life years lost per year from malnutrition, or about eight out of every thousand persons. More than half occur in children from birth to four years. The main deficiencies are protein, Vitamin A, iodine, and anemia. Diarrhea prevents nutrition absorption and is a primary cause of malnutrition. Malnutrition may have social causes (men eat before women, women before children; single working mothers serve junk food; colonially imposed crops like cassava are not balanced with complementary foods). *Diet and nutrition education* refers to the practice of educating, teaching, training, and imparting knowledge, ideas, and skills to people about diet, nutrition, additives, food safety, diseases related to undernourishment or malnutrition, exercise, eating habits, and lifestyle choices that affect health to improve quality and length of life.

nutrition, obesity, food choices, healthy eating, disease, exercise, health literacy, food intake, eating patterns, food contents disclosure, additives, vitamins, carbohydrates, fat, protein, dairy, fruit, vegetables, balanced diet, food deficits, World Food Program, malnutrition, food deficiencies, iodine, Vitamin A, diarrhea, junk food, slow food

FORESTRY
Forestry is the science, art, and practice of managing and using trees, forests, and their associated resources for human benefit. Forests can be managed to produce various products and benefits including timber, wildlife habitat, clean water, biodiversity, and recreation.

timber, stands, forests, trees, woodlands, commercial forestry, sustainable forestry, non-timber forest products, NTFP, forest ecosystem services, soil conservation, carbon dioxide sequestration, tree planting, clear cutting, thinning, silviculture, forest conservation, saplings, old growth forest, deforestation, afforestation, forest management

Agroforestry (285): Agroforestry is the practice of forest management where trees and shrubs are integrated with agricultural systems for multiple products and benefits. The integration of trees in farmland and rangeland diversifies and sustains production for increased social, economic, and environmental benefits for land users. In addition to commercial logging, developing nations rely on forests and woodlands for fuelwood, charcoal, rubber, browse, and other harvests. The concern is as much "sustainable forests" as sustainable forestry. Sustainable forests have been difficult to achieve because of the lack of clear property and access rights. In developing nations, the substitution of fossil fuels for fuelwood and land tenure and common property management rules are central issues, not resolved in developed nations which import both fossil fuels and many wood products.

multiple cropping, land use, food, fodder, wood, mulches, fuel, agriculture, tree, wildlife, farm, soil, watershed, reforestation, afforestation, crops, livestock, tree planting, soil erosion control, grazing, wood production, subsistence farming, forests, agrarian reform, wood use in developing nations, fuelwood, charcoal, rubber, forest products, land tenure, common property management, access rights, harvest rights, biofuels

Certified Timber Harvesting (4): In an attempt to make the global wood products industry more environmentally friendly, various "green certification" initiatives have begun to monitor tree harvesting and other forestry practices. When considered enviro-friendly, the trees or cut wood receives a green seal, ecolabel, or "report card" verifying enviro-friendly practices. The program can be third party, in-house, voluntary, or required; address the production methods; or simply certify the product. Forest biodiversity protection may or may not be included.

Forest Stewardship Council, ISO 14000, ISO 19000, ITTO, third-party certification, green certification, ecolabel, ecoseal, sustainable forestry, biodiversity, green buyers, chain-of-custody, green builders, NTFPs, nontimber forest products

Forestry Law and Policy (237): Sustainable forestry requires changes in many laws and policies on local, national, and international levels. Locally, any government agency or business can require paper to be bought with specific post-consumer waste or recycled paper pulp content. Nations can prohibit timber from legacy or old-growth forests or specific species considered threatened. Nations can rationalize land tenure rights, access to public forests for harvesting, set incentives or disincentives to encourage better forest management practices, stop illegal harvesting, and require safe minimum standards for logging. International organizations monitor phyto-sanitary and trade rules that impact cutting rates and debt-for-nature swaps. In addition, citizens can influence the wood products markets by boycotts or buycotts, media campaigns, and nonviolent direct actions.

forest protection, forest management, logging, forest conservation, timber products, forest economics, forest industries, pulp and paper, forest reserves, forest biomes, afforestation, Forest Management Act, U.S. Forest Service, recycled paper content, post-consumer paper content, ITTO, land tenure, common property management, common property resources, illegal tree harvesting, phyto-sanitary rules, debt-for-nature swap, boycotts

Global Wood Products Industry (3): The global wood products industry trades pulp, paper, softwood, and hardwood products as well as provides for international forest-based tourism (e.g. redwoods). Globally, natural forests have been in transition to managed natural stands and to tree plantations. Sustainable forests will be better conserved as their services are more accurately priced: erosion control, humidification, carbon-sinks, biodiversity support, groundwater recharge, and flood control. Acid rain continues to hurt productivity. Global warming has caused insect outbreaks and fires that can harm the industry. Tree farms with fertilizers, fire management, and pesticides will increasingly become sources of wood supply. Technology has increased efficiency by use of "minor species" and composites. The structural and appearance quality of softwoods will continue to deteriorate. Trade conflicts, especially an all-Asian market that does not respect sustainable forestry, may be the major challenge. Recycling and green products certification will take some pressure off logging. Deconsumption of wood products by developed nations has been contentious. Local sustainable forestry must deal with these global wood products pressures.

forests, forestry, wood products, chain-of-custody wood production, culture and wood, international trade, acid rain, plantations, tree farms, wood composites, paper recycling, post-consumer waste, wood recycling, green products, forest certification, U.N. Biosphere Reserves, ecotourism, boycotts, deconsumption, developing nations, developed nations, global/ local issues, ecosystem services, biodiversity protection

Logging (242): Logging is the practice of harvesting trees from a woodland or forest for commercial gain. The rate of deforestation, the management practices used, and the legality of the logging determine the sustainability of the activity and of the possible degradation to the forest environment. In certain forests, access roads have the most significant environmental consequences.

illegal logging, high-grading, unsustainable logging, forest management, timber products, erosion, clear-cutting, overcutting, silviculture, deforestation, harvesting, veneer, pulp, paper, construction materials, forest degradation, regeneration, reforestation, logging roads, forest disturbance, forests, shelterwood cut, clear-cut, selective cut, sediment, environmental impact, certified timber harvest, forest fragmentation

Plantations (13): Plantations or tree farms increasingly supply wood products to the world. They are typically grown after harvesting a more natural forest. Monterey pine, oil palms, acacia, and eucalyptus are common plantation species. Plantations provide more reliable harvestable yield and forest cover, can stabilize soils against erosion, and prevent leakage of nutrients. But they reduce biodiversity of the forest, encourage genetic impoverishment of tree varieties, and, if chemically treated, can cause water pollution. Plantations are only considered part of sustainable forestry when the creation of a plantation is mitigated by the protection of an equal area of natural or legacy forest.

plantations, tree farms, biodiversity, wood products industry, Monterey pine, eucalyptus, oil palm, acacia

Sustainable Forestry (1,411): Sustainable forestry is environmentally friendly, socio-economically productive, fiscally viable, and politically acceptable to the harvesters of forest products, the processors such as sawmills and pulp-paper makers, the wholesalers and traders, the retailers such as lumberyards, and the end users such as green buyers, green builders, and government agencies. Sustainable forestry is opposed to industrial forestry. Sustainable forestry custom designs forestry practices to bioregion, ecological web, and individual stands; performs green accounting at each stage of the chain-of-custody; considers watershed services and biodiversity as well as cultural values; and provides equitable rewards to individuals, communities, and other owners.

conservation, trees, forestland, reforestation, selective logging, watershed protection, runoff, soil erosion, forest products, economic productivity, environmental protection, regeneration, biodiversity protection, long-term management, ecosystem services, sustainable logging, forest ecosystem management, forest stewardship, nontimber products, forests, afforestation, forest management, green accounting, forest age, forest structure, patch size, patch isolation, fire regime, roads, sensitive species, forest preservation, forest fragmentation, clear-cut, genetic, resources

Urban Forestry (171): Urban forestry is the art, science, and practice of studying, cultivating, and managing trees, and their related natural resources, in the unique growing conditions of urban landscapes.

infrastructure, trees, city, street trees, urban ecosystem, urban landscape, forestry, shade trees, insulation, windbreaks, forests, afforestation, air quality, environmental planning, urban environment, greenways, greenbelts, urban park, flood control, stormwater management

GLOBAL CLIMATE CHANGE

Global climate change encompasses the long-term fluctuations in temperature, precipitation, wind, and all other aspects of the earth's climate. Global climate change is a natural process that has been exacerbated by human activity and is defined by the United Nations Framework Convention on Climate Change as "a change of climate which is attributed directly or indirectly to human activity that alters the composition of the global atmosphere and which is in addition to natural climate variability observed over comparable time periods."

global warming, science, biodiversity, carbon dioxide, pollution, emissions trading, stabilization, greenhouse gases, Kyoto, policy, regulation, fuel efficiency, international treaties, agreements, atmosphere, temperature, weather, energy production, coal, oil, gas, combustion, human activity, carbon budget, climate equity, climate justice, atmospheric carbon concentration

Climate Change (1,055): *Climate change* refers to the variation in the earth's climate or in regional climates over time. It describes changes in the variability or average state of the atmosphere—or average weather—over time scales ranging from decades to millions of years. These changes may come from processes internal to the earth, be driven by external forces (e.g. variations in sunlight density), or, most recently, be caused by human activities. In recent usage, especially in the context of environmental policy, the term *climate change* often refers to the ongoing changes in modern climate, including the rise in average surface temperature known as global warming. In some cases the term is also used with a presumption of human causation, as in the United Nations Framework Convention on Climate Change (UNFCCC). The UNFCCC uses "climate variability" for non-human-caused variations.

global warming, biodiversity, carbon dioxide, pollution, greenhouse gases, climate policy, regulation, fuel efficiency, atmosphere, temperature, weather, energy production, coal, oil, gas, combustion, global commons, carbon budget, climate equity, climate justice, atmospheric carbon concentration, climate change mitigation, low-carbon development, emissions trading, Kyoto Protocol, greenhouse effect, climate change and crop yields, climate change and impacts on weeds, diseases and insect pests, sea level rise, carbon cycle, climate change and water resources, agricultural emissions, climate change and ocean currents, wildlife responses to climate change, United Nations Framework Convention on Climate Change, Intergovernmental Panel on Climate Change (IPCC), general circulation models (GCM), climate change and fire, aerosols, climate change and melting ice, developed countries, developing countries, climate justice

Emissions Trading (53): Emissions trading is a market-based cap-and-trade program to reduce the emission of polluting gases into the atmosphere. Originally it was used to manage acid rain. The Kyoto Protocol targets greenhouse gases in order to reduce the overall emissions of gases that exacerbate global climate change. The trading mechanism allows participating polluting nations to either reduce their total emissions or to buy credits from less polluting nations to meet the limits set by Kyoto. As the price of credits rises, reducing emissions becomes more financially attractive to polluting nations or companies.

pollution, acid rain, greenhouse gases, air quality, sulfur dioxide, SO₂, mercury, carbon dioxide, offsets, cap and trade, carbon credits, emissions reduction, industrial pollution, climate change, carbon budget allocation, emission entitlements, fair allocation, atmospheric carbon concentration, public health, CO₂ equivalent, Kyoto Protocol, carbon taxes

Greenhouse Gases (210): Greenhouse gases are those gases that trap the heat of the sun within the earth's atmosphere, producing the "greenhouse effect," which warms the earth's surface. The main gases that contribute to the greenhouse effect are carbon dioxide and water vapor. Other greenhouse gases include methane, ozone, chlorofluorocarbons, nitrous oxide, and sulfur hexafluoride. The increase in the production of greenhouse gases has exacerbated heating within the atmosphere and the acceleration of global climate change.

emissions, global warming, carbon dioxide, methane, water vapor, nitrous oxides, aerosols, hydrofluorocarbons, HFCs, perfluorocarbons, PFCs, industrial pollution, burning fossil fuels, coal, oil, gas, biomass, international treaties, vehicle mileage standards, infrared radiation absorption, air, atmospheric carbon concentration, carbon sequestration, CO₂ equivalent, greenhouse effect, climate change and crop yields, climate change and impacts on weeds, diseases, and insect pests; carbon cycle, agricultural emissions, United Nations Framework Convention on Climate Change, Intergovernmental Panel on Climate Change (IPCC), climate change and melting ice, permafrost, sequestering carbon

GLOBALIZATION
Globalization is the process leading to increasing interdependence, integration, and interaction among people and corporations in disparate locations around the world. It is an overarching term that refers to a complex web of economic, social, technological, cultural, and political relationships.

economic internationalization, transnational corporation, multinational corporation, production chain, joint ventures, industrial cooperation agreements, licensing agreements, international subcontracting, "arm's length" transactions, national firms, state-owned enterprises, division of labor, geographical relocation, industrial environmental impacts, workers' rights, employment, environmental regulations, trade agreements, WTO, International Monetary Fund (IMF), North American Free Trade Agreement (NAFTA)

Currency Exchange (1): The exchange rate is the price at which one nation's currency can be converted into another. Most exchange rates float freely and change each trading day, while some are fixed. They play a crucial role in the alleviation of poverty and in creating trade balances and sustainable development. The poor do not have bank accounts, and the amount of cash they earn and save can be devastated by a change in exchange rates. Similarly, natural resource extraction often fluctuates with exchange rates. When, for instance, the dollar is worth less compared to the yen, Japan buys as much U.S. timber as possible. A weak dollar can override sustainable protections for forests. Sustainable development includes reinvestment in local or regional economy. If the exchange rate looks volatile, wealthier citizens tend to invest outside their own nation (capital flight), reducing local investment. With the global buying and selling of currencies over the Internet, governments can have a difficult time keeping their monetary policies sound and sympathetic to the poor.

currency, exchange rate, poverty, inflation, cash economy, savings, inflation, exchange control, exchange stabilization fund (U.S.), capital flight, International Monetary Fund, IMF, Latin American Monetary Union, Scandinavian Monetary Union, Tripartite Currency Agreement, euro

Fair Trade (324): Fair trade is an alternative trade system between small producers in developing nations and wholesalers in developed nations. Fair traders commit to social justice practices in which employees and farmers are treated and paid fairly, and sustainable environmental practices and long-term trade relationships are fostered that are equitable, less exploitative, and more secure.

commodities, wholesalers, retailers, producers, fair wages, employment, economic development, farmers, exploitation, certification, eco-labeling, group negotiating power, small growers, trading partnership, social justice, long-term trade relationships, cooperative association, marginalized producers, sufficient compensation, globalization, trade justice

Globalization Impacts (400): Globalization impacts involve the social, economic, and environmental outcomes of increased economic integration. The supply chain has been fragmented between nations, resulting in the extensive geographical relocations of each component (e.g., harvesting materials, parts manufacture, assembly, marketing and sales, distribution, research and development). Current globalization also includes a new international financial system, a new division of labor (e.g. unskilled and skilled workers, overseas buyers, traders, branchers/advertisers, retailers), and a new challenge to the nation-state to guide its economic development. Sustainability, on the global/local scales, tries to prevent transnational corporations from moving to nations where there are no or few enforceable standards for environmental and worker protection, and to prevent WTO regulations that would further remove regional and national authority over environmental and working standards.

economic internationalization, transnational corporation, multinational corporation, production chain, joint ventures, industrial cooperation agreements, licensing agreements, international subcontracting, "arm's length" transactions, national firms, state-owned enterprises, division of labor, geographical relocation, industrial environmental impacts, workers' rights, employment, material goods, corporations redefined and limited

International Debt (3): A debtor nation carries government, private business, and bank debts that exceed any expected income from foreign creditors. A debtor nation may find it difficult to obtain the capital needed to help the nation develop or to allocate government funds for sustainability projects. In some developing nations, the current account deficit is crippling and health care, roads maintenance, infrastructure, and education have suffered. To create more sustainable nations, there is a movement to forgive debts. Sometimes, debt forgiveness has strings attached (e.g. fiscal policies to lower inflation, cut imports, raise exports, cut government spending), which can increase citizen suffering and backfire (reducing government legitimacy and stability). Debt-for-nature swaps are one tool used by NGOs to reduce international debt and help save a nation's biodiversity heritage.

debt instrument, debt retirement, debt service, debtor nation, balance of payments, interest, principal, ability to pay, structural adjustment, external public debt, current account deficit, bank reserves, debt-for-nature swaps, International Monetary Fund, IMF, International Bank for Reconstruction and Development (IBRD aka World Bank), current account deficit, debt forgiveness

Trade Balance (2): The balance of trade is the difference (over a period of time) between the value of a country's imports and its exports of merchandise. If the trade balance is highly unfavorable, then many "peripheral" nations feel the "core" industrialized nations have turned imperial and that they have been trapped into a dependency relationship that hurts their development. Nations try to balance their trade deficits with trade barriers and export subsidies, which are contentious areas of trade negotiation between nations. The sustainability movement tries to level the playing field so that trade differences do not arise from one nation destroying its environment and exploiting its workers for a price advantage, while its trading partner suffers from practicing more environmentally friendly and better labor-related practices. NGOs and green companies may try to establish parallel trading systems and use ecolabeling to alert buyers (fair trade).

reciprocal trade agreement, trade barrier, nontariff barriers, rules of origin, antidumping measures, local content requirement, import quotas, import duties, protective tariffs, exchange control, sanitary and health regulations, trade bloc, trade deficit, balance of trade, balance of payments, current account deficit, fair trade, free trade, export processing zones, free trade zones, export credits, export guarantees, embargo, boycotts, GATT, WTO, ILO, NAFTA, ANCOM, AFTA (ASEAN), CARICOM, EFTA, EU, MERCOSUR, Trade Expansion Act

Transnational Corporations (14): Transnational corporations (TNC) are business firms with the power to control and coordinate operations in more than one nation, even if they do not own all the operations. A TNC profits by taking advantage of geographical differences in the factors of production (e.g. natural resources, capital, labor) and in state policies (e.g. taxes, subsidies, environmental regulations). Its additional advantage is its flexibility—its ability to switch and reswitch geographical locations at an international scale. TNCs, according to some viewpoints, exploit national economies or expand them; create jobs or exploit child labor; repress workers' rights; provide substandard conditions; spread technologies or preempt their use, creating external dependence on foreign technologies; interfere with the political process (including bribery) or help citizen well-being more than corrupted national officials; destroy a nation's natural capital or help organize its conservation and extraction. TNCs are a central focus of sustainability in the areas of employment, environment, national sovereignty, technology transfer, trade, and relative bargaining power of local communities, cities, states, and regions.

multinational corporation, production chain, natural resources, employment, natural capital, financial capital, geographic relocation, sustainable communities, anti-globalization, corporate responsibility, appropriate technology, fair trade, national sovereignty, export processing zones, TNCs, multinational companies, corporations redefined and limited, corporate concentration

GOVERNANCE
Governance is the practice of managing processes and systems within all institutions, be they nonprofit organizations, companies, state or national governments, or intergovernmental institutions.

good governance, government oversight, government reform, transparency, accountability, fair, open, impartial, administration, process, institutional accountability, democratic deficit, corruption, bureaucracy, world government

Global Governance (31): One earth, many worlds. Sustainability attempts to coordinate, network, correct, and govern those actions of the many worlds that may harm the one earth (the global commons) and incite destabilizing violence. Some believe in world government, others in markets, still others in a hierarchy of "federated" state governments, and others in a web of non-state-based civil society organizations as the best management practice. Global climate change, whaling, HIV/AIDS, open sea mineral harvests, acid rain, ozone, airspace for transport, near earth satellite and telecommunications infrastructure, planetary genetic resources (biodiversity), migratory birds and butterflies, waste disposal, weapons sales, Antarctica, nuclear testing, and chemical fallout have all been issues in which global governance and sustainability entangle. Sustainability, in part, rests on political stability and acceptance of government authority. The nation-state acts as a defender of its citizens' security, a container and nurturer for its citizens' economics and culture, and an internal legal and regulatory authority to preserve peace and justice. The nation-state authority has been challenged economically by globalization, its security by terrorism and climate change and other cross-boundary environmental issues, and its justice by special interests and electoral procedures. The nation-state has been challenged to preserve its natural capital and natural heritage for future generations by the sustainability movement.

world government, global environment, civil society, sovereignty, United Nations, World Trade Organization, World Bank, World Wildlife Fund, World Health Organization, International Monetary Fund, CSO, NGO, climate change, whaling, migratory birds and butterflies, waste disposal, weapons sales, Antarctica, nuclear testing, chemical fallout, biospheric commons, global commons, international convention, framework convention, common pool resources, natural resource management, nation-state

Good Governance (896): Good governance is the practice of decision-making and administering an organization, company, or public institution in an impartial, efficient, transparent, and fair way that is free of corruption and respects the rule of law.

organizational management, accountability, transparency, free from corruption, public office, public money, rule of law, fair, process, open, impartial, administration, decision-making process, policy implementation, public resource management, public service ethics, government, democratic deficit

Government Oversight and Reform (558): Government oversight is the practice of monitoring the activities and overseeing the performance or operation of government institutions with the intention of reforming and redressing any mismanagement of public funds and resources.

good governance, transparency, accountability, ombudsman, monitoring, oversight, government spending, public funds, policy, reform, abuse, corruption, government waste, awarding contracts, mismanagement, special interests, systemic abuse of power

Institutional Accountability (853): *Institutional accountability* refers to the obligation of institutions, be they nonprofit organizations, companies, corporations, or government bodies, to be responsible for their actions and for the appropriate management of resources that have been entrusted to them.

corruption, transparency, international conventions, oversight, reform, good governance, government agencies, mandate, fraud, federal, political, waste, international financial institutions, monitoring, ombudsman, responsibility, stakeholders, shareholders, auditing, public bodies, intergovernmental institutions, agreements, watchdogs, international law, democratic deficit, corporations redefined and limited

GREENING OF INDUSTRY

The greening of industry is the process of identifying the environmental cost of doing business, such as the use of finite natural resources for product manufacture and the wastes that are generated by the production process, and developing new ways of doing business that reduce environmental harm, make use of waste streams, increase product efficiency, and internalize the costs to the commons of industrial activity.

industrial ecosystems, industrial ecology, industrial symbiosis, Kalundborg, environmental externalities, pollution prevention, life cycle assessment, pollution remediation, waste cycling, gray water cycling, green chemistry, servicizing, resource efficiency, sustainable materials

Consumption and Green Consumers (523): *Consumption* refers to the rate and way that humans use both material and energy (natural resources). Many economists extend consumption to mean all material goods and services for the gratification of human desires. In sustainability, economists look for ways to reuse or recycle after use by humans; this is not a central concern of other economists. Nor are issues of "externalities" (environmental impacts that cost society), nor natural resource and manufacturing efficiencies. In general, citizens in developed countries are extravagant consumers and their use of resources is greatly out of proportion to their numbers. Sustainability tries to redress externality, efficiency, materials choice, inequitable use issues, and unfair labor practices. Green consumers try to purchase products that have been processed with a strong sense of environmental and social accountability. These include: ecologically harmless, no harmful additives, no irradiation of foods, high transparency in manufacturing process, honest labeling, fairness to employees, no sex discrimination, and high quality product assurance.

resource use, discard, product, consumerism, consumer power, consumer protection, resource waste, product choice, overconsumption, consumer education, green living, unsustainable consumption, environmental product choice, green products, environmental sustainability, ethical consumption, externality, life-cycle analysis, greenwash, customers, ecolabeling, certification, social accountability, quality assurance, green market, ecological footprint, environmental impact, industrial ecosystems, advertising, consumer fraud, additives, irradiation, social accountability, labor, socially conscious purchasing, industrial ecosystems, bioregional economies, local living economies

Ecolabeling and Certification (173): Ecolabeling and certification is the practice of assessing the environmental impacts of the harvest, manufacture, use, and discard of a product using defined environmental criteria. If an external organization can guarantee that the product has met the criteria, the manufacturer is able to package and market the product with the ecolabel to indicate the product's environmental credentials. Certification is the process of verifying that the product meets the criteria.

environmental standards, life cycle assessment, product differentiation, certification, external audit, validation, ecolabels, product manufacture, materials, product discard, sustainability standards, environmental impact, industrial ecosystems, material goods, products

Ecological Footprint (59): Ecological footprint is a measure of consumption and the impact on natural resources of a given population, in terms of the amount of land and resources required to support and to absorb the wastes produced by a group and their chosen lifestyle.

consumption, impact, monitoring, indicator, lifestyle, waste, resources, natural resource use, population resource consumption, reduce footprint, sustainable consumption, waste absorption, ecologically productive land, energy use, water use, building materials, consumables, land supporting human activity, environmental sustainability, ethical consumption, industrial ecosystems, material goods, products, ecological footprints

Environmental Monitoring (2,159): Environmental monitoring is the practice of quantifying the impact on the environment of various activities, from measuring a community's ecological footprint, to measuring the concentration of carbon dioxide in the atmosphere and monitoring the retreat over time of the polar icecap—an indicator of global warming.

environmental, natural, resources, impact, indicator, footprint, measure, continuous measurement, damage, quantify, monitor, environmental conditions, state of the environment, quality standards, metrics, comparative, evaluate environmental change, environmental exposure, analysis, parameters, baseline, environmental sustainability, research, study

Industrial Ecology (3,381): Industrial ecology (IE) is a systems approach to thinking about an industrial system as part of the environment, not in isolation from it, where the total quantity of materials and energy used from virgin material, to finished material, to component, to product, to obsolete product, and to ultimate disposal is optimized. It examines local, regional, and global uses and flows of materials and energy in products, processes, industrial sectors, and economies and focuses on the potential role of industry in reducing environmental impacts throughout the product life cycle.

industrial symbiosis, discard cycling, closed loop, manufacturing, life cycle assessment, waste as raw materials, sustainable materials, material use reduction, waste reduction, waste cycling, industrial ecosystem, energy efficiency, pollution reduction, material flows, energy flows, dematerialization, decarbonization, ecodesign, design for the environment, extended producer responsibility, product stewardship, ecoindustrial parks, product-oriented environmental policy, ecoefficiency, industrial pollution, environmental sustainability, true cost accounting, cradle-to-cradle, cradle-to-grave, material substitutions, waste mining, emissions, by-products, remanufacturing, residual products, water conservation, energy conservation, reuse, solid waste, liquid waste, gas waste, aerosol waste, toxicity or ecotoxicity, noise, odor, occupational health, material good, waste recovery, water efficiency

Life Cycle Assessment (39): Life cycle assessment (LCA) has become a major tool for sustainability accounting among businesses and corporations. LCA evaluates the environmental burdens of each product, process, and activity of a business. It quantifies materials and energy used, wastes produced, and their impacts on the environment. It goes from "cradle-to-grave" or "cradle-to-cradle" if recycling and reuse are part of the process. LCA includes extraction, processing, manufacturing, transport, distribution, use, reuse, recycling, and final disposal. LCA complements risk assessment and technology assessment in the green business tool kit.

product stewardship, resource and environmental profile analysis (REPA), cradle-to-grave analysis, cradle-to-cradle analysis, ecobalance (Europe), waste audits, environmental impact assessment, technology assessment, risk assessment, LCA, material goods

Natural Resource Management (329): Natural resources are any material part of the environment used to promote the welfare of humans or other species (e.g. freshwater, forests, soils, minerals, huntable species). Rarely, the term is used for crops or domesticated animals. Natural resources can be renewables (wind, water) or nonrenewable (oil, coal, some kinds of groundwater). Careful natural resource management from harvest to extract through manufacture and processing to use and recycle/waste management is a central theme of sustainability.

natural resources, water, air, land, soil, minerals, coal, petroleum, iron, plants, forests, ecosystems, consumption, monitoring, supply, natural systems, environmental protection, resource scarcity resources, groundwater, wind, rivers, wildlife, ores, timber, wood, nonrenewable, renewable, resource extraction, resource harvest, resource processing, waste, recycling, common-pool resources, watershed management, industrial ecosystems

Recycling and Reuse (4,346): *Recycling and reuse* refers to the process of collecting and separating waste or a discarded product, and recovering materials that can be used again in the same form (reuse), remanufactured into other products (recycling), or used as a fuel to generate power (refuse-derived fuel). It also includes increasing repair services rather than discard and replacement.

conservation, reuse, compost, resource, recycle, salvage, building materials, preowned materials, discarded materials, reprocessing, closed loop recycling, raw materials conservation, resource recovery, waste cycling, pre-consumer waste, post-consumer waste, environmental sustainability, industrial ecology, waste reduction, solid waste, landfill, life cycle assessment, repair, back-hauling, incinerator, collecting, sorting, decontaminating, ISO 14000, solid waste disposal, contaminants, reverse logistics, value recovery, asset recovery, reconditioning, retrofitting, industrial ecosystems, material goods, waste recovery

Sustainable Materials (178): Sustainable materials are materials used for building, product and parts assembly, and other applications that are used in manufacturing. Sustainable materials are often renewable, energy efficient, and nontoxic, and are chosen for the low impact of the product over its entire lifetime. The sustainability of materials production, use, and discard is often determined using an LCA. Life cycle assessment (LCA) is the practice of evaluating the environmental impact of a product throughout its lifetime from the extraction of resources for product manufacture, through the use of the product, to the method of product disposal.

LCA, environmental impact of material production, materials use, materials discard, recycling, energy efficient, waste cycling, industrial ecology, renewable, reuse, green procurement, greening supply chain, environmentally responsible purchasing, natural resources, sustainable production, clean production, materials production, energy use, renewable products, packaging, product life, human toxicity, durable, locally sourced, biodegradable, easily recycled, low LCA, low embodied energy, incorporate recycled content, salvage, environmental sustainability, bio-based resource productivity, green design, transportation, material selection, reclamation, industrial ecosystems, material goods, sustainable materials design

Sustainable Production (258): Sustainable production is defined as the creation of goods and services using processes and systems that are nonpolluting, conserve energy and natural resources, are economically viable, are safe and healthy for employees, communities, and consumers, and are socially and creatively rewarding for all working people. As opposed to older uses of the term, it adds an intergenerational ethic and incorporates environmental consequences.

working conditions, sustainability, resource use, waste reduction, sustainable development, clean manufacturing, energy conservation, sustainable materials use, economically viable, industrial ecology, environmental sustainability, renewable materials, recycling, reuse, employee welfare, water conservation, toxics, pollution-free production, industrial ecosystems, material goods, products

HEALTH

Health refers to the total state of physical, mental, and social well-being of a person, including the absence of disease.

well-being, physical, mental, social, absence of disease, treatment, diet, nutrition, access, equity, human rights, poverty, disease prevention, public health, HIV, AIDS, health education, sanitation, environmental health, endocrine disruptors, alternative medicine

Alternative Medicine (148): *Alternative medicine* refers to treatments and therapies that are used instead of, or complementary to, conventional allopathic medical practices. Alternative medicine treatments include acupuncture, ayurvedic, tribal, traditional, homeopathic, herbal medicine, and chelation therapy.

therapy, health, medicine, herbal, homeopathy, healing, chiropractic, acupuncture, naturopathy, massage, healing tradition, indigenous healing, complementary medicine, holistic, ayurvedic, tribal-traditional medicine, photo therapeutic, nonallopathic medicine, ecological medicine, treatment, conventional Western medicine, earth medicine

Asthma (33): Asthma is a chronic respiratory disease that causes difficulty in breathing, including wheezing and shortness of breath, which is often triggered by an allergic reaction to factors including environmental pollutants such as chemicals and smoke.

environmental health, pollution, children, quality of life, air quality, triggers, breathing, environmental pollution, industrial pollution, respiration, environmental allergens

Cancer (20): Cancer covers a wide-ranging series of diseases characterized by the uncontrolled development of cells that have the ability to invade tissues of the human body. Cancer is the lead cause of death in developed nations. Although cancers have a genetic component, many are associated with environmental causes. These include tobacco, excessive alcohol, obesity, too much bright light, UV exposure, radiation exposure, and specific persistent organic pollutants. The reduction of certain cancers can be helped by removal of the pollutants and education of the public (e.g. teen smokers) on the dangers as well as control of advertising and product development (e.g. asbestos board).

malignancy, metastasis, carcinogens, public health, immunotherapy, chemotherapy, radiation therapy, alternative medicine, viruses, radiation, tobacco, alcohol, obesity, ultraviolet light, visible light, asbestos, melatonin, persistent organic pollutants, environmental health

Ecological Change and Emerging Diseases (29): Ecological changes have increased the probability and severity of many diseases. Changes in the atmosphere, increased population densities, mobility and travel, toxins introduced into the nutrient cycles, the spread of still waters (reservoirs, ponds), the addition of drugs to the food web, megacities, and land use changes such as soil pollution and deforestation have all altered or added important vectors, hosts, and transmissions of disease. Dams, for instance, increase the transmission of water-related schistosomiases and malaria. Air conditioners can spread Legionnaires' disease. Reforestation can increase Lyme disease. Ozone-layer changes can increase cancers. The abusive use of DDT can produce pesticide-resistant malarial mosquitoes. Solutions may require global/local actions: heat waves, air pollution, and disease transmission by human migration (especially air travel) bring about new challenges to public health.

Legionnaires' disease, hemorrhagic fevers, Lassa fever, Lyme disease, cryptosporidiosis, AIDS/HIV, Escherichia coli *O157:h7, hantavirus, human granulocytic ehrlichiosis (HGE), schistosomiasis, malaria, helminthiases, cholera, dengue, leishmaniasis, oropouche, red tide, Rift Valley fever, ozone hole, radioactive fallout, global warming, carcinogens, irritants, air pollutants, radiation, infectious disease, dose, disease transmission, exposure, international convention, global climate change, ozone, health hazard, host, herd immunity, disease, morbidity, pathogen resistance, precautionary principle, disease reservoir, vaccine, vector, environmental health*

Endocrine Disruptors (4): Endocrine disruptors are chemical substances that interfere with hormone function in animal species, affecting the body's ability to regulate functions such as growth, maturation, and development. Chemicals that are known endocrine disruptors include dioxins, PCBs, DDT, and other chemicals found in some insecticides, herbicides, and fungicides. Animal species, including humans, are exposed to endocrine disruptors through direct contact with pesticides and other chemicals and through the ingestion of contaminated water, air, and food.

health, environment, exposure, hormone, xenoestrogens, environmental estrogens, semen quality, sperm count, breast cancer, testicular cancer, cryptorchidism, feminization, embryonic development, environmental pharmaceuticals, estrogen mimics, androgen mimics, hormone antagonists, hormone synthesis, hormone metabolism, hormone receptors, environmental health

Environmental Health (2,123): *Environmental health* refers to those factors in the environment that can affect human health and well-being, from air pollution to water contamination to the impacts of industrial waste. The environmental health movement began in the 1860s with the rise of industrialization. Along with the conservation movement, it is the most solid citizen-based sector of the modern environmental and sustainability movement. The environmental health movement extended self-protection of the body to concern for the environment that, many times, contributes to or causes disease. The environmental health movement works on global issues such as the health impacts of Chernobyl and the international sales of harmful products, on networked issues like Minimata disease, and on local issues of toxic pollution. It is deeply involved in monitoring advertising based on misleading claims that targets the vulnerable with unhealthy products and withholds information important to health decision-making.

alternative medicine, ecotoxicology, global public health, epidemiology, genetically modified foods, factory farms, consumer protection legislation, green chemistry, bioremediation, sanitation, food and workplace safety, product safety, organic food standards

Environmental Toxicology (47): Environmental toxicology encompasses classical toxicology and ecotoxicology. It is concerned with the toxic effects, caused by natural or synthetic pollutants, on the components of ecosystems (microbes, plants, animals, including humans). It tries to reduce harm by reducing releases into the environment, understanding transport into living creatures (with or without transformations), understanding the risk of exposure to target organisms, and the hazard or risk to different species. It encompasses risk perception, risk assessment, and risk management as well as regulatory needs and product substitutions. Ecotoxicology is the study of the harmful effects of chemical compounds and toxic substances on species, populations, communities, and ecosystems.

ecotoxicology, toxicology, food web, dose, exposure, risk management, risk assessment, toxic pollution, pollution source, species, population, ecosystem, bioaccumulation, biomagnification, pesticides, fungicides, plasticizers, solvents, polychlorinated biphenyls, halogenated aryl hydrocarbons (HAH), endocrine disruptors, environmental estrogens, polynuclear aromatic hydrocarbons (PAH), petroleum hydrocarbons, organotins, anti-fouling agents, molluscicides, imposex, environmental impact, carcinogens, teratogens, radionuclides, metalloids, precautionary principle, genotoxins, environmental pharmaceuticals, transfer pathways, interactions, concentration, toxicologic testing, poisonous, biomagnification

Green Hospital Movement (13): Sometimes called the "healthy hospital movement," this is a worldwide effort of over 440 organizations in over 55 nations to "green" health care facilities and to ensure that medical care does not further contribute to diseases caused by environmental contaminants. Typical projects include stopping the manufacture and use of mercury-based thermometers (U.S.), eliminating hospital incinerators that emit toxic dioxin, improving hospital food quality, green purchasing of materials, green building to save energy and water and eliminate volatile pollutants, and elimination of PVC in hospital equipment. The movement has been spearheaded by nurses, doctors, and concerned NGOs.

sustainable health care systems, green health care, PVC, waste incineration, medical waste disposal, dioxin, environmental health, ecologically sustainable health care, sustainable health care, sustainable medicine, disease prevention, raising awareness, health literacy, health promotion, health care, medical facilities, clinics, emergency care, public health, green building, green purchasing, food ingredients, food safety, hazardous substances, persistent organic pollutants, volatile organic pollutants, body burden, biomonitoring, ecotoxicology, nutrition, mercury pollution, public health, epidemiology, workplace safety, food safety, product safety, toxicology, food web, dose, exposure, risk management, risk assessment, toxic pollution, pollution source, bioaccumulation, biomagnification, pesticides, fungicides, plasticizers, solvents, polychlorinated biphenyls, halogenated aryl hydrocarbons (HAH), endocrine disruptors, environmental estrogens, polynuclear aromatic hydrocarbons (PAH), petroleum hydrocarbons, precautionary principle, environmental pharmaceuticals

Health Care Access (4,035): Access to health care refers to the rights of all people to have equal access to basic health care and to information about health, regardless of financial status, ethnicity, gender, ability, or other factors. Universal health care is achieved when all citizens have equal access to health care and patients pay by their income level through the tax system, insurance policies, or co-pay fees.

human rights, poverty, health care, medical facilities, clinics, emergency care, public health, social justice, health equity, health insurance, health payment fees, public health taxation, medical treatment, willingness-to-pay policies, health policy

Health Education (2,201): Health education is the practice of educating, teaching, training, and imparting knowledge, ideas, and skills about health, including public health, disease prevention, lifestyle choices, and medical conditions and health care.

disease, prevention, knowledge, information, exercise, promotion, healthy lifestyle, diet, nutrition, active lifestyle, training, teaching, raising awareness, well-being, health literacy, health promotion, community health, health equity, health policy

HIV/AIDS (2,754): Human immunodeficiency virus (HIV) is the virus that causes acquired immunodeficiency syndrome (AIDS), a fatal disease. HIV is contracted through blood or other bodily fluids; by a transfusion of infected blood; from a mother to her baby; through the use of contaminated hypodermic needles; or through sexual contact with a person who has the disease. The HIV/AIDS epidemic is having widespread social and economic impacts on communities, countries, and regions, and tackling the HIV/AIDS epidemic involves public health and health education, as well as addressing the equity of access to treatments and the human rights of those affected directly and indirectly by the disease.

orphans, epidemic, treatment, equity, medicine, prevention, protection, educating, raising awareness, safe practices, retrovirus, immune system, generic drugs, access to medicine, health justice, public health, economic impact, social impact, policy, funding, budgeting, programs, discrimination, health equity, health policy

Infectious Diseases (204): Infectious diseases are different from noninfectious diseases because a transmission process (contagion) determines who becomes ill. Contagion can occur through waterborne, foodborne, airborne, or vectorborne transmission. Sustainability brings a more holistic view of infectious diseases than simply a triangle of Agent, Host, and Environment. It understands that social policy, for instance, between environment and host may reduce infection rates and include better nutrition, hygiene, preventative treatment, and housing (e.g. TB, HIV/AIDS, diarrheal diseases, acute respiratory infections). It investigates the life cycle and ecology of disease agents (e.g. cholera, hantaviral disease, hookworm, schistosomiasis), and understands the centrality of environmental disruption (e.g. rabies, Lyme disease, malaria, cryptosporidiosis). Sustainability works to improve both human and natural environmental conditions as part of its management of infectious disease.

contagion, host organism, host habitat, disease agent, disease vector, disease reservoir, environmental disruption, sanitation, exposure, hygiene, nutrition, housing conditions, shantytowns, pathogens, immune responses, host specificity, tissue responses, public health, preventative treatment, disease transmission, herd immunity, anthroponotic diseases, zoonotic diseases, emerging disease, vaccine, human ecology, resistance to treatment, environmental health

Malaria (48): Malaria, which kills 1 to 2 million people each year (90 percent in Africa) and affects hundreds of millions, is the world's most important vectorborne disease. Malarial epidemics are complex: rain and temperature affect the number of mosquitoes and the development of parasites in mosquitoes. Human landscape changes such as farm clearing, deforestation, and dam and housing construction also affect populations of mosquitoes. Some mosquitoes and parasites have become resistant to insecticides like DDT or drugs like chloroquine. Human movement can spread the disease. Public health infrastructure is crucial to managing malaria, and most recent outbreaks have occurred in areas of civil strife. Malaria is a major challenge to public health and the creation of sustainable communities in sub-Saharan Africa and other infected areas.

Plasmodium falciparum, Anopheles mosquitoes, public health, drug resistance, DDT resistance, ecosystem change, vectorborne disease, water management, climate change, deforestation, civil strife, Blue Nile Health Project (Sudan), economic development and disease, human migration and disease, bednets, pyrethroid insecticides, indoor spraying, sub-Sahara Africa, vector control, environmental health, public health

Medical Biotechnology (26): Medical and pharmaceutical biotechnology can speed diagnosis, prevention, and certain therapies. Biotech medicine includes the creation of new vaccines, nutraceuticals, cosmetics with active biological ingredients, and medicines from transgenic animals and plants. Although not as controversial as food biotechnology, ethical concerns have been raised about stem cells, genetic enhancement, cell cloning, testing of new drugs in developing nations, controls of transgenic crops, and international regulation and enforcement.

biotechnology, vaccines, nutraceuticals, pharmaceutical companies, transgenic organisms, stem cells, genetic enhancement, HIV/AIDs, Ebola virus, SARS virus, flu virus, Chagas disease, Parkinson's disease, regenerative medicine, reproductive cloning, medical diagnosis, biological patents, health policy, biopiracy, intellectual property, genetic resources, human genomics

Pesticides (321): Worldwide an estimated 5 billion tons of pesticides are applied annually. Most are used in agriculture to destroy, prevent, or control unwanted plant or animal pests (e.g. herbicides, fungicides, insecticides, and rodenticides), but a few are used as anti-fouling agents in cooling fluids, ships, and metal machinery. Sustainability concerns include harm to nontarget species such as birds, oysters, and beneficial insects, human health impacts, and alternative farm and industrial methods to control damaging pests. Most pesticides are chlorinated organics, organophosphates, carbamates, and pyrethroids.

exposure, inhalation, toxic pesticides, farmworkers, agricultural pollution, weed killers, pesticide drift, chronic exposure, organophosphates, carbamates, biomagnification, health impacts, chlorinated organics, pyrethroids, persistent organic pollutants, POPs Treaty, bioavailability, autoimmune suppression, dermal toxicity, ingestion, Yusho, U.S. Toxic Substances Control Act, cancer promoters, aerial transport, Agent Orange, lindane, DDT, phthalate, 2,4,5-T, bisphenol, DES, DDE, PCB, aldrin, dieldrin, heptachlor, chlordane, endrin, HCH, mirex, kepone, parathion, malathion, carbaryl, carbofuran, dichlorvos, 2,4-D, diquat, paraquat, atrazine, PCDD, PCDF, HAH, environmental health

Public Health (3,214): Public health is at the center of the human side of sustainability. It works on all scales (individual, household, community, state, region, global), and becomes entangled in issues of urbanization, migration, refugees, industrialization, modern agriculture, climate change, energy use (especially fuelwood), as well as air, land, and water pollution. It must involve itself in human adaptations to, for instance, climate change and infectious disease, nutrition, sanitation, vector breeding and migration, pathogen transmission, and virulence. Public health officials must deal with uncertainties, benefit/cost analysis, policy disputes, technology access of patients, and funding priorities as well as triage when multiple crises occur. Public health is sometimes viewed as any medical aid to a population regardless of funding sources, and sometimes as medical aid that comes through government or philanthropic funding (versus private health care).

disease, virus, epidemic, public health awareness, access to services, preventive health care, environmental health, sanitation, environmental hazards, hygiene, health equity, GIS, malaria, tuberculosis, cancer, HIV/AIDS, infectious disease, World Health Organization, epidemiology, EPA, U.S. Department of Health and Human Services, Centers for Disease Control and Prevention, National Institutes of Health, CIESIN, Pan American Health Organization, Food and Agriculture Organization (U.N.), public health policy

Sanitation (924): *Sanitation* refers to maintaining clean and hygienic living conditions to prevent the spread of disease by properly disposing of sewage and other waste, correct handling of food and water supplies, and the control of disease carriers such as mosquitoes and rats.

clean, water, disease, hygiene, rural, developing countries, sewage, garbage, purification, pest control, waste disposal, wastewater, water treatment, spread of disease, public health, access to water, water supply security, microbial contaminants, purity, environmental health, air, waste, stormwater

Tuberculosis (54): Tuberculosis (TB) is one of the top four infectious diseases of humans. It is transmitted in airborne droplets during coughing and sneezing. In 2004, there were 14.6 million humans with active TB (with 9 million new cases) and about 1.7 million deaths. About one-third of the world population has latent TB. About 10 percent of these cases will lead to death. TB occurs mostly in developing nations but can appear anywhere with immune-compromised patients. TB is resurging, in part, from the spread of HIV/AIDS, in part from drug-resistant strands of the bacterium, and in part from deteriorating environmental conditions in various parts of the world. The existence of a public health infrastructure for early diagnosis and quarantine is essential to limiting the spread of TB.

mycobacterium tuberculosis, infectious diseases, global health emergency, HIV/AIDs, immuno suppression, drug resistance, living conditions, public health, public health infrastructure, vaccine, consumption, environmental health

HUMAN RIGHTS AND SOCIAL JUSTICE

Human rights are those basic rights that belong to people because they are human beings, regardless of their nationality, race, ethnicity, gender, or religion, and without which people cannot live in dignity. The human rights that are considered by most societies to belong to all people include the rights to life, justice, freedom, and equality. Social justice is the concept that community and state activity should be based on just and equitable treatment of all people regardless of color, race, socioeconomic class, gender, age, or sexual preference.

dignity, basic rights, innate rights, inalienable rights, inherent, right to education, life, right to adequate standard of living, freedom of movement, freedom from torture, prohibition of genocide, right to self-determination, right to participate, freedom of religion, equality, right to health, right to education, right to shelter, right to employment, right to food, civil rights, freedom of expression, water as a human right, environment as a human right, social justice

Climate Justice (87): Climate justice is the environmental justice concept that the negative environmental and social impacts of global climate change should not disproportionately affect populations of people in developing countries and poorer nations while predominantly being fueled by wealthier industrialized nations. Climate justice seeks the acknowledgment that all human beings belong to an atmospheric commons and that equity within and between nations is necessary as a precursor for tackling climate change.

social equity, economic equity, climate change, solution, basis, ecological debt repayment, equal access to resources, global inequality, resource use inequality, access to energy, significant carbon emission reduction, grassroots action, inequitable impacts of climate change, inequitable consumption, inequitable production of emissions and waste, weather pattern change, rights, human impacts of climate change, North-South inequity, climate equity, environmental racism

Disability Equality (1,350): *Disability equality* means that people regardless of their disability have the same opportunities to fulfill their human rights and potential, and are not denied equal access for reasons related to their disability to services, to jobs, to education, to economic development, to the allocation of resources or benefits, and in all aspects of life.

disability, equality, equal access, equal rights, services, facilities, physical access, human rights, access to jobs, institutional inequalities, bias, discrimination, attitudes of nondisabled, inclusion, social justice, learning difficulties, physical disability, mental disability, discrimination, political rights, social exclusion

Distributive and Economic Justice (2,935): Distributive justice tries to ensure that the benefits and burdens of public policy are fairly and equally distributed. Distributive justice concerns itself with past injustices, such as compensation of Holocaust victims who lost property, as well as treaty violations. It tries to ensure that the benefits of infrastructure development (e.g. water, roads, schools, health) will reach the least well-off and that burdens are not disproportionately laid on the most politically powerless (dam or highway construction). Distributive justice raises issues such as: What is fair? What is equal? Since exposure to environmental risks and access to resources cannot be distributed equally, how should they be shared? Economic justice is that part of distributive justice that is concerned with fairness and equity in economic affairs.

public policy, past injury, treaty violations, confiscated property, fairness doctrine, natural rights, distributive justice and ethics, corporations redefined and limited

Environmental Justice (2,064): Environmental justice is a broad concept to define political actions to stop the victimization of disadvantaged communities of color and poverty by projects that cause environmental harm. It began in North Carolina (1982) over a PCB landfill in a poor Afro-American town. It became part of federal policy in 1994 when President Clinton signed an executive order charging all federal projects to incorporate environmental justice into their operations. Environmental justice became internationalized at the United Nations Rio Summit on the Environment (1992). It is a multifaceted movement that includes laws and policies from civil rights, distributive justice court decisions, public participation equality issues, social justice ethics, and actions by the global sustainability movement. Sustainability interacts with issues of justice. It attempts to eliminate harmful impacts such as pollution rather than redistribute them. It attempts to find ways to ensure access to resources such as water and energy that are equitable but do not harm the environment. It promotes the precautionary principle as a way to avoid inequitable environmental burdens. Environmental justice and sustainability also deal with very large economic justice issues: Can all consumers reach U.S. standards of living? Will this cause too much environmental damage? What is an equitable distribution of resources? What kinds of limits (e.g. taxation of luxuries and high incomes) should be imposed on the wealthy? What is the minimum lifestyle that should be guaranteed all humans?

environmental racism, civil rights, Executive Order 12898, Title VI, distributive justice, sustainability, social justice, environmental ethics, public participation, racial inequality, poverty, disadvantaged communities, environmental hazards, human rights, indigenous peoples, race, social class, power, environmental democracy, toxic waste, environmental enforcement, North-South equity, economic equality, economic discrimination, fairness, equity, economic affairs, international financial institutions, transnational corporations, economic inequalities, economic access, social justice, debt relief, gender, North-South solidarity, inequity, economic solidarity, social exclusion, inequity of power, justice and sustainability, enviro justice, race, poverty and the environment

Ethnic Equality (4,492): *Ethnic equality* means that people, regardless of their race and ethnicity, have the same opportunities to fulfill their human rights and potential, and are not denied equal access on the basis of their ethnicity to services, to jobs, to education, to economic development, to the allocation of resources or benefits, and in all aspects of life.

human rights, racial equality, ethnicity, equality, discrimination, minorities, social justice, racial justice, access to jobs, economic development, fair treatment, equal opportunities, embedded institutional discrimination, laws, policy, rights, tribes, tribal equality, indigenous, political rights, social exclusion, inequality of power

Gay Rights and Equality (1,610): Basic human rights and equality have been denied to gay, lesbian, bisexual, and transgender people on the basis of their sexual orientation. Equality means that people regardless of their sexual orientation have the same opportunities to fulfill their human rights and potential, and are not denied equal access on the basis of their sexual orientation to services, to jobs, to education, to economic development, to the allocation of resources or benefits, and in all aspects of life.

human rights, transsexual, transgender, gay, lesbian, bisexual, tolerance, sexual orientation, gender identity, gender expression, equal rights in law, discrimination, prejudice, civil rights, social justice, sexuality, political rights, sexual minorities, social exclusion

Human Rights and Civil Liberties (8,052): Human rights are the legal and written recognition of the dignity and equality of all persons. They are considered indivisible, inalienable, and universal. It is recognized that some may be compromised during periods of armed conflict and disturbance. Those rights that cannot be compromised are called "fundamental guarantees" or "nonderogable rights." Fundamental guarantees prohibit torture; cruel, inhuman, or degrading punishment or treatment; slavery, slave trade, or servitude; jail for those who cannot fulfill a contract without trial; and provide that all humans should have the freedom to choose their religion, thoughts, and beliefs (including to manifest these beliefs). Nations that have not adopted the death penalty include the right to life. Other conventions include freedom from retroactive laws and various guarantees for children and families. Civil liberties are those basic human rights that are protected by law against arbitrary interference particularly by governments. These fundamental individual rights include the right to life, freedom from torture, freedom from slavery and forced labor, the right to liberty and security, the right to a fair trial, the right to privacy, freedom of conscience, freedom of expression, freedom of assembly and association, and the right to marry and have a family. These rights and liberties are usually created and protected by a constitution.

protected by law, civil rights, equality, human rights, freedom, law, freedom of speech, thought, action, protection, justice, inclusive, right to life, freedom from torture, freedom from slavery and forced labor, right to liberty, security, right to a fair trial, right to privacy, freedom of conscience, freedom of expression, freedom of assembly and association, right to marry, right to have a family, race, gender, sexuality, disability, asylum, discrimination, international law, political rights, identity rights, democratic deficit, freedom of conscience, Universal Declaration of Human Rights, International Covenant on Civil and Political Rights, International Covenant of Social, Economic and Cultural Rights, African Commission of Human and People's Rights, European Court on Human Rights, Inter-American

Court on Human Rights, fundamental guarantees, nonderogable rights, genocide, children, women, refugees, torture, interfaith tolerance

Human Rights Education (3,612): Human rights education is the practice of teaching, imparting knowledge, and raising awareness of basic individual human rights, of human rights law, and of those civil liberties protected by law, and of areas that some believe should be considered as human rights, such as the right to water and the right to a healthy and sustainable environment.

raising awareness, teaching, training, publicizing, universality, indivisibility, interdependence, equality, human dignity, respect, nondiscrimination, information, knowledge, human rights law, protected civil liberties, justice, race, gender, sexuality, disability, international law, political rights, identity rights

Human Rights Monitoring (2,053): Human rights monitoring is the practice of identifying, recording, and monitoring human rights violations and abuses to bring attention to those abuses in order to protect and defend individual human rights from erosion by governments and institutions. There are judicial and nonjudicial monitoring actions that can be initiated by conventions, individuals, states, or NGOs. They are usually diplomatic and are more than legal mechanisms. They rarely prevent human rights violations; they can only identify them.

human rights violations, international human rights law, fact-gathering, fact-finding, procedure, international law, conventions, political rights, identity rights, Human Rights Committee, Committee Against Torture, Committee on the Rights of the Child, Committee on the Elimination of Discrimination Against Women, Committee on the Elimination of Racial Discrimination, African Commission of Human and People's Rights, European Court on Human Rights, Inter-American Court on Human Rights

Human Rights and Natural Law (220): Natural law is linked to moral and ethical principles. It states that humans have certain inalienable and irrevocable rights, even if they are not written into texts. Natural law has been used by citizens of conscience to oppose written law. The most famous example has been the abolishment of slavery. What should be included in natural law remains controversial. Natural law pertains to both times of peace and situations of armed conflict.

natural law, humanitarian law, positive law, jus cogens, international law, customary law, inalienable rights, inviolable rights

Human Rights Protection (7,352): Human rights protection is the practice of preserving and defending those rights which all humans possess because they are human, such as the right to life, liberty, dignity, justice, and equality.

abuse, violations, torture, protection, raising awareness, recording, publicizing, defending, rights, legal protection, public pressure, law, policy, observance, advocacy, intervention, state responsibility, justice, legal challenges, international declaration of human rights, race, gender, sexuality, disability, discrimination, international law, conventions, political rights, identity rights, prosecution

Human Trafficking and Slavery (573): Human trafficking is the illegal practice of using force, coercion, or deception to move people across national or state borders for sexual and labor exploitation, and subjecting them to involuntary acts such as slavery. Traffickers deny their victims basic human rights and use coercive tactics including intimidation, isolation, the threat and use of physical force, or debt bondage to control their victims.

human rights, slavery, child trafficking, abduction, servitude, prostitution, social justice, debt bondage, bonded labor, international law, conventions, identity rights, sex industry, Convention on the Abolition of Slavery, the Slave Trade, and Institutions and Practices Similar to Slavery (U.N.)

Social Justice Education (8,250): Social justice education is the activity of educating, teaching, training, and imparting knowledge, ideas, and skills to people about the concept that community and state activity should be based on just and equitable treatment of all people regardless of color, race, socioeconomic class, gender, age, or sexual preference.

social justice, raising awareness, education, skills, understanding, problem-solving, information, knowledge, teaching, advocacy, economic justice, environmental justice, rights, poverty, climate justice, access to resources, food justice, trade justice, quality of life, basic needs, human rights, direct action, discrimination, social exclusion, community cohesion, inequality of power

INDIGENOUS PEOPLE AND RIGHTS
Indigenous people are those people and their descendants who are native to a geographical area and were the first people to live there, prior to colonization. *Aborigines* and *the first nations* are alternative names for indigenous people. Indigenous rights are those human and legal rights that have been denied to indigenous people because of their heritage.

native, first nations, aboriginal, peoples, autochthonous, human rights, indigenous culture, historic, sacred, protection, traditional knowledge, land rights, tradition, teaching, tribes, cultural heritage

Indigenous Lands (121): *Indigenous lands* describes the geographical area that is home to a particular indigenous community and may have sacred, cultural, agricultural, or economic value for that community. The rights of indigenous people are those human and legal rights that have been denied to indigenous people because of their heritage, particularly in relation to their land and homes.

land rights, protection, ownership, aboriginal, livelihoods, subsistence, economic development, sense of place, environmental protection, land seizure, land sovereignty, natural resources management, cultural easements, biodiversity knowledge, self-determination, tribe, tribal, local, community, culture, aborigines, native peoples, first peoples, autochthonous, education, minority, racism, cultural diversity, sovereignty, human rights, land claims, self-government, autonomy, political rights, identity rights, racial healing

Indigenous People and Culture (1,341): Indigenous culture encompasses the knowledge, arts, traditions, medicines, values, habits, lands, economic, political, and social systems of indigenous communities. Indigenous people can be subject to discrimination, inequality, and threats to their culture, and are working to preserve their culture and fully realize their human rights.

culture, arts, traditional knowledge, protection, conservation, oral history, skills, tradition, family life, community, stability, cultural heritage, tribe, tribal, aborigines, native peoples, first peoples, autochthonous, education, minority, racism, cultural diversity, cultural freedom, population, traditions, healing, ethnobotany, human rights, discrimination, autonomy, social exclusion, earth medicine, herbal medicines

Indigenous Rights (997): Indigenous rights are those human and legal rights that have been denied to indigenous people because of their heritage, particularly in relation to their land and homes.

land rights, protection, ownership, aboriginal, livelihoods, subsistence, economic development, sense of place, environmental protection, land seizure, land sovereignty, natural resources management, cultural easements, biodiversity knowledge, self-determination, tribe, tribal, local, community, culture, aborigines, native peoples, first peoples, autochthonous, education, minority, racism, cultural diversity, sovereignty, human rights, land claims, self-government, autonomy, political rights, identity rights, environmental racism

INLAND WATER ECOSYSTEMS

Inland waters make up a minute portion of the planet's water resources (less than one hundredth of one percent). Nevertheless, they are crucial to human survival, hold a disproportionate number of species (more threatened than either oceans or land species), and are central to sustainability issues. Inland water ecosystems include rivers, lakes, streams, ponds, standing water, and wetlands.

inland waters, freshwater, lakes, rivers, streams, ponds, water resources, dams, channelization, floodplains, lentic, lotic, aquatic ecosystems

Inland Aquatic Ecosystems (565): Inland aquatic ecosystems include running waters (e.g. rivers, streams, thermal springs), standing waters (e.g. lakes, soda and saline lakes, ponds, even minute pools in some plants like bromeliads), and transitional wetlands. Freshwater environments are central to sustainability as inland waters provide drinking water, recreation, irrigation, industrial water, hydropower, fishing, wetland food production, waste discharge and foodplain hazards. Inland water uses are contentious because of dams, channelization, downstream pollution, invasive species, flood-control systems, sediment from hillslope impacts, and water level depletions.

inland waters, freshwater, lakes, rivers, streams, ponds, water resources, dams, channelization, floodplains, lentic, lotic

Lakes and Ponds (23): A lake is a still or standing body of water that is deep enough to prevent rooted plants from growing all the way across the bottom. The largest are sometimes called "seas" (e.g. Aral Sea, Caspian Sea), even though their waters are fresh. The oldest lakes (4 to 30 million years old) have very high species richness and endemics (e.g. Lakes Baikal, Tanganyika, Malawi, Victoria). Some lakes are saline or alkaline (e.g. the Dead "Sea" or Great Salt Lake, Utah). Because of the relatively low turnover of water flow, lakes can become easily polluted by fertilizers and industrial toxics. They are especially sensitive to invasive species and water losses from agricultural diversions (Aral Sea).

lakes, ponds, soda lakes, salt lakes, freshwater seas, closed basins, lentic ecosystems, water pollution, eutrophication, water transport, invasive species, Societas Internationalis Limnologiae

Riparian Ecology and Conservation (1,741): *Riparian habitat* (sometimes called *riverine*) refers to the environment found on banks of streams and rivers (and sometimes lakeshores). Riparian systems feature a "hydroperiod pulse"—a seasonal spreading of water laterally from the channel and increase in speed and volume in an upstream/downstream direction. Riparian zones, which connect the channel to the hillslope in most watersheds, support patchy, diverse mixes of plants and animals. Riparian zones act as buffers and filters between urban areas and agricultural zones. They supply nutrients and shade to migratory fish. They support unique and large numbers of species. Many riparian areas have been subject to heavy use by subdivisions, agriculture, livestock, recreation, fishing, in-channel transport, road development, and invasive plants. Riparian restoration may require land purchase, artificial flooding, removal of exotic species, watershed management, and revegetation.

seasonal flood, 100-year floodplain, dikes, levees, channelization, ectone, edge effects, riverine mangroves, river swamps, bottomland hardwoods, bosques, watershed management, Bureau of Reclamation, Federal Water Pollution Control Act, National Environmental Policy Act, floodplain protection, grazing, fire, invasive species, National Flood Insurance Act, Bureau of Land Management, U.S. Forest Service, riparian zones, setbacks, stream banks, flooding, flood meadows, watercourses, buffer zones, stream restoration

River-Lake Ecology and Biodiversity (60): *Limnology* is the technical term for the study of inland, freshwater lakes, rivers, and other water bodies. Inland waters, although a tiny fraction of liquid water on the planet, support 40 percent of all known fish species and almost all the 5,000 known amphibians (e.g. frogs, salamanders). Two predominantly aquatic reptile orders (turtles and crocodilians) and over 150 duck and swan species are uniquely adapted to ponds and lakes. A large number of "waders" (e.g. cranes, screamers, coots, egrets, ibises) require freshwater habitats. A few mammals are wholly confined to inland waters (the river dolphins and two manatee species). A few must require aquatic habitats but use the land (e.g. hippopotami, beaver, capybara, Nile lechwe). Because of competing demands, inland water ecology has received exceptional attention. Duck hunters have sponsored perhaps the best population monitoring of wildlife in the world. Nevertheless, nearly 20 percent of inland water species appear to be threatened, especially freshwater turtles (60 percent).

conservation, education, aquaculture, freshwater, seawater, brine, limnology, habitat, aquatic ecosystems, fish, aquatic species, lakes, ponds, streams, rivers, water, biodiversity, inland waters, fish, amphibians, birds, ducks, stream restoration

Rivers and Creeks (42): A river is a surface drainage channel that has a large annual discharge of water. Most water flow ultimately discharges into the ocean, but some rivers discharge into closed basins. They have three zones that are important to sustainability: the headwaters, which are typically the sources of water and sediment; the middle reaches, where sediment can be both transferred and stored; and the delta or alluvial fan, where sediment is deposited and freshwater mixes with saltwater. Each zone supports different economic activities (e.g. mines are often in headwaters; cities in delta regions). There are many variations. Sustainability concerns itself with upstream/downstream, in-stream/off-stream, and aboveground/below-ground sharing of the seasonal water resource.

rivers, creeks, watersheds, catchment basins, drainage basins, drainage net, headwaters, river reaches, mainstem, tributary, delta, alluvial fan, closed basin, estuary, river channel, erosion, sediment, river discharge, floodplain, floods, in-stream flow, riparian vegetation, lotic ecosystem, Ca, Cauvery, Chao Phraya, Columbia, Gambia, Ganges-Brahmaputra, Godavari, Indus, Irrawaddy, Krishna, Ma, Magdalena, Mahanadi, Mekong,

Narmada, Niger, Nile, Pahang, Paraná, Paraíba, Penner, Perak, Salween, São Francisco, Senegal Sitang, Song Hong (Red), Tapti, Tambesi-Hari, Volta, stream restoration

Wetlands (875): Wetlands are areas of land that are regularly or continually saturated with shallow water, and dominated by large aquatic plants and patchy diverse microhabitats. The shallow waters are determining factors in soil development, ecology, vegetation growth, and associated animals within the ecosystem. Bogs, fens, marshes, and swamps are all examples of wetlands. Some are very extensive (e.g. Everglades, Florida or Sudd swamps, Sudan). Humans harvest food (e.g. frogs, crocodiles, waterfowl, beavers, muskrats, manatees, seeds, tubers, watercress) and grow rice and sago palms in wetlands. Wetlands can mitigate flood peaks.

wetlands, bogs, fens, marshes, swamps, peat, forested freshwater wetlands, water table, zone of saturation, perched water, wells, pumps

LAW, POLICY, AND PROPERTY RIGHTS
Law and policy refers to the legal and criminal justice systems, legal and legislative institutions, and processes of making and changing laws, including lobbying institutions for policy change, within organizations, regions, or countries.

institutions, protect, sustainable, society, legal, counsel, research, analysis, advocacy, education, training, lobbying, public policy, law reform, legal advocacy, representation, legal aid, policy change, environmental law, legislation, government, enforcement, crime, policing, prison, legal services, rule of law, corruption

Biological Patents (22): *Biological patents* refers to the ownership and legal rights to intellectual property that can be granted by a government to an applicant for a limited amount of time for discoveries that relate to living organisms, such as gene sequencing or biotechnology products. The scope of biological discoveries and materials that can be patented is subject to ethical questions particularly in relation to patenting human genes.

intellectual property rights, human rights, patents, property, ownership, ethics, corporatization, environment, biotechnology, genetically modified organisms, GMOs, gene patenting, globalization, trade, gene sequences, research, development, public ownership, biological material, standards, limited patents, biopiracy, international law, genetic resources, human genomics

Conservation Easements (902): Conservation easements are used to restrict development on private property in order to protect the natural resources and conservation value of the land. The easement is either a voluntary donation or is sold by the landowner and is a legally binding agreement that limits the use of the land, or prevents development from taking place on the land in perpetuity while the land remains in private hands. The landowner receives tax relief and retains property rights from the easement.

development easement, land conservation tool, voluntary legal agreement, landowner, restrict land development, tax benefits, limits development rights, purchasing development rights, sustaining rural economies, enabling conservation and farming, transfer of selected property rights, conservation covenant, environmental protection, sustaining agriculture, farmland preservation, land trust, conservation easement

Crime and Policing (516): *Crime and policing* refers to the criminal justice system, the treatment and rights of suspects and prisoners, the approach and action of police to suspects and members of the public, including police harassment, aggression, and racism, within communities, regions, and states.

racism, protection of rights, abuse, police harassment, persecution, law, rule of law, respect for law, racial profiling, social justice, equality, rights, violence, police aggression, rights of the police, prisoners, criminals, suspects, criminal justice system, arbitrary arrest, detention

Environmental Law and Policy (2,394): *Environmental law and policy* refers to the practice of creating new regulations, laws, and practices, and of changing existing ones through campaigning, lobbying, advocacy, and direct action, which have been adopted by a government or an organization, and which govern activity related to the environment, such as laws and policies relating to land-use planning or toxic dumping.

environmental law, reform, public interest, protection, land, pollution, externalities, lobbying, legislation, environmental protection, enforcement, access to legal defense, community representation, information, skills, resources, policy change, advocacy, analysis, resource management solutions, environmental law enforcement, environmental policy, toxic waste, chemical

International Humanitarian Law and War Crimes (507): Only parts of human rights conventions apply during times of armed conflict, civil unrest, emergency, siege, or other extraordinary circumstances. At these times, governments have the right to restrict some human rights and freedoms. During these times, it is best to distinguish between human rights guaranteed during peacetime and international humanitarian laws that protect the "fundamental guarantees" agreed upon by treaties and conventions. The central international humanitarian law for times of tension is the Geneva Conventions. Rules pertain to the wounded, sick, shipwrecked, prisoners of war, and civilians. During ambiguous periods between peace and war, and even during armed conflict, nongovernmental humanitarian groups play a critical role in bringing violations to light and pursuing redress. For over fifty years, since the Nuremberg and Tokyo trials after World War II, the United Nations has attempted to set up an international tribunal (the International Criminal Court) to judge and punish crimes against humanity. Some progress has been made and ad hoc courts have looked into Yugoslavia and Rwanda war crimes. But, because of arguments about what constitutes a war crime and the fear of giving up national power to an international court, the process is not complete. It has been agreed that war crimes are not subject to statutory limitations. War crimes, in general, include genocide, rape, sexual slavery, enforced sterilization, conscripting children under the age of fifteen, intentional starvation of civilians, the use of civilians as human shields, and the use of weapons of mass destruction.

fundamental guarantees, nonderogable rights, international law, natural law, armed conflict, detention, prisoners of war, children, women, wounded and sick, civilians, noncombatants, Geneva Conventions, Hague Conventions, international criminal tribunals, children, women, international humanitarian law, penal sanctions, statutory limitations, individual recourse, universal jurisdiction, international fact-finding commission, immunity, amnesty, International Criminal Court (ICC), genocide, crimes against humanity

Land Reform (51): Land reform is the process of redistributing ownership of land, usually implemented by a government authority backed by the military. Land reform can distribute large land holdings to peasants, privatize common property holdings, or take private lands for government use.

land ownership, landless, landholdings, restructuring, land ownership, smallholdings, land tenure, plantations, agribusiness, ranches, collective ownership, agrarian reform, compensation, individual ownership, private property, large estates, land division, equitable land distribution, land titles, rights of occupancy, land rights, privatization, property takings, eminent domain, property confiscation, public domain, property rights, farmers' rights

Land Tenure (8): *Land tenure* is a general term to describe the holding of land by an individual or group of persons. It has rules of access, use, extraction, and intergenerational inheritance. In many developing nations, land tenure relies on kinship and lineage. These customary rights have confronted modern interventions from colonial administrations, state legislation, oligarchical tribalism, and international requirements for loans and aid. In many developing nations, clear and enforceable communal and/or private rights are much desired by peasants, farmers, and indigenous peoples as a form of security for food production and freedom from politically influenced loss of homeland. Sustainability initiatives in these locales requires mapping, surveying, education as to ownership options, acceptance by the state, and the help of NGOs.

rural communities, developing nations, common property resources, food security, homeland security, kinship, tribe, clan, customary rights, colonialism, tribalism, structural adjustment, private property, communal property, land-use policy, property rights, farmers' rights

Land Trusts and Land Conservation (1,194): A land trust is an organization that enables private landowners and communities to find ways to conserve and to protect their land from economic pressures and overdevelopment, enabling people to use the land while also protecting the land from development, preserving the biodiversity and ecological integrity of the land for the long term. Land trusts use a variety of tools to conserve land, including purchasing or leasing land for conservation and creating conservation easements for private landowners.

conservation, management, conservation easements, development, buffer zone, purchase, gift, donation, land trust, subdivision, parcel, consolidation, wildlife corridor, ecosystem, parcel, legal protection, preservation, land conservation, conservancy, farmland preservation, land acquisition for protection, property rights

Land Use Policy (657): *Land use policy* refers to the practice of creating new regulations and practices, and of changing existing ones that have been adopted by a government or an organization, and that govern activity related to land use and planning, through electoral politics, bond issues, referenda, private land purchase, campaigning, lobbying, advocacy, and direct action.

land use, zoning, sustainable land development, urbanization, permitting, environmental impact, commercial activity, industry, right of way, public access, fragmentation, planning, land rights, land reform, agrarian reform, neighborhood renewal, ordinances, permits, bonds, referenda, property rights, greenbelt expansion, wildlife corridors

Law and Policy Reform (2,594): *Law and policy reform* refers to the practice of creating new regulations and laws, and of changing existing laws that have been adopted by a government or an organization, and that govern activity related to laws, legislation and policy-making, through campaigning, lobbying, advocacy, and direct action.

new policy, reform, law, legal system, equality, equity, rights, justice, advocacy, lobbying, policy making, legislative change, government, state, enforcement, organization, legislation,

Legal Services and Representation (2,458): *Legal services* refers to those services provided by lawyers or people with legal authority to assist individuals in seeking or obtaining legal help in civil or criminal matters, assisting them to secure and exercise their legal rights, and to meet their legal responsibilities. Legal services include legal education, advice, and representation, enabling often minority or low-income people to define, assert, promote, and enforce the full range of their legal rights. *Legal representation* refers to the ability, opportunity, and right of people in need to have access to adequate legal representation when needed, regardless of financial circumstances, education, gender, ethnicity, or any other factor that might prevent adequate legal representation being available.

low cost, referral, economic access, advice, legal issues, representation, information, legal system, minority access, knowledge, legal rights, defending rights, access to services, equal access, pro bono, legal aid, civil, criminal, law, low-income, access to legal services, legal education, legal procedure, availability, access to legal representation, right to representation, courts, advocacy, law, legal defense, legal services, legal support, discrimination, right to a fair and public hearing

Precautionary Principle (70): Precautionary principle is the principle of taking pre-emptive action to forestall long-term environmental damage despite scientific uncertainty of such damage occurring. Where the potential damage is severe and irreversible, as in the case of climate change, a lack of scientific proof is an insufficient reason to justify inaction to prevent such damage occurring.

uncertainty, mitigation, long-term, prevention, preemptive action, averting irreversible damage, prevent harm, risk management, potential harm, tipping point, toxic substance combinations, restrict use, known to be safe, anticipate environmental damage, environmental health, cumulative harm, ethical technology use, potential negative consequences, environmental sustainability, environmental health

Prison Reform and Policy (416): *Prison reform and policy* refers to the practice of creating new regulations and practices, and of changing existing ones that have been adopted by a government or an organization, and that govern activity related to prisons, through campaigning, lobbying, advocacy, and direct action, including prison conditions, the treatment of convicted criminals, access to due process from inside prison, and the treatment and rights of suspects who should be presumed innocent until proven guilty.

human, alternative sentencing, prevention, prison, labor, prisoners, rights, overcrowding, juvenile, education, legal services, law, criminal justice, injustice, racism, abuse, humane conditions, political prisoners, prisoners of conscience, access to legal counsel, representation, penal offense, arbitrary arrest, unfair detention, rule of law

Property Rights (12): Private property rights have been part of societies for hundreds of years. They are contested arenas of power. Private property provides private goods with closed access to others. Private property can be contrasted with international, state, and communal property, which provide common goods and have controlled access. The extreme contrast is open access lands with neither management nor rules. Economists speak of the "jointness" and "exclusivity" of public goods (e.g. land, water), while some advocates believe in total exclusivity; anything can be extracted or modified on their private property. Most nations believe that property rights are granted by the government and can be modified for the greater good. In the American West, for instance, the government may hold subsurface mineral rights, while the property owner holds surface rights. Sustainability reinforces the legitimacy of the common goods approach to property because land and water should be protected for future generations. Sustainable land ownership is more a usufruct right (use right) than an absolute right. Throughout the world, property rights remain contentious.

property rights, rural communities, land ownership, government lands, state lands, crown lands, usufruct rights, Magna Carta, land tenure, food security, land use regulation, common property management, common property resources, closed access, open access, controlled access, private property, state property, international property, communal property, joint property management, exclusive property management, land security, partial subtractibility, natural resources, customary law, land rights

MEDIA
Media refers to all of the different technologies, materials, and means of communicating information to people, such as through radio, television, and video. *Media* is a term that also applies to those people whose profession it is to discover, report, and communicate information and news to the public.

mass media, communication, information, source, independent, freedom of expression, tools, education, knowledge, raising awareness, radio, television, film, video, publishing, newspapers, magazines, press, journalism, audio-visual, uncensored, accountability, censorship, progressive, radical, alternative media, Internet activism, independent media

Advertising (6): Advertising is closely linked to sustainability. Its purpose is to persuade consumers to buy more products. This can lead to overconsumption, health problems, and environmentally destructive consumption. However, sustainability tries to persuade those in wealthy nations to limit their rate of material and energy consumption and to avoid health problems or ingredients that may injure their bodies. Advertising has special financial incentives, especially in the United States where it is a tax-deductible business expense. Advertising to children is strictly regulated and, in some European nations, banned altogether. The monitoring and regulation of advertising is a crucial aspect of limiting overconsumption, consumer fraud, health protection, and protecting the environment.

consumerism, overconsumption, advertising messages, commercials, product packaging, ecolabels, media, commodity fetishism, children's advertising, product promotion, cigarette advertising, alcohol advertising, environmental impacts, consumer fraud, consumer safety, Federal Trade Commission, Federal Communications Commission, Children's Television Act (1990, U.S.), tax policy, independent media

Film (160): Film is the medium that uses documentaries, feature films, and cinema as tools for communication.

film, filmmaking, documentaries, independent, uncensored, alternative, freedom of expression, media, communication tool, raising awareness, environmental films, environmental, educating, alternative media, independent media

Internet (734): The Internet is the global network for communication using computers to publish, share, generate, and find information. It has become a crucial tool for civil society and sustainability, connecting global and local projects and tasks, creating both networks and virtual communities.

communication, tool, networking, connecting, Web, computers, access, equality, network, global, freedom of information, empowering, access to information, publicizing, raising awareness, educating, media, independent, uncensored, information and communication technology, ICT, Internet activism, alternative media, digital divide, independent media

Journalism and the Press (708): *Journalism and the press* refers to the profession of collecting, verifying, reporting, and analyzing information regarding current events, including trends, issues, and people. Issues that are central to this profession and to the communication of information include journalistic standards, the independence of reporting, censorship, bias, corporate ownership of the media, and the safety and human rights of the press.

alternative press, independent, full disclosure, uncensored, censorship, access to information, freedom of the press, freedom of expression, freedom from intimidation, media worker rights, press freedom, journalism, ethics, reporting, news, watchdog, media ownership, progressive media, alternative media, independent media

Media and Communication (1,575): Media is a crucial aspect of social change. It spreads ideas, it inspires groups fighting for power, and it tries to change organizations through which power is exercised. Media has informational, educational, commercial, and entertainment values. Media is the vehicle of one of the central principles of democracy—what was called "freedom of the press" but now (with television, radio, movies, and the Internet) might be called "freedom of the media." The media as a communications form is amoral and can work for or against sustainability on a local, regional, or global scale. Spreading ideas, ethics, stories, and information about sustainability has become a major task.

full disclosure, information, communication, advertising, marketing, public relations, media, watchdog, alternative media, progressive media, radical media, independent media, First Amendment rights, rights to free speech, freedom of the press, commercial media, public media, television, radio, Internet, movies, newspapers, magazines, zines, comics, books, reporting

Photography (48): Photography is the medium that uses still pictures and prints as tools for communication and documenting events through photojournalism.

photojournalism, images, communication tool, nature, representation, environmental damage, educating, reporting, journalism, visual information, media, independent, freedom of expression, uncensored, alternative media, independent media

Publishing (1,260): *Publishing* refers to those forms of printed material, such as journals, magazines, newspapers, and newsletters, which are used as tools for communication, including printed media that is also published in electronic form.

journals, newsletters, newspapers, books, magazines, reporting, journalism, information, knowledge, access to information, communication, freedom of expression, independent, editorial control, bias, media, uncensored, online journals, alternative media, independent media

Radio and Audio (239): Radio and audio are the auditory media that use recordings and the broadcasting of live programs as tools for communication.

communication, music, information, broadcasting, knowledge, independent, sound, editorial control, recordings, access to information, news, uncensored, remote, community, source, audiovisual, media, freedom of expression, alternative media, independent media

Television (146): Television is a communication tool that uses a telecommunication system for broadcasting and receiving moving pictures and sound over a distance, broadcasting a wide variety of news, discussion, and other programs.

communication tool, news, uncensored, independent, editorial control, information, quality of information, access to information, money, funding, program content, freedom of expression, public access, community, broadcasting, media, ownership, alternative media, independent media

Video (139): *Video* refers to the means of recording and playing back audiovisual programs on demand, as well as a method of distributing information and a tool of communication. *Video* also refers to the personal technology that allows anyone with a video camera to make a film or program and to share the media in a variety of ways, including making it available on the Internet.

audiovisual, communication, tool, information, access to information, program, educating, knowledge, raising awareness, on demand, dissemination, media, independent, freedom of expression, uncensored, Internet broadcasting, alternative media, independent media

MINING
Mining is the practice of extracting minerals from the earth. Minerals are elements or compounds that occur naturally in the earth's crust. They are major driving forces in human history from the Roman exploration of Britain for tin to the lure of gold in Amazonian and Indonesian rain forests. Minerals are metallic (metals such as iron, aluminum, copper), or nonmetallic (minerals such as sand, silicon, stone, salt, and phosphates). Minerals are a nonrenewable resource. Each mineral has an estimated mineral reserve and a "life index" after which there will be no more economically available sources. For instance, copper reserves may be depleted by 2050. Sustainable mineral practices—keeping minerals available for future generations—are a major challenge. The future may see minerals extraction from the ocean floor (manganese modules) and by concentrating seawater.

minerals, mineral reserves, metals, nonmetals, rocks, ores, aluminum, borax, chromium, cobalt, copper, gold, gravel, gypsum, iron, lead, magnesium, manganese, mercury, molybdenum, nickel, phosphate, platinum, potassium, salt, sand, silicon, silver, sulfur, tin, titanium, tungsten, zinc, deep-sea mining, seabed mining, mountaintop removal

Fossil Fuels (76): Fossil fuels are the partially decayed bodies of organisms formed millions of years ago. They are hydrocarbons (mixtures of carbon and hydrogen) and include methane, coal, oil, and natural gas. Hydrocarbons also can be produced as synthetic fuels from tar sands, oil shales, gas hydrates, liquefied coal, and coal gas. They pose two problems to sustainability (besides the fact that they are nonrenewable). There are the consequences of combustion and the impacts of production and transport. Land damage occurs, especially when coal is strip-mined. Water pollution is a common indirect impact of production. Oil spills occur in transport, and greenhouse gases and acid rain can result from combustion.

fossil fuels, nonrenewable resources, Strategic Petroleum Reserve, methane, coal, lignite, bituminous coal, soft coal, anthracite, hard coal, Surface Mining Control and Reclamation Act, acid mine drainage, dragline, mountaintop removal, mine reclamation, acid rain, scrubbers, clean coal technologies, petroleum, crude oil, liquefied petroleum gas, petrochemicals, cogeneration, oil spill, greenhouse gases, global warming, proved reserves, shale oil, speculative oil resources, tar sands

Minerals Law and Policy (24): The minerals industry is one of the most powerful globalized industries on the planet. It has been the focus of some of the most tense and violent conflicts between environmentalists, indigenous peoples, industrial promoters, sympathetic government politicians, and, at times, the military. Minerals are so precious as raw materials for goods and items of trade that the industry has been awarded many subsidies, exempted from many environmental laws, and rarely concerned itself with sustainable mineral value chains. Antarctica is the best example of international cooperation to protect the mineral commons.

General Mining Law, Surface Mining Control and Reclamation Act, Madrid Protocol, Environmental Protection Protocol to the Antarctic Treaty

Mining and Refining Ores (151): There are many kinds of mines, depending on the depth and concentration of the mineral ore. Iron, copper, stone, and gravel are usually extracted by open-pit mining or strip mining. The overburden may be 60 to 99 percent of all the earth and rock moved. Subsurface mines are usually called shaft mines. They disturb the earth less but can be more hazardous to miners. Open-pit mining disturbs huge areas of land and wastes huge amounts of encountered groundwater. Water flowing through waste rock or tailings dissolves or suspends toxic substances that have killed waterfowl and fish. The most damaging mining to rivers is called "hydraulic mining," in which pressure hoses take apart the riverbanks in search of gold. The extracted ores are refined at smelters which, without pollution control devices, emit dangerous gases such as sulfur and mercury. Sustainable mineral industries eliminate these human and ecosystem hazards but at high cost.

coal, lead, water, pollution, contamination, mine waste, strip mining, extraction, environmental impacts, erosion, energy use, open mine, quarrying, auguring, alluvial dredging, ore, by-product, tailings, salt mining, sand extraction, acid drainage, stream pollution, abandoned mines, Superfund sites, smelters, scrubbers, electrostatic precipitators, open-pit mines, subsurface mines, shaft mines, strip mines, overburden, toxic pollutant, heavy metal pollutants, hydraulic mining

Mountaintop Removal (26): Mountaintop removal (MTR) is a type of coal mining where the vegetation and topsoil are removed from a mountaintop and dynamite is used to blast away eight hundred to a thousand feet of mountaintop to reach a seam of coal for mining. The large quantities of debris removed from the mountaintop are dumped into nearby valleys with significant impacts on the environment.

mining, coal, flooding, landslides, clear-cutting, deforestation, dust, valley fill coal mining, pollution, water quality, tailings, subsidence, sediment, debris, sediment ponds, erosion, recontouring, land reclamation, resource extraction, mountaintop

Sustainable Minerals Industry (6): To make a mineral useful for humans, the mineral deposit must be located, extracted, and beneficiated, processed, or refined by concentrating it and removing impurities. The purified mineral can then be used to make a product. Mining and smelting can damage land, and disposal can pollute the air, soil, and water. Industrial mining is infamous for its impacts on human safety and health. The environmental costs of extraction, processing, and disposal of wastes have not been captured in the actual price of the final products, especially in developing nations. The inability to capture "externalities" in mineral costs has saddled taxpayers in developed nations with the cost of reconstruction and restoration of mined mountains and polluted waters. "Sustainable" mineral flows seek ways to minimize environmental impacts, maximize the health and safety of workers and citizens of nearby watersheds, extend the useful lifespan of processed minerals by reuse, and rehabilitate the derelict lands left by mining operations. Sustainable industries also look for materials substitution and new product designs to minimize the need for nonrenewable minerals (ceramics, plastics). Future trends include: at the extraction stage, microorganisms "biomine" low-grade ores to increase concentrations of metals like gold; at the refining stage, technological advancements can extract more minerals per ton of country rock; at the goods and factory step, scraps can be returned to the smelter for reuse; at the finished product and consumer step, recycling can return used goods to become secondary materials (scrap); and at the disposal level, certain tailings and landfills can be "mined" for recycled products.

mineral flow or value chain, reserves, proved reserves, industrial ecosystems, mineral conservation, recycling, reuse, solid waste management, sanitary landfills, durable goods, reparable goods, beverage deposits, dematerialization, mineral substitutes, biomining, reserves, proved reserves, acid mine drainage, tailings, watershed conversion, toxic minerals, derelict lands

PEACE, WAR, AND SECURITY

Peace and *security* refer to seeking peace, resolving conflict, disarming military regimes and groups, and establishing a society where violence is not the solution to conflict and where all people are safe from threats to their lives and livelihoods from war and violent conflict. Approximately fifty international and internal wars have taken place since World War II. About twenty were in progress in the year 2000. The "rules of war" attempt to reduce civilian (noncombatant) casualties, prevent gratuitous massacres and extermination of civilian populations, limit the types of weapons employed, prevent the ill-treatment or murder of hostages and prisoners of war, limit the gratuitous destruction of villages, towns, and cities, and prevent the rape of civilians by the military and the plunder of private homes and public buildings. While the spirit of the rules of war is clear (to avoid unnecessary suffering), the effectiveness of the rules has been questioned.

peace, nonviolent, disarmament, conflict resolution, security, human rights, social justice, values, absence of war, freedom from threats to life, injury, person, property, arms trading, land mines, peacemaking, nuclear disarmament, militarism, peacebuilding, Geneva Conventions, Hague Conventions, U.N. Security Council, peacekeeping force, belligerents, combatants, international humanitarian law, internal conflicts, international armed conflict, genocide, women, children, pillage, duty of commanders, terrorism, warfare

Arms Trading (218): Arms trading is the practice of manufacturing, buying, or selling weapons, armaments, military equipment, or the raw materials to make weapons. The five major exporters are: the United States, the Russian Federation, France, Germany, and the United Kingdom. Internal conflicts are fueled by the weapons trade. Over 500 million weapons are in circulation and 40 percent are traded illegally. Besides national embargoes, few international laws exist regulating small-arms trade. Arms trading includes both illegal and legal direct arms trade and arms brokering, where sales are made through an intermediary.

illegal trade, black market, weapons, legal arms trading, conflict promotion, access to weapons, arms industry, terrorism, peace, security, internal conflict, weapons embargo, war, transactional crime

Conflict Resolution (3,208): Conflict resolution is the process of resolving a dispute or a conflict by consulting each side and adequately addressing their interests so that they are satisfied with the outcome without resorting to violence. Negotiation, mediation, arbitration, and conciliation are common methods of conflict resolution.

nonviolent, negotiation, stakeholder, mutual agreement, mutual respect, peace, alternatives to violence, tolerance, peacemaker, peacekeeping, reconciliation, environmental conflict, warring faction, conflict prevention, armed conflict, international law, parties to conflict, ceasefire, collective security, diplomacy

Land and Naval Mines (102): Land mines are explosive devices that are buried in the ground and are triggered by people or vehicles putting pressure on the mine. Mines are used in military conflict and present significant and disproportionate danger to civilian populations during and after armed conflict, preventing the use of land for agricultural and other uses, deterring economic development, endangering people and communities, and damaging the environment.

weapons, land mines, arms trading, land mine removal, mine clearance, disarmament, illegal trade, black market, peace, international ban, injury, amputation, maim, civilian casualty, international law, Mine Ban Treaty, antipersonnel land mines, antitank land mines, naval mines

Militarism and Violence (931): Militarism is the ideology that maintaining and using military strength and aggression is an appropriate method of defending or promoting national, tribal, ethnic, religious, or other interests. A culture of militarism promotes and fosters the use of violence and power over another person as behavior that can be acceptable to solve conflicts or to advance conquest.

antiwar, pacifism, aggression, oppression, criminal justice, war crimes, enforcement, policy, assault, armed conflict, international law, genocide, torture, ethnic violence, attacks, siege, ethnic cleansing, displaced persons, genocide, perfidy, pillage, persecution, terrorism, crimes against humanity, mercenaries, weapons

Military Disarmament (1,619): Military disarmament is the act of reducing military forces, military equipment, weapons supplies, nuclear weapons capability, or the complete abandonment and abolition of the armed forces and use of weapons by states and by other military groups. Disarming states and other groups is a prerequisite for building a just, equitable, peaceful, secure, and sustainable world.

military disarmament, peace, demilitarization, security, militarism, nonviolence, limitation, military equipment, arms control, mutual disarmament, biological weapons, chemical weapons, nuclear weapons, international law

Nuclear Disarmament (154): There is no international law banning nuclear weapons. But nuclear weapons come under the restrictions imposed on all weapons of mass destruction. They should be banned but possessors of nuclear weapons, except the Russian Federation and the United Kingdom, refuse to sign any convention limiting nuclear weapons production. The Treaty on the Non-Proliferation of Nuclear Weapons has been suspended since 1995. The Comprehensive Nuclear Test Ban Treaty has not been adequately ratified. Inspection and enforcement as well as "threshold" nuclear powers are major barriers to nuclear disarmament.

nuclear weapons, weapons of mass destruction, nuclear disarmament, Treaty on the Non-Proliferation of Nuclear Weapons, Comprehensive Nuclear Test Ban Treaty, ratification of treaties, inspection teams, nuclear bomb tests, enriched uranium, peace, security

Peace and Peacemaking (7,916): Peace is sometimes defined as the absence of war. More often it is a state of society when differences are resolved without the use of armaments and when public order is respected by the parties in conflict. Peace does not mean that conflicts do not exist, only that they are resolved by pacific means—arbitration, mediators, courts, regional organizations, diplomacy, or the United Nations. In times of peace, both national and international humanitarian laws are in force. Peacemaking is the act of building and maintaining peace through consensus, mediation, and conciliation, with the consent of all parties in a conflict. Peacemaking includes pacifism and campaigns to prevent military action, and can involve diplomatic and military action, with the intervention of the United Nations, to further the aims of peace and security.

peace promotion, understanding, education, communication, mutual, cooperation, disarmament, nonviolent, reconciliation, antiwar, anticonflict, dialogue, peace building, pacifist, end conflict, peacekeeping, conscientious objection, international law, antiviolence, ceasefire, collective security, human rights, international humanitarian law, parties to the conflict, public order, peacekeeping, Security Council (U.N.), war, United Nations, conflict resolution, nonviolent communication

Protected Areas, Individuals, Objects, and Property (77): In an attempt to minimize unnecessary suffering and superfluous destruction, international humanitarian law has attempted to define who and what should *not* be the targets during armed conflict. Protected areas include undefended locales, hospitals, natural environments, demilitarized areas, declared "safety areas," and areas claiming neutrality. Protected objects and property include medical units, places of worship, cultural buildings, infrastructure essential to the survival of the civilian population, infrastructure that can cause excessive danger (e.g. nuclear power plants, dams). Protected persons should include the wounded, sick, medical and religious personnel, parliamentarians, relief organizations and civilians. Of course, in actual armed conflicts, these rules can be very difficult to apply.

fundamental guarantees, relief personnel, civilian population, protected persons, medical services, safe areas, peacekeeping, duty of commanders, protecting powers, Red Cross, journalists, protected areas and zones, wounded and sick persons, protected objects and property, war, conflict, protected zones, international humanitarian law

Weapons (153): Humanitarian law may prohibit the use, stockpiling, production, or selling of certain weapons that cause unnecessary suffering or superfluous injury and harm (e.g. biological and chemical weapons and to some extent land mines). Specific weapons are covered by numerous treaties and conventions but are more generally covered by the Geneva Conventions, which is considered customary law. Certain conventions try to prevent the use of weapons that will have long-term harmful impacts on the environment, connecting weaponry to sustainability.

edge weapons, firearms, chemical weapons, biological weapons, nuclear weapons, bacteriological weapons, incendiary weapons, weapons of mass destruction, WMDs, land mines, Convention on the Prohibition of the Development, Production, Stockpiling and Use of Chemical Weapons, Convention on the Prohibition of the Development, Production, Stockpiling and Use of Bacteriological (Biological) and Toxin Weapons, Convention on Conventional Weapons, Hague Conventions, Geneva Conventions, the Mine Ban Treaty, Treaty on the Non-Proliferation of Nuclear Weapons, biological warfare, terrorism

PLANTS

Plants are members of kingdom Plantae, which are multicellular, autotrophic organisms that create their own food through photosynthesis and lack the power of locomotion. The four other kingdoms are Animalia, Fungi, Protista, and Monera.

flora, management, protection, conservation, ecology, ecosystems, photosynthesis, biodiversity, sun, energy, primary producers, botany, trees, shrubs, organisms, kingdom Plantae, native, indigenous

Endangered Plant Species Protection (658): Endangered plant species are those plant species whose ability to survive and reproduce has been jeopardized by human activities, such that the size of the population is so low that the species is at risk of extinction. The World Conservation Union lists 34,000 plants as threatened with extinction. Although hard to calculate, about 400 plant species appear to have gone extinct from human activities. Plants become endangered predominantly through habitat loss and degradation and competition with exotic species. Climate change may accelerate these losses.

danger of extinction, conservation, biodiversity, flora, plants, vascular plants, climate change, endemic, plant biodiversity, invasive species, land conversion, Endangered Species Act, World Conservation Union Red List of Threatened Plants

Endemic Plant Species Protection (1,062): Endemic plants are exclusively and uniquely native to a place, biome, or region and not naturally found anywhere else. Regions or places with high endemism include the tropical moist forests of Hawaii, the Fynbos in South Africa, and lowland forests of Madagascar. Endemic plants are susceptible to clear-cutting, agriculture (both modern and swidden), and grazing by domesticated animals.

danger of extinction, conservation, biodiversity, flora, plants, vascular plants, climate change, endemic, plant biodiversity, invasive species, land conversion, Endangered Species Act, World Conservation Union Red List of Threatened Plants, native plant species, conservation, restoration, protection, invasive plants, botany, plant, horticulture, education, science, population, ecosystem, management, indigenous, naturally occurring, threats, climate, niche, range, adaptation, introduced species, seeds, preservation, genetic diversity, nonnative species, ethnobotany

Ethnobotany (69): Ethnobotany is the study of the relationship between people and plants, including how different cultures make use of indigenous plants for uses such as medicine, religious ceremonies, food, housing, and clothing, and how people view nature. Currently, sustainable economies are involved in bioprospecting for new medicinal plants, creating seed banks to maintain genetic diversity, and tackling the issues of genetic property rights.

aboriginal botany, indigenous plant use, traditional medicine, medicinal plants, spiritual plant use, native plants, plant knowledge, economic botany, intellectual property rights, cultural heritage, food plants, human diet, agriculture, cultivars, varieties, landraces, plant domestication, nutrition, cereals, bioprospecting, seed banks, ethnozoology, agricultural diversity, botanical gardens, benefits sharing, earth medicine, herbal medicines

Plant Ecology (593): About 270,000 plant species are known on earth. It is estimated that there are about 320,000 in all. Plants are the source of almost all food in terrestrial ecosystems, capturing sunlight in photosynthesis. Plant ecology is the scientific study of the interactions between plants and their environment, from the cycling of nutrients through plants, to photosynthesis, to the impact of air pollution on plant morphology and growth.

botany, vegetation, plant biomes, terrestrial ecosystems, flora, nutrient cycling, photosynthesis, plant community, population, adaptation, range, niche, herbivores, grazing, agriculture

POLLUTION

Pollution is the introduction by humans of substances or energy into the environment that are liable to cause hazards to human health, species, and ecosystems; to damage structures, ecosystem services, or amenities; or to interfere with economic uses of the environment. As a sound bite, pollution is something in the wrong place at the wrong time in the wrong quantity. Pollution can come in the form of energy (noise, light, heat, or radioactivity); or in the form of a substance (a vast number of chemical elements and compounds, smog, or soot). Pollutants are viewed in many ways: by their organic or inorganic chemistry; by their state as liquid, gas, or solid; by their biodegradability or persistence; by the environment where they can be found (marine, soil, freshwater, air); by their source (industry, agriculture, power plants); or by their targets (crops, wild species, buildings, infants, the elderly, etc.).

hazards, toxics, human health, ecosystem health, noise pollution, light pollution, heat pollution, thermal pollution, radon pollution, radioactive pollution, nucleotides pollution, dust, soot, smog, smaze, smoke, water pollution, air pollution, soil pollution, industrial pollution, agricultural pollution, household pollution, military pollution, microbial pollution, fungal pollution, pollution sources, pollution target, chemical pollution, point source, nonpoint source

Chemical Pollution (22): Most chemical elements and many compounds are absolutely necessary for life. But, at certain concentrations, they become toxic to living organisms. Toxicology attempts to determine safe concentrations, exposures, and risks to health of various chemicals. Discovering when a substance is hazardous can be difficult because toxicologists must extrapolate from animal studies, discover the life stage (particularly fetal, perinatal, or elderly) when a human has a "window" of sensitivity, take into account genetic variability, and determine if the effects are cumulative, take a long time to show, or work synergistically with other chemicals. Important chemical pollutants are endocrine disruptors, mutagens, teratogens (impacting embryos), and carcinogens. Suspect substances of particular concern have been persistent organic pollutants (POP) such as PCBs and chlorinated hydrocarbons (dieldrin) and heavy metals such as mercury.

pesticides, fungicides, herbicides, agricultural pollution, industrial pollution, pollutant concentration, exposure, pollutant persistence, toxic substances, toxicology, synthetic organic molecules, endocrine disruptors, mutagens, teratogens, carcinogen, persistent organic pollutants (POP), heavy metals, chlorinated hydrocarbons, dose-response relationship, median lethal dose, phase-outs, PCBs, dieldrin

Energy Pollution (11): Energy pollution comes from the discharge of energy during some human activity that harms or interferes with human health or ecosystems. Typical forms of energy pollution are: noise pollution from subsonic testing by the navy or too many decibels from heavy traffic or large machines, thermal discharges from power plants, radioactivity from building materials with concentrated radon or from nuclear power plants, light that interferes with astronomy or bird migration, and increased ultraviolet ray exposure from depletion of the ozonosphere. Climate change can be considered a form of energy pollution.

sky glow, illumination, safety, photopollution, ecosystem impacts, potentially physically harmful, distracting ambient sound, high intensity sound, hearing impairment, irritating noise, road traffic, aircraft, unwanted man-made sound, subsonic noise pollution, thermal discharge, heat pollution, thermal pollution, radioactivity, radon, bird migration, ultraviolet radiation, climate change, power plant pollution, industrial pollution, electromagnetic emissions, power line emissions, light pollution, noise pollution

Global Pollution (21): Specific pollutants do not respect international boundaries. Some nations try to use the international "commons" as a dump. The most obvious global pollutants are volatile organic pollutants, acid rain, radioactive chemicals, and greenhouse gases that move with air currents; ocean dumping and outfall pipe pollutants that may move in ocean currents and through the food chain; and freshwater pollutants in rivers that border or cross into multiple nations. The pollutants of the global commons require international negotiations and rules to protect each nation's citizens. The difficult issues include harmonizing standards for emissions and enforcement.

globalization, global commons, volatile organic pollutants, radioactive pollutants, greenhouse gases, marine pollution, ocean dumping, watershed management, International Atomic Energy Agency, International Commission on Radiological Protection, International Registry of Potentially Toxic Chemicals, international standards, international monitoring, international enforcement, UNEP International Monitoring Program, international agreements, pollution

Hazardous Solid Waste (24): Solid wastes become hazardous wastes when they pose an immediate or future threat to human health, are persistent and nonbiodegradable, can be amplified by the food chain, or cause other impacts on the landscape. Hazardous wastes can be toxic, corrosive, explosive, flammable, radioactive, or biologically harmful. Sustainable approaches to solid waste include: reduce the volume, separate out the harmful from harmless components, detoxify the harmful portion, and recover useful materials.

land disposal, incineration, resource recovery, recycling, product substitution, product change, dematerialization, military waste, landfill, ocean dumping, biological warfare, chemical warfare, biomagnification, pollutant, medical waste, toxic waste, industrial waste, sewage sludge, heavy metals, illegal trade, dumping, toxicity

Light and Noise Pollution (74): Light pollution is the wasted light from streetlights and other sources that is created by humans, which lights up the night sky unnecessarily and can disrupt ecosystems and obscure the stars at night. Noise pollution is any unwanted man-made noise that penetrates the environment and is irritating and potentially harmful to humans and animals, such as traffic noise or noise from heavy machinery.

sky glow, illumination, safety, light, loss of contrast, artificial light, direct light intrusion, upwards spill, photopollution, luminous pollution, ecosystem impacts, noise, potentially physically harmful, distracting ambient sound, high intensity sound, hearing impairment, irritating noise, road traffic, aircraft, unwanted man-made sound, persistence, recurrence, interference, energy pollution

Petroleum in the Environment (56): *Petroleum in the environment* refers to the polluting effects of petroleum extraction, refining, and transportation, and encompasses crude oil and any refined oil products such as kerosene or diesel. Examples of petroleum polluting the environment include oil spills that result from accidents in oil transportation, spills from loading and unloading oil, and discharging oil into the ocean when washing tanks in oil tankers.

pollution, prevention, contamination, health, safety, petroleum, crude oil, spill, gaseous, liquid, solid, animals, environmental impact, eradication, persistence, water supply, cleanup, remediation, bioremediation, fractions, gasoline, benzene, naptha, diesel, kerosene, aviation fuel, asphalt, tar, coke, bitumen

Pollution Prevention and Reduction (1,195): Pollution prevention and reduction can occur at any stage of the chain-of-custody from harvesting the materials, to processing them, using them, and disposing of the "wastes." Pollution prevention starts at the source by reducing the amount of material required or by substituting a nonpolluting ingredient. It occurs by using "end-of-pipe" solutions such as wastewater treatment or smokestack scrubbers before release into the environment. Pollution can be reduced by recycling the polluting substance or isolating it from the environment by designing a safe waste disposal site. It can also be reduced by lowering demand for the product (product substitution) or using social fixes such as cap-and-trade emissions permits and emissions standards. Who pays (polluter or citizen through taxes) has been a major barrier to cleaning up heavily polluted sites (e.g. Superfund in the United States, asbestos mines).

end-of-pipe solution, pollution controls, emission controls, cap-and-trade emissions management, demand reduction, emissions offsets, process modifications, wet scrubbers, electrostatic precipitators, bag filters, heat recovery, emissions standards, pollution control legislation, pretreatment of wastes, landfill management, sanitary landfill, wastewater treatment, water treatment, dechlorination, sludge, residual products, regulation, hazard insurers, Kalundborg, preemptive Clean Air Act, Clean Water Act, Environmental Pesticide Control Act, Federal Insecticide, Fungicide, and Rodenticide Act, Federal Water Pollution Control Act, Hazardous Transportation Act, National Motor Vehicle Emissions Standards, Solid Waste Disposal Act, Toxic Substances Control Act

Pollution Remediation (239): Pollution remediation is the practice of repairing and removing the damage to the land, air, and water that is caused by any pollution. Many remediation activities involve the removal of toxic contaminants produced by industrial pollution on brownfield sites. There are many methods of remediation, from removing contaminated soil, to capping and sealing the source of contamination, to using living organisms for bioremediation of the pollutants.

groundwater, bioremediation, in situ, ex situ, contaminants, pump and treat, mitigation, contamination, chemical binding, bio-slurping, air sparging, soil vapor extraction, total fluids extraction, air stripping, filtration, ion exchange, flocculation, solvent recovery, catalytic oxidation, industrial waste, soil pollution, chemical pollution

Toxic and Hazardous Substances (619): Hazardous substances are polluting materials and wastes that are harmful to public health and to the environment, and are toxic, corrosive, ignitable, explosive, or chemically reactive.

chemicals, pollution, hazardous, waste, materials, contamination, safety, health, cleanup, incineration, burning, reduction, biomagnification, pollutant, lead, mercury, dioxin, PCB, PVC, plastic, waste, toxicity, disposal, pollutants, solid waste, medical waste, spill, environmental impact, health impact, treatment, use, persistence, industrial pollution, toxic waste, chemical, illegal trade, dumping

Water Pollution (154): In the past, the most common phrase was "the solution to pollution is dilution" and engineers tried to calculate what volume of contaminated liquid they could dump into a river or estuary. Ecological studies have proven this phrase inadequate. Some pollutants persist and bioaccumulate, causing harm to health; some microorganisms thrive and multiply when released into an estuary; other polluted effluents change the long-term acidity, the oxygen concentrations, and the temperature of rivers. Multiple releases of pollutants interact and cause problems that overwhelm a watershed's ability to self-cleanse. The reverse ideal is now "zero discharge" of any pollutants. The point sources (like an outfall pipe of an industry) of pollutants are more easy to correct than nonpoint sources such as runoff from a feedlot or city street.

sewage outfall, river aeration, biological oxygen demand, coliform count, water quality standards, discharge standards, watershed management, point source pollution, nonpoint source pollution, runoff, acid rain, groundwater pollution, eutrophication, acid mine drainage

POPULATION
Population refers to the number of people in a specific geographic location or region. The understanding of population helps measure the environmental, economic, and social impacts on local, regional, and global scales.

migration, growth, consumption, rights, impact, refugees, death, birth, quality of life, food, requirements, carrying capacity, natural resource use, sustainable population, migrants, demography, family planning

Demographics (47): Demographics is the quantitative study of population size and structure by age groups and sex as well as geographic distribution and is general to all animals. Demographics is important to sustainable futures. Industrialized nations are top-heavy with citizens beyond working age. Nonindustrialized nations have many unskilled young citizens. Reconciling the flow of human migrants to meet labor and elder-care needs as well as to support families in developing nations is a major concern. Demographics are also crucial to wildlife management policies.

age structure, sex structure, birth rates, death rates, fecundity, mortality, age of first reproduction, survival pattern, post-reproductive age, human migration, women's issues, children and youth, elder care, migrants

Family Planning (676): Family planning is a service to help families with birth control and child health to help people have the number of children they desire.

family planning, contraceptives, abortion, child spacing, birth rate, in-union women, contraceptive pill, oral contraceptives, progesterone implant, spermicides, contraceptive diaphragm, RU 486, abortion pill, rhythm method, coitus interruptus, vasectomy, tubal ligation, birth control pill, IUD, condom, sterilization, reproductive health services, one-child families, reproductive rights, population

Global Migration (864): One out of every forty or fifty persons dwells in a nation that was not the nation of his/her birth. The best predictor of where global migrants go is not poverty or war but the dispersed networks of humans with similar language, origins and ethnic worldviews. Global movement takes four forms: professional transients (e.g.

business, tourism, students, visitors), illegal and legal job seekers (maybe 100 million transnationals each year), and refugees fleeing violence (14.5 million in 2000). The refugees must meet the criteria of the Geneva Conventions. Perhaps the 20 million or so internally displaced persons (IDP) should be added, although because they do not cross borders, they are not counted by the United Nations. They usually need as much aid as official refugees. Finally, there are uncounted "environmental" and "development project" "oustees" forced from their homes by natural disasters or dams and shantytown upgrades. The extreme form of refugee is called the "permanent stateless nonperson." They are neither refugees nor citizens.

migrants, refugees, Geneva Conventions, professional transients, internally displaced persons, IDPs, environmental refugees, development refugees, permanent stateless nonperson, illegal migrants, mojados, contract workers, mail order brides, human trafficking, dislocation, population

Human Population Growth and Impacts (522): *Human population growth and impacts* refers to those environmental, economic, and social impacts that result from the growth of population on local, regional, and global scales. Such impacts include the environmental impacts of increasing demand from a growing population for energy that exacerbates climate change, and of the degradation of natural resources from overfarming resulting in food scarcity. A simple model has been used to sketch population consequences to sustainability: $I = P \times A \times T$, where I is environmental impact, P is the number of people, A is the affluence per person (a measure of consumption), and T is a measure of the effects of technologies on the extraction and use of resources.

overpopulation, scarcity, resource depletion, environmental impacts, carrying capacity, food supply, energy use, natural resource consumption, demographic alarmism, birth rate, death rate, rate of natural increase, population doubling time, infant mortality rate, total fertility rate, population under age 15, dependent population, life expectancy, percent urban population, age structure, sex structure, age at first reproduction, post-reproduction age, age classes, Ehrlich-Holdren Model, nonrenewable resources, marriage age, educational opportunities and fertility, family planning, U.N. International Conference on Population and Development, pronatalist,

Refugees, Internally Displaced Persons, and Migrants (1,882): Refugees and IDPs are humans who do not choose to migrate for jobs or family. They have been forced from their homes by civil war, governance collapse, border conflicts, famine and other natural disasters, restructuring of the economy, as development "oustees," and by persecution. Many cannot return (repatriate) because of land mines. The rights of nonnationals differs widely from nation to nation. They range from those seeking "asylum rights" because their life would be in jeopardy if repatriated to illegal job seekers seeking medical care to legal migrants trying to establish citizenship.

immigrant, migrant, human rights, legal rights, access to legal services, naturalization, working conditions, legal status, undocumented immigrants, living conditions, racism, civil rights, refugee status, trafficking, immigration law, asylum, discrimination, identity rights, internally displaced persons, IDPs, repatriate, nonnationals, asylum, population movements, inter-state conflict, civil war, security, genocide, atrocities, social security)

POVERTY ERADICATION
Poverty is a broad term that describes many circumstances where people lack resources such as money, housing, food, clothing, jobs, and suffer physically, socially, and

emotionally from this material deprivation. *Absolute poverty* describes a standard of living where a person is unable to afford even a basic diet. International definitions of poverty rely on the classification of the poverty level, which is a level of income below which a person cannot afford to buy all the resources required to live. Poverty eradication encompasses the will and strategies to stop all people from living in poverty.

poverty alleviation, sustainable livelihoods, absolute poverty, material possessions, destitute, hunger, homelessness, being without, hardship, deprivation of essential goods and services, Millennium Development Goals, extreme poverty, income poverty, human poverty

Affordable Housing (2,849): *Affordable housing* refers to the provision of housing and shelter as a basic human right, including housing for the homeless and for those of low income. Affordable housing occurs only in nations that can support such projects or in locales that have substantial foreign aid. Affordable housing is a range of both subsidized and nonsubsidized housing designed for those whose low or moderate incomes make them unable to purchase or rent housing on the open market. The provision of affordable housing—where the homeowner or renter spends no more than 30 percent of their gross income on housing including utilities—reduces the financial burden of housing costs and provides opportunities for people to move out of poverty.

affordable housing, low-income, economic, community, development, housing conditions, environmental health, available housing, homelessness, social justice, access to housing, standard of living, equity, policy, access to financing, shelter, human right to housing, slums, favelas, barrios, housing rights, compensation for evacuation, landlord, right to adequate housing

Crises and Disaster Aid (883): The poor of the world suffer most in times of crisis. The disasters include tsunamis, hurricanes, tornadoes, earthquakes, floods, dam breaks, famine, drought, plagues, and transmitted diseases. Part of sustainability includes helping people to recover from these crises through humanitarian aid, volunteer work, food-for-work programs, donations, and microinsurance.

tsunamis, hurricanes, tornadoes, earthquakes, floods, dam breaks, famine, drought, plagues, transmitted diseases, humanitarian aid, donor aid, medical aid, food-for-work programs, microinsurance, crisis aid, humanitarian relief

Poverty Alleviation (9,240): *Poverty alleviation* (or *reduction*) describes strategies to eradicate poverty. It is any process that seeks to reduce the level of poverty in a community or among a group of people or countries. Poverty alleviation programs may be aimed at economic or noneconomic poverty. Some of the popular methods used are education, economic development, and income redistribution. Poverty alleviation efforts may also be aimed at removing social and legal barriers to income growth among the poor. Economists such as Hernando de Soto see improvement in property rights as being instrumental in poverty reduction. Other economists also highlight government corruption as a chief problem in reducing poverty in the developing world.

unemployment, poverty eradication, hunger, food, homelessness, income, education, job, training, health, credit, relief, housing, disenfranchised, marginalized, advocacy, rights, low-income, quality of life, disease, standard of living, public health, social justice, human rights, equality, material deprivation, poverty line, barriers to education, barriers to health, property rights, North-South inequity, social exclusion, environmental health, poverty reduction

Squatter Communities (46): In many developing nations, sustainable building is a long-range goal. In squatter towns, with no property rights, homes are first built with any materials available (e.g. scrap wood, mud, pieces of tin, cloth) and then upgraded as the homeowner finds better materials or earns enough to purchase building materials. The goal may be cinder block with rebar and cement. At first, open sewers, then neighborhood latrines, then inside-the-house toilets. At first, water is hauled; then pirated water from the city mains for a neighborhood of homes; then internal water. Over a billion humans live in squatter homes (one out of every six people). These homes are considered sustainable by the squatters as they keep them sheltered in the city (in nations where there is little government or donor help) until they can improve upon their dwellings.

squatter town, favelas (Brazil), colonias *(Mexico)*, kijiji *(Kenya)*, johpadpatti *(India)*, gecekondu *(Turkey)*, aashiwa'i *(Egypt)*, barriadas *(Peru)*, kampongs *(Indonesia)*, mudukku *(Sri Lanka)*, penghu *(Shanghai)*, shantytown, slum, common property, evictions, sanitation, pirated water, rebar, open sewer, sewer lines, self-building, cinder block, brick, squatter homes, corrugated plastic, tin, bamboo, tile, caulk, neighborhood latrine, communal water source*

Sustainable Livelihoods (2,754): A sustainable livelihood is one that encompasses meaningful work that fulfils the social, economic, cultural, and spiritual needs of a person, and safeguards cultural and biological diversity. Livelihoods are environmentally sustainable when they maintain or enhance the local and global natural resources on which livelihoods depend. Livelihoods are socially sustainable when they can cope with and recover from stress and shocks, and can provide for future generations. Creating sustainable livelihoods is a development strategy for poverty eradication that focuses on more than income generation alone.

sustainable living, meaningful living, development strategy, poverty eradication focus, quality of life, human capital, natural capital, financial capital, social capital, livelihood building blocks, protection from shocks, sustaining rural livelihoods, access to credit, choice in education, health, access to assets, basic living requirements, local community living systems, food, water, clothing, self-reliance

RELIGION, ECOLOGY, AND SUSTAINABILITY
Religion, ecology, and sustainability refers to the spiritual and metaphysical aspects of valuing the earth and working toward sustainability.

human-earth relations, interrelatedness, interconnectedness, understanding, inner peace, faith, spiritual, beliefs, ethics, religion, sacred, divine, intuitive, philosophy, outside science, metaphysical, transcendental, earth-honoring, values, interfaith, faith-based, environmental ethics, ecopsychology, green organized religion

Ecopsychology (138): Ecopsychology is the study of the mind and the synergistic relationship between environmental and personal well-being. Ecopsychology seeks to bring ecological thinking to psychotherapy and to fostering lifestyles that are both ecologically and psychologically healthy.

environmental philosophy, psychology, ecology, synthesis, interdisciplinary, study, human-nature relationships, connection, human mind, human relationship to the natural world, environmental sanity, counseling, holistic, psychotherapy, spiritual, metaphysical, ethics, earth values, ecotherapy

Environmental Ethics (842): Environmental ethics explores the moral values and ethical relationship between human beings and the environment in which they live. In sustainability projects, players ask: What do we want (if no barriers existed to our desires)? What do we know? What will we accept (given the situation)? Asking what we want and what we will accept always incorporates environmental ethics as it may be difficult to clearly and practically perceive one's desires and the future consequences of actions. The line between ethical and spiritual or religious judgment is not clear.

moral values, ethics questions, ethics of resource consumption, extinction, future generations, stewardship, human rights, deep ecology, rights of biotic entities, conservation ethic, ecological interdependence, environmental utility, anthropocentrism, bioethics, environmental ethic, discrimination, utility value, existence value, option value, nonconsumptive use value, production value, ideals, praxis, practices, action, nonviolence, conscience, conscientious objector, ecoethics, earth values, earth respect, green values, earth appreciation, earth defense, long-term survival

Religion and Ecology (16): The environmental crisis represents a serious challenge to contemporary religion. In order to save nature, religions are re-evaluating their relationships to nature both in practice and scripture. In the past twenty-five years, especially after the 1986 Assisi meeting of religious leaders, many churches, synagogues, temples, and informal congregations of organized religions (e.g. Christianity, Judaism, Hinduism, Islam, Buddhism) have returned to their scriptures and rituals to better understand how faith and religious practices work together with caring for the earth. Many have encouraged projects from environmental cleanups to working in crisis areas as part of their religious obligations.

Christianity, Judaism, Hinduism, Islam, Buddhism, Shintoism, Confucianism, Bahai, Daoism, Jainism, Earth care, environment and religion, faith, interfaith tolerance

Sustainability, Religion, and Spiritual Issues (920): Certain sustainability issues rest on deep ethical and spiritual commitments, in senses of faith, salvation, karma, truth, and beauty. Spiritual commitment contemplates the beauty of landscapes, the sacredness of specific places (archeological, historical, and architectural), and the soulfulness of nonhuman species as worthy of care and protection. Spiritual and metaphysical understanding of what animates life (the soul) may define an individual's or group's attitude toward family planning, birth control, abortion, and other aspects of population management. Spiritual and religious attitudes toward the "good life" or "happy life" will influence their giving to sustainability NGOs, volunteer time, attitudes toward governance and political officials, call to service, "necessary" material gain and accumulation, desire for upward mobility, etc. Spiritual realities—the soul, the sacred, the religious, and the divine—are considered realities that are outside those of reason and science.

inspiration, inner peace, faith, spiritual, beliefs, ethics, religion, sacred, divine, intuitive, philosophy, outside science, metaphysical, transcendental, earth-honoring, material satisfaction, materialism, voluntarism, volunteer time, tithe, gift, religious service, worship, soul, sin, beauty, satyagraha, compassion, bodichitta, long-term survival, interfaith tolerance

Sustainable Living (5,627): Sustainable living focuses on individual and community responsibilities for sustainability and focuses on choices, values, ethics, and the way in which human beings interact with the natural world. Sustainable living is a lifestyle choice that considers a person's relationship within the community and the natural environment and seeks harmony with both.

voluntary simplicity, conservation, self-sufficiency, quality of life, simple living, lifestyle, choices, community responsibility, individual responsibility, natural resource use, footprint, values, consumption changes, social and ecological harmony, earth stewardship, nature's limits, duty of care, conservation-based development, nature's carrying capacity, environmental sustainability, earth values, ecoliteracy, conservation economy, local economy, local self-reliance, economic relocalization, environmental literacy

SENIORS

Seniors are those members of the population who are older adults and may be retired from working life. The seniors area of focus refers to those rights and issues that concern older people, such as seniors' rights, health, and the valuing of their experience and knowledge in the community.

senior, mentors, retired, mentoring, participation, health, education, old age, pensioners, skills, elders, wisdom, knowledge, experience, retirement, rights

Seniors' Health (207): Seniors' health refers to the health issues of the older members of the population, from access to health care, to health education, to quality of life as people grow older.

health care, access to health, equality, services, basic operations, quality of life, dignity, healthy lifestyle, activity, education, information, community, seniors

Seniors' Rights and Participation (339): Seniors' rights are those human rights that are denied to older people on the basis of their age, such as equal opportunities to rent housing and to employment. Seniors' civic participation addresses the ability of seniors to actively participate in community, local, and national institutions, as well as removing the social, cultural, and economic barriers to civic participation experienced by seniors on the basis of their age. The effective civic participation of seniors strengthens institutions and empowers seniors to defend their rights.

human rights, respect, dignity, advocacy, lobbying, community participation, active participation, social justice, discrimination, ageism, knowledge, civil disobedience, direct action, retirement

Senior Volunteerism and Mentoring (89): Senior volunteerism is the practice of seniors' giving their time to impart their knowledge and offer their skills, talents, and efforts to contribute to the community through mentoring and in other ways.

activity, giving back, involvement, participation, volunteering, mentoring, experience, sharing, skills, activity, service, community work, mentor, sharing, knowledge, guiding, active teaching, learn by example, professional development, learning, positive influence, ethics, guidance

SUSTAINABLE CITIES AND DESIGN

Sustainable cities are those cities, towns, and urban areas that use their resources to meet current needs while ensuring that adequate resources are available for future generations. Sustainable cities seek community development that enhances the local environment and quality of life as well as developing a local economy that supports both thriving human and ecological systems. Sustainable cities are characterized by improved public health and a better quality of life for all the residents by limiting waste, preventing pollution, maximizing conservation, promoting efficiency, and developing healthy regional economic development and vibrant communities. Integrated planning and design are key elements of developing sustainable cities and communities.

sustainable development, efficient resource use, waste cycling, solid waste, energy use, transportation, services, sprawl, integrated planning, livable cities, neighborhoods, walking, ecological system health, economic health, sustainable livelihoods, sense of place, communities, towns, cities, urban areas, vision, locally based economy, whole community participation, economic self-reliance, green roofs, ecosystem protection, broad-based citizen participation, urban ecology, urbanization, participatory budgeting, PB, social exclusion, community cohesion, neighborhood renewal, bioregional economies, green cities, green city movement, design, green architecture, holistic design, green design, sustainable design, ecocity

Ecovillages (150): Ecovillages are intentional communities, ideally of 50 to 150 people, that live sustainably and are dedicated to creating and demonstrating ecological, social, economic, and spiritual sustainability. The ecovillage movement is a global movement for small community living where consensus-based decision-making is possible and where the communities minimize their environmental impact and are as self-reliant as possible.

community, environmental, sustainable living, socially, economically, environmentally sustainable social network, community size, incubator, lifestyle, cooperation, collective independence, low-impact living, supportive social environment, 50–150 people, resource recycling, green building, self-sufficient, consensus decision-making, renewable energy, sustainable cities, ecoplanning, cooperatives, ecoliteracy, common vision, economic relocalization, environmental literacy, ethical lifestyle, sustainable communities, local self-reliance

Green Roofs (117): Green roofs are created when the rooftops of buildings are designed or reengineered to be gardens planted with vegetation. Green roofs improve a building's thermal management by providing insulation in the colder months and by keeping the building cool in warmer months, with vegetation that produces oxygen, absorbs carbon dioxide, filters air pollution, provides wildlife habitat, and slows stormwater runoff by absorbing much of the rain that falls on the roof.

water recycling, insulation, habitat, rainwater catchment, increased cooling, shade, stormwater runoff reduction, species, sustainable cities, urban, rural, soil, sod, plant species, water efficiency

Infrastructure (940): All modern communities must move people along streets, bridges, and sometimes through tunnels and underground subways; many move freight by rail, ships, and airfields from central transport centers such as harbors and airports. Modern communities must be powered by electricity and usually natural gas and steam; they must be kept clean by water and wastewater treatment facilities as well as garbage (solid waste) disposal and reuse. And, all modern communities have mechanized modes of communication such as telephone by landlines, cell phones by airwaves, as well as postal services. Each of these services requires many components configured and connected into a network. The sum total of all these working parts and the system itself is called infrastructure. Infrastructure is considered by many to be the most important driving force in community economic development. Infrastructures shape housing and commercial density, green spaces, health, and sustainability of communities. In developing nations, roads, electricity, and water and wastewater infrastructures are the highest priorities.

transport, streets, bridges, tunnels, subways, freight, rail, ships, air cargo, highways, power, power plants, transmission lines, electricity, water treatment, wastewater treatment, sewerage, watershed management, stormwater, urban runoff, garbage, solid waste, landfill, garbage dumps, telephone, cell phone, telecommunications, mail, community development, sustainable cities, stormwater

Sustainable Building (424): Sustainable green building is the practice of constructing buildings using environmentally responsibly produced materials and techniques that minimize the environmental impact of the building throughout its lifetime. For example, a green building can be built using energy-efficient and nontoxic materials and can use renewable power generation to provide electricity for the building.

environmental impact, sustainable building materials, house, energy efficient, nontoxic, reuse, recycle, green roofs, ecoroof, stormwater recycling, buildings, cities, efficiency, harvest energy, harvest water, renewable energy, PV, natural daylighting, improved indoor air quality, productive indoor environments, LEED certification, light, air, heating, water, waste, life-cycle assessment, construction, operation, end of life, deconstruction, building reuse, straw building, straw bale, passive solar, resource efficient housing, sustainable design, natural ventilation, urban environment, low-energy cooling, bioclimatic housing, passive cooling, soil as a malleable material, water efficiency

Sustainable Transportation (884): Sustainable transport planners concern themselves with moving people and moving goods. They attempt to integrate both so that humans can have efficient access to jobs, stores, and community activities, while freight and products efficiently move in and out of the city. Sustainable transportation includes land use zoning such as bike paths or truck-free neighborhoods, infrastructure development such as subway systems, and socioeconomic incentives such as carpool lanes. In metroplex areas, sustainable transportation is complex and highly political because of the overlay of local, state, and regional transport and commerce authorities and bureaucracies. The general goals of sustainable transportation are: reduce pollution, increase efficiency of people and cargo movement, maintain safety, reward low-energy vehicles and transport modalities, discourage development that separates workers from their place of work, and harmonize vehicle fees, taxes, and tolls to favor these goals.

pollution, urban, planning, bicycling, commuting, cycling advocacy, road safety, non-polluting buses, biodiesel vehicles, pollution reduction, integrated transportation planning, walking, access by proximity, car-free, pedestrian areas, trams, trolley cars, monorail, light rail, affordable transportation, cities, bicycle friendly, bike, transportation energy security, mass transit, bicycle freeway, green cities, public transit, freight, cargo, car pool, bicycle paths, land use zoning, parking fees, parking meters, alternate side-of-the-street parking, green taxes, tolls, hybrid vehicles, bus lanes, subways, transportation authorities, harbors, ports, railyards, air cargo facilities, traffic-calming, traffic cameras, street surface textures, pedestrian traffic, noise abatement, greenways, speed humps, raised crosswalks, pedestrian refuges, bus bulbs, neckdowns, pedestrian-phased red lights, mobility

Sustainable Urban Environmental Services (67): Cities must provide certain services in order to sustain citizen well-being and especially health. These include the provision of clean drinking water, removal and treatment of wastewater and solid wastes, and protection of lives and property from flood damage. Sustainable environmental services consider the natural environment that helps provide urbanites with these benefits—the watershed of the water source which may be hundreds of miles from the city, the rivers and oceans where wastewater is released, the preservation of floodplains to manage hundred-year floods. During the design process, planners integrate ecosystem services with human infrastructures so as not to significantly harm the environment and to balance the multiple benefits of the environment (flood management, biodiversity, modifying the heat island effect, a sink for greenhouse gases produced in the city, water purification by soils and streams, etc.).

environmental infrastructure, reservoirs, aqueducts, water delivery system, clean water, water treatment plant, fire protection, water storage, water leaks, water conservation, waterborne disease, wastewater treatment, sewerage, sludge, cogeneration, heavy metals, water pollution, ocean dumping, effluent discharge, combined sewers, flood waters, storm drains, on-site stormwater management, garbage, solid waste, recyclables, sanitation equipment, garbage export, landfills, incineration, street cleaning, snow removal, salt pollution, heat island, urban landscaping, stormwater, wastewater recycling

Sustainable Urban Power (11): Cities require electricity to power their lights, computers, telephones, elevators, mass transit, heating, and cooling. Since urban electricity cannot be stored (in batteries), the amount arriving and amount used must always be equalized or "power quality" mishaps occur. In addition, cities must have a reserve to generate for peak demands such as early morning or extremely hot days. Almost all cities are interconnected through regional power grids, so "sustainable management" includes regional coordination to prevent brownouts and blackouts. Cities have an increasing and enormous role to play in sustainable energy policies because they can require a certain percentage of renewable energy in electricity contracts, require energy-efficient buildings, set rates that encourage conservation, purchase power from nearby sources (reducing transmission losses), install metering devices, and allow decentralized production by encouraging cogeneration facilities.

electricity, power infrastructure, electricity consumption, electricity conservation, power plants, cogeneration, decentralized power generation, distributed power, power grid, peak demand, power quality, brownouts, blackouts, alternative energy sources, green energy, power distribution, wholesale power purchasers, high-voltage transmission lines

Sustainable Urban and Regional Planning (1,110): Sustainable urban and regional planners attempt to find ways of organizing the structure and function of cities, including land use, buildings, and infrastructure (e.g. water, wastewater, flood management, transport, etc.), in order to bring them into greater harmony with their natural or original surroundings. Sustainable planning attempts to reduce the ecological impacts of both the urban footprint, such as sprawl, and the source areas from which resources are imported to the city. Sprawl is the spread of suburban development, usually into rural and subrural landscapes. It may also mean the densification and spread of industrial and urban footprints. In either case, the sprawl is minimally constrained by land use zoning and is characterized by significant land consumption, centralized infrastructure for water and wastewater, automobile dependence, disregard for ecosystem services such as wetlands as components of floodplain management, and minimal concern for amenities such as parks, greenways, and wildlife. Sustainable planners include both social and economic impacts in order to plan an energy- and materials-efficient economy with minimal pollution and equitable distribution of benefits.

land use, efficiency, infrastructure, urban sprawl, open space, preservation, materials, proximity to services, walking distance, integrated planning, holistic, sense of place, access to proximity, car-free areas, land use mix, brownfield development, greenbelt, open space, quality of life, designing for people, environmental planning, smart growth, neighborhood renewal, ecoplanning, car-free streets, greenbelt expansion, urban vitality, mass transit, sustainable design, greenways, urban parks, brownfields, urban renewal, green cities, urban footprint, urban resource imports, new urbanism, commuting, smart growth, building development, greenbelt development, environmental impact, unplanned development, land consumption, fragmentation, transportation, sustainable cities, urban, rural encroachment, ecosystem services, land-use zoning, housing density, infrastructure, road density, urban footprint, suburban growth, housing mix, farmland, commuting, bicycle friendly

Urban Communications (29): Communications infrastructures have complex entanglements with sustainable cities. They allow families to remain in touch even when kids and parents are in separate places; facilitate banking, financial, and health services; keep markets running more efficiently; allow global-local information networks; provide community information and entertainment and much more. Urban telecommunications requires hidden (e.g. switching stations, underground cables, satellites) as well as visible infrastructure (post offices). The capacity, quality, and location of each communication system (e.g. telephone, mail, airwaves) balances private for-profit companies with government pressure to provide equal and fair access to all citizens. Major concerns for sustainability include access to emergency services, traffic information, local share of TV and radio broadcast time, equitable access to the World Wide Web, special phone and Internet rates for the elderly, incapacitated, and the poor, and identity theft. Sustainable cities try to limit telecommunications by encouraging face-to-face meeting places and conversation.

communication, telecommunication, telephone, cell phone, postal service, airwaves, television, radio, cable communications, satellite communications, World Wide Web, Internet, equitable monthly service rates, face-to-face conversation, communications infrastructure

Urban Ecology (220): Urban ecology is an applied use of science to better understand urban and urbanizing watersheds (including their weather patterns, solar inputs, geology, soils, plants, and animals) with the goal of bringing urban lifestyles, land use practices, and infrastructures into greater harmony with the environment.

urban areas, urban habitats, urban environments, enhancement, sustainable environmental design, greening the built environment, ecology, restoring natural systems, urban water cycling, nature in cities, healthy cities, sustainability, cities, urban watersheds, infrastructure, stormwater, greenways, urban parks, street trees, urban landscape, floodplain preservation, wetland preservation, nonpoint pollution, street runoff, heat island, impervious surfaces

Urban Revitalization (273): In 2006, for the first time in human history, more people lived in urban than rural areas, placing increasing pressure on urban areas and urban infrastucture. Urban revitalization is the process of rebuilding thriving economically, environmentally, and socially sustainable urban areas and populations in areas that have been in decline and in those urban areas that are stressed from the continuing influx of people to urban areas.

urbanization, economic revitalization, urban population growth, rural to urban migration, land development, sprawl, land use change, impervious surfaces, heat island, infrastructure, quality of life, economic development, job availability, living conditions, working conditions, housing, sustainable cities, slums, shanty towns, neighborhood renewal, greenbelt encroachment, megacities, favelas, colonias, informal economy, urban vitality, community gardens, urban farms, livable cities

Waste Management (1,440): *Waste management* refers to the methods of collecting, transporting, processing, treating, storing, and disposing of municipal, industrial, and commercial solid, liquid, and gaseous waste. The focus of waste management is increasingly to reduce the impact of waste on the environment as well as on human health, from reducing the quantity of waste produced to recycling waste materials to using landfill gas as a fuel.

solid waste, liquid waste, management, disposal, incineration, recycling, reuse, reduction, remediation, pollution, toxic waste, hazardous waste, solid municipal waste, power plant waste, fly ash, landfill, waste-to-energy, construction debris, commercial refuse, sludge, semisolid waste, reduction, sustainable waste management, cities, urban, rural, waste cycling, chemical, green cities, waste recovery, waste disposal, biodegradable, material goods

SUSTAINABLE DEVELOPMENT
Sustainable development encompasses economic and social development. It takes full account of the environmental and social consequences of economic activity and is based on the use of resources that can be replaced or renewed, meeting the needs and improving the quality of life of current generations without compromising the ability of future generations to meet their own environmental, social, and economic needs. Sustainable development practices do not allow for "externalities" that exist in "economic development." Sustainable development broadens the accounting system to include green accounting, equity, and intergenerational issues. Its goal is not maximum economic growth but more balanced development of environmental, social, political, and economic resources.

environmental sustainability, social, economic activity, healthy environment, triple bottom line, social conditions, economic sustainability, sustainable economies, job creation, community capacity building, internalizing externalities, equality, justice, improved standard of living, poverty reduction, income, sustainable wealth creation, improving prosperity, economic and social well-being, renewable resource use, social capital, natural capital, human capital, economic capital, integrated value creation, ecocentric, ecological economics, low carbon development, progress as if survival matters, long-term survival

Biological Development (21): *Biological development* describes the differentiation of cells and tissues into organs to form a plant, fungus, or animal. Biological development can be roughly contrasted with biological growth, in which cells of the same type reproduce themselves to increase the size of a tissue. Biological development is important because humans can now modify the genetics of development to create new cells and tissues. There is also argument about when the development of embryonic tissues of a fetus should be called "human." There are many health concerns about stopping the development and growth of cancers and the impacts of toxic chemicals on embryos and sperm development.

cells, tissues, genes, genetically modified organisms (GMO), biotechnology, abortion, family planning, cancer, endocrine disruptors, teratogens, mutagens, toxics, pregnancy, industrial pollutants, sterile seeds, "Terminator" technology

Economic Development (5,678): *Economic development* refers to the sustained increase in the economic standard of living of a country's population, improving the quality of human life through increasing per capita income, reducing poverty, and enhancing individual economic opportunities by developing technology, making more productive and efficient use of physical capital, and increasing human capital. The concept, as written above, is controversial because of past economic imperialisms, the spectacular failure of many multilateral development projects, the harmful social and environmental impacts of various development projects, the demand for structural adjustments, the increasing gulf between rich and poor in many developing nations, and its confusing understanding of the difference between economic development and economic growth. Sustainable development, in part, arose as an answer to Western-generated development theories, especially those of the World Bank and International Monetary Fund. In general, critics of Western concepts of economic development have pointed out that: rapid GNP growth may not be the best indicator of well-being; economic well-being is not the source of all other ingredients that make up a good life (peace, family stability); state and multilateral manipulation of the economy does not guarantee equitable distribution of wealth; and technology (especially technology for greater resource extractions) may harm future generations more than it helps them. They have proposed alternative development, which usually includes a higher prioritization of self-empowering the poorest, and custom-designing any economic aid and financial restructuring to local traditions and desires. It prohibits megaprojects characteristic of the earlier economic development investments.

economy, improved economic standard of living, job creation, poverty reduction, wealth creation, improving prosperity, economic well-being, developing human capital, technological development, sustainable development

Rural Development (3,842): *Rural development* refers to the economic and social development of rural areas and communities to improve the quality of life of rural populations. Local, state, international, and/or private funds may work to help generate off-farm employment, new sources of economic growth, foster value-added enterprises and new commodities, encourage more sustainable use of resources, and subsidize the delivery of services, such as transport and health care. Resistance movements may go further and try to reorganize the urban/rural market system to provide greater profits to producers, reduce middlemen, or offer alternative high-paying crops (e.g. cocoa, opium, marijuana). Maintaining a viable rural population has been difficult in all nations because of price fluctuations paid for rural commodities, lack of adequate price supports and disaster relief or insurance, lower wages for service workers (e.g. doctors, teachers), rural social mores that demean women, fewer educational opportunities for children, and the feeling of isolation brought about by the media's glorification of urban life.

rural mobility, natural resource protection, sustainable resource use, land management, economic development, family farmers, sustainable rural economy, rural job creation, poverty eradication, services to rural communities, appropriate technology, farmers' rights

Social Development (10,283): *Social development* refers to the improvement in qualities of life and human well-being by organizing human governance and affairs to accomplish such tasks as the eradication of poverty, the reduction of income disparities, the elimination of violence, the guaranteed right to clean water and health services, the increased respect for nonhuman creatures and their ecosystems, and the structuring of a just legal system and system of representation.

quality of life, well-being, standard of living, poverty eradication, equality, eradication of violence, social justice, education, social entrepreneurs, social welfare, social responsibility, reciprocity, respect for cultural differences, traditions, social benefit, fundamental guarantees, ethics, representation, universal health care, environmental ethics, human rights, social and moral development, social accountability, social security, social enterprise

TECHNOLOGY
Technology can broadly be defined as the development and application of tools, machines, materials, and processes that help to solve human problems.

nanotechnology, environmental technology, green technology, appropriate technology, engineering, science, high technology, technology transfer, technological solutions, information technology, biomimicry, biotechnology, information and communication technology, digital divide, ICT

Appropriate Technology (539): Appropriate technology (AT) is an applied engineering science suitable to the level of economic development of a particular group of people. Ideally, AT is decentralized, can be used and operated by most of the concerned citizens (i.e. does not require outside operators), uses local or regional fuels and materials in an efficient manner, and involves machines that can be locally repaired. It is sometimes called "alternative technology" and sometimes used for the "best choice" of a technology no matter how complex ("green technology").

green technology, energy efficient, environmental health, efficiency, recycling, energy production, materials, biomimicry, clean technology, environmental impact, efficient resource use, manufacturing methods, sustainable production, bio-based materials, technology transfer, alternative technology, green engineering, green chemistry, renewable resources, green building, culturally appropriate technology, biofuels

Biomimicry (29): Biomimicry is a technical discipline that takes inspiration from natural organism shapes and forms to develop new engineering solutions and industrial designs that imitate nature's designs.

mimicking nature, science, model, design, ecological principle, product, process, feedback system, sustainable, ecological standard, biomimetics, design by nature, biological design, environmental sustainability, industrial design, economics, business, technology

Biotechnology (189): Biotechnology researches and manufactures products based on the modification of DNA, proteins, cells, and tissues. Some of the products have been compatible with sustainability goals, such as the production of insulin by genetically modified bacteria, which has saved thousands of laboratory rabbits. Others, such as the use of genetically modified soil bacteria to speed the cleaning of toxics, have raised concerns about "escaped" microbes. Most concern has been raised over genetically modified crops and forest trees. Enzymes to save energy, microbes to replace toxic solvents, and artificial skin all bring promise and fear to the sustainability movement.

safety, agriculture, food, genetic engineering, biological processes for product manufacture, biology in industrial processes, pest resistance, plant modification, animal modification, agricultural products, food science, medicine, food production, transgenic, altered genetic makeup, gene manipulation, cloning, mutagenesis, resistance, DNA, foreign genetic material, bioelectronics, tissue engineering, biomaterials, enzymatic engineering, environmental remediation, agricultural biotechnology, PCR, GMOs, "Terminator" technology, sterile seeds, traitor technology, synthetic biology, human genomics

Information and Communication Technology (640): *Information and communication technology (ICT)* refers to the ability to have the skills, and physical and financial means, to access information particularly using computer-based communications technology, such as the Internet. The ability to access and use information and communication technology is seen as a way of overcoming the "digital divide"—the gap in access to technology that exacerbates economic, social, and developmental inequalities.

ICT, information and communication technology, computer skills, gathering, storing, retrieving, processing, analyzing, transmitting, information, information technology, data networking technologies, telecommunications, access to information, network, digital divide, lack of access, capability to use information technology, access equality, equity, computer-based information systems, Internet, World Wide Web, skills to use ICT, community development, community resources, training, lifelong learning, vocational training, economic justice, hardware, software, computer infrastructure, equipment, intellectual property, digital commons, access to technology and information

Sustainability and Technology (133): Technological change, the development of new production methods and new products, is one fundamental driving force for enhancing economic growth. At the moment, research and development has been dominated by transnational corporations (TNC), which rarely contract these activities to "third world" countries. In addition, TNCs may or may not transfer technologies to developing nations or produce new technologies addressing the needs of nonindustrialized peoples. In contrast, NGOs and socially responsible entrepreneurs have tried to find ways to transfer and develop appropriate technologies (e.g. pumps, irrigation pipes, stoves), that the poorer peoples of the world can purchase and maintain, that reduce work burdens and increase incomes (e.g. cell phones to determine market prices). Education on technology use has focused on extension agents or worker-to-worker teaching.

manufacturing methods, technology transfer, alternative technology, appropriate technology, NGOs, education

Technology Transfer (16): Technology transfer can be crucial to the improved livelihood of citizens. Technology transfer is part of sustainability when: 1. a local firm is given the production contract and technology and provides paying jobs in a "third world" nation; 2. a local firm forms a strategic agreement (usually a licensing agreement) to acquire a technology, modify it, and market a product in a "third world" nation; 3. a local NGO searches and finds an appropriate technology and finds an entrepreneur to finance its development and use in a needy nation; and 4. NGOs or a government agency does extension work (either through its own employees or by sponsoring local citizens) to demonstrate and spread a new technology to other citizens with the same need. Three types of technology transfer have been controversial in sustainability projects. First, some technology has been inappropriate for the people involved; big dams are an example. Second, copyright laws can prevent development of local industries and lead to pirated and illegal copy products (e.g. medicines, CDs, watches, software). Third, transnational pharmaceutical companies can search out and utilize plants from third world nations but refuse to pay appropriate fees for intellectual property rights.

transnational corporations, research and development, R&D, foreign direct investment, appropriate technology, biotechnology, alternative technology, fair trade, intellectual property rights, copyright laws

TERRESTRIAL ECOSYSTEMS

Terrestrial ecosystems are those ecosystems—a community of plants, animals, and microorganisms that are linked by energy and nutrient flows and that interact with each other and with the physical environment—that are land-based. The main broadly defined terrestrial ecosystems are tundra, taiga (boreal forests), tropical forests, temperate forests, deserts, and grasslands.

savanna, steppes, temperate grasslands, coastal deserts, semiarid deserts, hot and dry deserts, cold deserts, taiga, boreal forests, rain forests, arctic tundra, alpine tundra, biomes, bioregions, ecological function, tropical moist forests, tropical dry forests, rangelands

Deserts and Semideserts (239): Nearly 10 million square kilometers of land are hyperarid or true desert with low and erratic rainfall. The Sahara covers 70 percent of the hyperarid areas. Semideserts have more rain but plants cover no more than 80 percent of the ground. Temperate deserts and semideserts (cold deserts) cover another 6 million square kilometers. It is not true that all deserts have low biodiversity—all terrestrial groups are represented and many are endemic (unique) to the deserts. Many larger vertebrates are threatened by hunting and habitat loss. Domestic livestock and introduced species, off-road vehicles and irrigation also cause harm. True deserts should be distinguished from grasslands that have been desertified.

deserts, desertification, reforestation, ecosystems, biodiversity, arid climates, erosion, evapotranspiration, desert encroachment, semideserts, temperate deserts, polar deserts, drylands, cold deserts, overgrazing, off-road vehicles, invasives

Forest Ecology and Conservation (1,097): Forest ecology inventories, monitors, and models the changes in species mix, the forest's structure, and watershed consequences (water cycle, nutrient cycle, erosion/sediment cycle). Forests are dynamic communities and forest ecologists have had to focus on forest change from climate change, to acid rain, to fire policies, invasive species, forest diseases and pathogens, wildlife (e.g. beavers, woodpeckers, elephants), as well as logging practices. Forest ecologists may have an influence on conservation practices and sustainable forestry: methods of logging, appropriate machines and other technology, road development, riparian buffers, rotation cycles, preservation of species, and watershed management.

native species, forest, forestry, preservation, woodland, regeneration, habitat preservation, forest organisms, woodland edges, old growth, fire, prescribed burns, thinning, watershed management, biodiversity, forest structure, climate change, forest health, acid rain, invasive species, forest diseases, forest pathogens, forest wildlife, logging practices, threatened species, riparian forests

Grasslands and Savannas (246): Grasslands are lands dominated by grasses (excluding bamboos) and other herbaceous plants, with few woody plants. Drought, fire, grazing, freezing temperatures, and soil fertility combine to produce grasslands. About 20 percent of the earth's surface is covered by grasslands. The tropical grasslands (savannas) of Africa support a remarkable diversity of antelopes and other herbivores. Other grasslands have many fewer species. Larger vertebrates and specialty birds are threatened by grassland conversion and hunting. Grasslands are major sources of meat and milk. Anthropogenic grasslands now dominate the earth.

temperate prairies, pampas, campos, steppes, tropical grasslands, savannas, grazing, fire ecology, livestock, cropland conversion, livestock, grazing, invasives, rangeland

Shrublands (1): Shrub communities with woody plants, usually adapted to fire, occur in all parts of the earth with about 200 to 1000 millimeters annual rainfall. They cover 2.5 million square kilometers on the west side of continents facing cold ocean currents. The vascular plant diversity is high for the amount of land covered. This flora is largely threatened by human activities, invasive plants and insects, and changed fire regimes.

fynbos, Mediterranean shrubland, maquis, chaparral, mallee, Australian health, matorral, fire ecology, grazing, invasives, agricultural impacts

Sparse Trees and Parklands (14): In areas where forest turns to nonforest are parklands with trees whose canopies shade out about 10 percent of the land. They are common in the boreal regions and seasonally dry tropics. The African wooded savannas are species rich. In other parklands, the species numbers and variety are lower. These areas can be threatened by woodcutting and agricultural conversion.

boreal forest, taiga, forest tundra, indigenous people, Inuit, Sami, open woodland, cerrado, caatinga, wooded savannas, forest plantations

Temperate and Boreal Needleleaf Forests (2): The cold-climate needleleaf forests cover a larger part of the earth's surface than any other forest type (13.1 million square kilometers). They include the tallest and biggest organism on the planet (redwoods) but also many small-tree pine forests (pygmy forests). Short growing days, cold, and competition with deciduous forests limit needleleaf forests. Wildlife has a strong influence. About 22 percent of the planet's 630 conifers are considered threatened by logging for softwoods, paper pulp, and replacement by plantations. Specific species are in danger because of logging (e.g. the northern spotted owl). These forests have high recreational and cultural value. Needleleaf forests store about one-third of the carbon stored by forests.

conifers, needleleaf forests, pine forests, redwoods, spruce forests, firs, larches, Douglas firs, hemlocks, logging, threatened species, recreation, softwoods, paper, plantations

Temperate Broadleaf and Mixed Forests (15): Temperate broadleaf and mixed forests cover 7.5 million square kilometers. They include the mixed deciduous forests (China, United States, Japan), the broadleaf evergreen forests (Japan, China, New Zealand), and the hard-leaf (sclerophyll) forests (Mediterranean and California). Many are impacted by fire, and few old-growth forest groves remain because of human use. They supply humans with hardwoods, chestnuts, and mushrooms, watershed protection, genetic reserves for tree plantation species, and recreation.

temperate deciduous forests, broadleaf evergreen forests, eucalyptus, watershed management, plantation forests, recreation, deforestation, logging

Tropical Dry Forests (15): In tropical areas with seasonal drought are underappreciated tropical dry forests, now reduced to less than 4 million square kilometers. They are seasonal (dropping leaves in dry periods) with smaller trees. In Africa and South America, fire and termites help shape the forests. Mexico and Bolivia harbor the most diverse tree species with great endemism of both plants and mammals. They are highly threatened by hunting, wildlife trade, and land conversion. The most exploited tree has been teak. Tropical dry forests conserve soil, provide for pollinators during temperate winters and dry seasons, and produce honey.

deciduous forest, seasonal dry forest, thorn forest, short tree forest, wildlife trade, hunting, land conversion

Tropical Moist Forests (270): Tropical moist forests (rain forests) cover more than 11.5 million square kilometers (6 percent of the earth's surface) and are themselves highly various in types (e.g. cloud forests, *vareza*, etc.) and species. Global terrestrial biodiversity of species is concentrated in these forests. They have tall stature, many vines, epiphytes, and, in the neotropics, many palms. They have few restrictions on seasonal growth, though climate change and droughts may become important. Severe logging of tropical hardwoods, clearing for pasture and crops, as well as natural catastrophes (hurricanes) threaten some moist forests and highly charismatic species like the bonobo, Sumatran rhino, the indri, and the recently rediscovered Edward's pheasant. These forests are key to regulating global climate and part of the carbon cycle. They provide rubber, Brazil nuts, rattan, watershed protection, and other services and products.

lowland tropical broadleaf rain forests, vareza forests, igapo forests, peat forests, cloud forests, mangrove forests, primates, inland water fish, insects, ecosystems, biodiversity, deforestation, climate change, carbon cycle, cattle ranching, logging, rain forest expansion

Tundra (21): Tundra is a type of ecosystem characterized by lichens, mosses, grasses, and dwarf woody plants, which is found at high latitudes (polar tundra) and high altitudes (alpine tundra). Arctic tundra is characterized by permafrost and, like alpine tundra, is mostly treeless, and has a short growing season. Alpine tundras can have hot summers. Thawing of permafrost from global warming and loss of habitat for tundra-breeding mammals and birds are of major concern.

arctic tundra, alpine tundra, permafrost, climate change, carbon sink, carbon cycle, polar tundra, montane ecosystems

WATER
Water refers to all freshwater ecosystems and functions, and encompasses the management and conservation of those aquatic environments, the flow and supply of freshwater resources on land, rights to water, water and industry, policymaking, and other activities affecting water resources.

conservation, supply, purity, quality, health, reuse, wastewater, pollution, sustenance, security, salination, desalination, water as a human right, freshwater, aquatic ecosystems, water use, wetlands, watersheds, management, water policy

Dams (314): Dams are barriers that are placed across flowing water, which divert or retard the flow of water. There are over 45,000 large dams in over 140 nations. Europe and North America have slowed or stopped dam building as favorable locations no longer exist. China, India, and South Korea still build large dams. Some dams in North America and Europe have been decommissioned and removed to return more natural flow regimes. The impounded water can create a reservoir or lake to supply water for drinking, for irrigation, or to generate electricity using hydroelectric turbines. Large dams modify river flows, fragment and transform aquatic and riparian ecosystems, may cause saltwater intrusion at the river's mouth, threaten aquatic species, increase waterborne diseases, produce greenhouse gases, displace thousands of local residents, and alter water rights and working rules for water use. Sustainable water developers favor microdams, conservation practices, transparency, accountability, public participation, and protected local residents' rights over macroeconomic goals.

dam, barrage, benefit sharing, decommissioning, displaced people, environmental management, export credit agency, externalities, flood control, greenhouse gases, impoundment, indigenous peoples, mitigation measures, multipurpose dams, performance bonds, precautionary principle, reservoir, World Commission on Dams (WCD), river basin, run-of-river dams, irrigation, domestic/commercial use, hydropower, World Bank, costs and benefits, recreation, pumped storage, small hydropower, fish migration, fish ladders, fish spawning

Groundwater (58): Groundwater is water located beneath the surface of the earth. Groundwater is an attractive water supply source because it is available year-round, does not lose water by evaporation, and usually is of a high quality. Groundwater can be renewed by rain and snowmelt seepage, if it is not overpumped, but the groundwater that was accumulated thousands of years ago (fossil groundwater) is a nonrenewable resource. A disadvantage of groundwater can be the cost of energy to pump it from the ground. Sustainable groundwater management balances withdrawals with recharge. Unfortunately, because of the complex and invisible nature of groundwater supplies, and the lack of laws addressing groundwater extraction, very few aquifers have been managed in a sustainable manner.

aquifer, artesian well, cone of depression, confined aquifer, consumptive use, drawdown, groundwater, infiltration, recharge, overdraft, ground subsidence, unconfined aquifer, produced water, oil and gas extraction, agricultural consumption use, groundwater commons, conjunctive use right, dewatering, groundwater right

Hydrology and the Global Water Cycle (105): Hydrology is the study of the water cycle by watershed, catchment, drainage basin, region, or worldwide. The cycle includes precipitation; interception; water returns to the atmosphere by evaporation; water losses from seepage and changes in quality; water use by vegetation; water in the soil; groundwater; and resulting water balances. Applied hydrology focuses on humans and runoff, flood hazards, erosion, water uses, and water quality changes. Recent concerns about climate change have spurred attempts to understand how global warming will shift rainfall, snowfall, permafrost, and evapotranspiration patterns. Major human consumptive uses include agriculture, industrial and energy, and commercial/residential withdrawals from natural systems. Major human impacts on the water cycle include greenhouse gases, dams, aqueducts, deforestation, and urbanization.

water cycle, hydrological cycle, infiltration, percolation, precipitation, evaporation, evapotranspiration, water table, groundwater, atmospheric water, oceanic water, reservoirs, lakes, runoff, soil moisture, river flows, freshwater reserves, glaciers, permafrost, river basins, watersheds, drainage basins, catchments, climate change

Water and Energy (45): Water and energy are intimately intertwined. It takes energy to move water sometimes hundreds of miles in aqueducts, sometimes hundreds of feet in elevation from aquifers. It takes energy to achieve a harmless water quality. The greatest operation costs of sewage plants is the energy to clean water. The most expensive way to make water is desalinization because it is so energy intensive. In addition, every energy recipe says "add water." Water is required in mining fuels, in cooling power plants, and in boilers in industry. Water is lost from reservoirs in the production of hydropower. Biomass energy sources like corn need water to grow. Where water flow or energy flow is limited, infrastructure fails to provide life support.

water transfer projects, water pumps, groundwater, desalinization, fuel cycle, coolant water, boiler water, thermal electric generating facilities, cooling ponds, cooling towers, once-through cooling, water consumption, water withdrawals, slurry pipeline, mine reclamation, oil recovery, synthetic fuels, flue gas desulfurization, hydropower, reservoirs, evaporative losses, geothermal, water treatment, biomass, irrigation, dust control

Water Law and Policy (318): *Water law and policy* refers to the practice of creating "working rules" for the volume and timing of withdrawal of water, its shared use, the required water quality of return flows, as well as other aspects such as the minimum flow required in a creek (instream flows) or quantity of water reserved by the government (reserved rights). Water policy may attempt to coordinate consumptive users in one watershed, in one groundwater basin, or in a major river basin that spans multiple jurisdictions. Water policy and laws are made by legislative bodies, by government regulations, by litigation and the courts, by direct actions that sway voters and elected officials, as well as by local "informal" agreements. In nondemocratic nations, water policy is made by technocrats and the political leadership. Members of the sustainability movement have argued that a minimal volume of clean water for household use is an inherent human right.

water right, water law, water regulations, prior appropriation right, riparian right, Pueblo right, Indian water right, federal water right, reserved right, conjunctive use right, Clean Water Act, dispute resolution, water as a commons, water ownership, grandfathered right, reasonable use doctrine, instream water right, rule of capture, water allocation, water diversion, return flow, water quality, water claim, perfected right, water trust, private water right, public water right, clean water standards

Water Quality and Health (1,902): Water quality is a key indicator of the ability of a nation to achieve sustainable development. Water quality has two broad forms: microbial and industrial chemical contamination. Industrialized nations have managed waterborne bacterial and viral diseases well, though global warming may revive a need to manage diseases associated with vectors like mosquitoes (avian flu, malaria). Industrialized nations now deal with toxic chemical compounds (metals, radionuclides, and synthetic organics). Developing nations must deal with both waterborne microbial diseases and toxics without much funding and with dense concentrations of humans in burgeoning cities. Sustainable development addresses protection of the sources of drinking water by, for instance, better agricultural practices (no nitrates, no pesticides) and elimination of toxics in the industrial process. While wealthier nations can afford water treatment, poorer nations do not have the ability to build the protective infrastructure. Comprehensive watershed management and water quality monitoring are still rare government functions in all nations.

water quality, contamination, waterborne disease, bacterial disease, viral disease, water-related insect vector, toxic pollutants, endocrine disruptors, polluting metals, polluting synthetic organic molecules, polluting organics, polluting radionuclides, nitrates, dissolved oxygen, eutrophication, sanitation, water treatment plants, wastewater treatment plants, urban infrastructure, fungicides, pesticides, fertilizers, dyes/pigments, pharmaceuticals, radon, halogenated hydrocarbons, phenols, phthalates, organochlorines, megacities

Water Rights (216): Water rights are the legal rights that define ownership of water and water sources (surface and subsurface), the use of water, and the priority of water use. Water rights allocate water to different users and can be contentious in areas where water supplies are insufficient for the demands upon the supply, and where people are denied or deprived of access to water. The right to water is increasingly considered a basic human right that has to be reconciled with legal water rights already in existence.

resource, drinking water, privatization, policy, access, equity, provision, clean water, irri-

gation, agriculture, industry, ownership, freshwater, water as a human right, claims, legal rights, allocation, surface water, subsurface water, aquifers

Water Supply and Conservation (1,630): The water supply is the total quantity of water available for consumption by humans and for other uses. Water conservation is the practice of using water more efficiently, reducing water use, and capturing all available sources such as rainwater harvesting. This includes pricing water to promote conservation and reusing water (recycling) according to its quality. Water conservation may also include leaving water or returning water to a stream to ensure adequate instream flows. Water conservation techniques help to secure a sustainable water supply.

scarcity, protect, groundwater, gray water, wastewater recycling, drought, aquifers, household use, commercial use, urban water conservation, rainwater harvesting, cloud seeding, cisterns, low flush toilets, low flow appliances, drip irrigation, xeriscape, evaporative losses, surface flow, stormwater, water efficiency

Water and Sustainable Development (143): The water-related goals of sustainable development are: adequate clean drinking water for all citizens, adequate waste disposal systems and the elimination of waterborne diseases, adequate water flow to maintain wetland and other water-dependent ecosystems, water laws that harmonize with our scientific understanding of hydrology (especially groundwater and rainfall variability), and an equitable share of water resources to allow nations that share a water source to supply agriculture and industry. These goals have been severely challenged by population growth, migration to megacities, lack of mutually agreed upon rules for water sharing, the conflict between privatizing water facilities and the inherent "right" to clean water, climate change and increased uncertainty, as well as the costs for water-related infrastructure. Hope can be seen in smaller scale solutions (small dams, proper crop choice), efficiencies (drip irrigation, etc.), recycling, consideration of "externalities" such as environmental and social costs in plans for new water projects, increased sharing of data and hydrological understandings, and a priority focus on affordable, clean water supply for shantytowns.

sustainable development, equitable utilization of water, population growth, megacities, clean water, water efficiency, international water law, waterborne disease, wetlands, hydrology, privatization, water recycling, externalities, environmental impacts, social impacts

Watershed Management (2,638): Watershed management is the practice of analyzing, protecting, developing, and managing the land, vegetation, and water resources within a single drainage basin—watershed—for the conservation of all its resources and for the benefit of both humans and the ecosystem.

water, quality, river, stream, protection, restoration, land, management plan, conservation objectives, habitat, ecosystem management, drainage basin, vegetation, development, runoff, ecology, deforestation

WOMEN
Recognizing women's equality with men and working toward addressing existing inequalities is a basic requirement for the development of a sustainable and equitable world. *Women* refers to those rights and issues that concern women and girls, such as women's rights, women's health, gender equality, and the empowerment of women.

trafficking, rights, education, economic development, vocational training, equality, violence against women, participation, equal wages, family, children, human rights, slavery, health, environmental health, girls, female circumcision

Female Circumcision (107): Female circumcision is the practice of removing a portion of the external female genitalia in procedures such as the clitoridotomy, clitoridectomy, and infibulation. These procedures on girls and young women are frequently carried out for cultural reasons and, in other cultures, are considered to present health risks, to be illegal, and to be an infringement of human rights.

clitoridotomy, clitoridectomy, infibulation, women, girls, health, infection, infertility, tradition, human rights, choice, consent, genital circumcision, initiation, indigenous, culture, social practice, control over own bodies, sexuality, FGM, gender

Gender Equality (4,836): *Gender equality* means that people, regardless of their gender, have the same opportunities to fulfill their human rights and potential, and are not denied equal access on the basis of their gender to services, to jobs, to education, to economic development, to the allocation of resources or benefits, and in all aspects of life.

female, male, women, men, equal, conditions, human rights, social, equity, empowerment, sexuality, gender justice, advocacy, vulnerable, voice, equitable, freedom of expression, feminism, inclusive, equal pay, discrimination, economic equality, gender neutral, social exclusion, inequality of power, gender equity

Trafficking of Women (476): Trafficking of women is the illegal practice of using force, coercion, or deception to move women and girls across national or state borders for both sexual and labor exploitation, and subjecting them to involuntary acts such as prostitution. Traffickers deny their victims basic human rights and use coercive tactics including intimidation, isolation, the threat and use of physical force, or debt bondage to control their victims.

women, girls, slavery, abuse, forced prostitution, violence, coercion, labor exploitation, sexual exploitation, abduction, domestic service, vulnerable, human rights, gender

Women and the Environment (382): *Women and the environment* addresses the relationship between women and their environment, including the role of women in environmental protection, environmental leadership, and environmental education; environmental health and the impacts of pollution on women's health; and the empowerment of women through environmental restoration and conservation work.

ecofeminism, gender perspective and sustainable development, environmental decision-making, environmental health, environmental health and women, women's empowerment through conservation, women's environmental leadership, Wangari Maathai, Margaret Mead, Rachel Carson, environmental justice

Women's Civic Participation (2,046): *Women's civic participation* addresses the ability of women to actively participate in community, local, and national institutions, as well as removing the gender-based social, cultural, and economic barriers to civic and democratic participation. Effective civic participation of women strengthens institutions and empowers women to be leaders within their communities.

engagement, voice, diversity, expression, empowerment, organizing, democratic institutions, representation, expanding civil society, women's leadership, strengthening civic institutions, gender, civil disobedience, direct action

Women's Economic Development (3,901): Women's economic development focuses on enhancing the opportunity for women to generate and to increase their income and to improve their standard of living.

jobs, wages, economic independence, development, self-sufficiency, equal pay, economic justice, standard of living, quality of life, access to jobs, increasing income, access to credit, microfinancing, gender, economic empowerment

Women's Education (2,416): *Women's education* refers to those issues surrounding education that are influenced by gender. These include access to education, access to educational resources, equality in teaching practices, training and skills sharing, and overcoming cultural, social, and economic barriers and discrimination that deny women an opportunity to learn.

training, skills, access to education, literacy, basic education, enabling economic independence, equal access to jobs, rights to education, lifelong learning, leadership, barriers to education, socioeconomic, cultural barriers, discrimination, gender, empower women

Women's Health (2,144): *Women's health* refers to improving women's access to health care, educating women about health, and working on issues of health that are gender specific, such as reproductive issues including general gynecology, gynecologic surgery, conception, and pregnancy.

health education, environmental health, reproduction, children, access to health care, toxics, well-being, fragmented care, treatment, medical profession, equality, access to information, health rights, gender, reproductive rights, health equity, maternal mortality

Women's Leadership (2,522): *Women's leadership* refers to the ability of women to transform economic and social development when empowered to fully participate in the decisions that affect their lives through leadership training, coaching, consulting, and the provision of enabling tools for women to lead within their communities, regions, and countries.

leadership training for women, social revolution, empowerment, education, economic development, social development, self-reliance, responsibilities, agents of change, traditional prejudices, subjugation, sustainable community development, coaching, consulting, gender

Women's Rights (4,721): Women's rights are those basic human rights that tradition, prejudice, and social, economic, and political interests have combined to relegate to "special interest" status within the human rights framework, which women are working to redress. These include changing the understanding of human rights to reflect more gender-nuanced definitions, as well as promoting gender-specific rights such as reproductive rights, and advancing the protection and respect for women's rights.

human rights, empowerment, economic rights, dignity, female, status, equality, abuses, monitoring, violations, gender, inalienable, universal, indivisible rights, humanity, quality of life, social justice, defense of rights, discrimination, violence against women, advocacy, gender-based violations, women, girls, sexual rights, reproductive rights, political rights, social exclusion, empower women, gender equity

Women's Safety from Violence (2,281): *Violence against women* refers to any act of gender-based violence that results in physical, sexual, or mental harm or suffering to women. Violence against women affects the health and human rights of women and includes trafficking, slavery, rape, abuse, domestic violence, and harmful cultural practices and traditions that damage women's reproductive and sexual health. Women have the right to be safe from violence.

women, girls, domestic violence, antiwar, pacifism, rape, violence against women, aggression, oppression, physical, vulnerable, advocacy, criminal justice, war crimes, enforcement, policy, assault, abuse, transforming a culture of violence, gender-based violence, armed conflict, international law, genocide, torture, domestic violence, violence against women

Women's Vocational Training (1,681): Women's vocational training is training in which women learn skills and knowledge that are specific to a particular job, profession, trade, or vocation, which will enhance their ability to do that job.

training, empowerment, employment, financial independence, skills, jobs, vocation, career development, occupational skills, knowledge, teaching, learning, on-the-job training, work preparation, technical training, lifelong learning, gender, empower women

WORK
Work refers to workers' rights, employment, the informal economy, and all of the issues relating to labor and the working environment, such as the right to organize, the right to a safe working environment, and the existence of laws to protect workers.

working conditions, living wages, labor rights, health, safety, environmental health, standards, exploitation, sweatshop, workplace, occupational hazards, standards, regulations, enforcement, violations, procedure, risk management, employment law

Employment (1,387): *Employment* refers to a job for which a person is paid, generating income for the employee and giving him or her the opportunity to be self-reliant and to achieve a better standard of living. Generating jobs and employment within a community, such as through community enterprise, contributes to community development.

work, employees, job creation, economic activity, income, community enterprise, community jobs, remuneration, benefits, employment law, standards, conditions, workforce

Global Labor (101): Globalization has encouraged the free movement of goods, services, finances, and information. It has not encouraged the free movement of labor. Nation-states have tried to balance citizen-worker needs, new economic development, and movement of corporations to lower wage-earning nations. Global issues include: the brain drain of skilled workers from developing nations; guest worker and illegal worker jobs in developed nations; outsourcing of jobs and job loss in developed nations; worker remittances and national balance of payments; sweatshops in both developed and developing nations; equal worker conditions and benefits in developed versus developing nations; and training programs in developed nations for workers who have lost jobs due to corporate movement to a lower-wage nation.

ILO, brain drain, guest worker programs, illegal workers, outsourcing, worker remittances, sweatshops, worker rights, working conditions, worker training

Informal Economy (9): The informal economy can be the largest employer within developing nations. Many NGO microfinance and aid programs focus on this sector. These are jobs not recorded by the government and not taxed. Their work and markets are usually local. They include jobs from hawking goods on streets, sales at nonregistered kiosks, markets, or shops in squatter towns, the economies of indigenous people, barter trade, as well as unreported employment in drug cartels, weapons sales, gambling, and other illegal economic activities. Employment statistics collected by governments and agencies may not accurately reflect the true state of an economy with a large informal sector.

drugs, squatter towns, nonregulated shops, weapon sales, gambling, barter trade, black market, underground economy, gray economy, unregulated market, shadow economy, labor

Living Wages (705): A living wage is a wage that allows a person to afford a basic standard of living. A living wage is often higher than the minimum wage and the definition of the standard of living enabled by a living wage is culturally dependent. A living wage description often encompasses the ability to afford housing, food, utilities, transport, health care, and some recreation, working an average of forty hours per week.

standard of living, quality of life, afford, basic needs, housing, food, transport, utilities, health care, dependents, minimum wage, workers, employees, labor equality, equal pay, poverty, low income, advocacy, raising awareness, ordinances, legislation, employment law

Vocational Training (2,317): Vocational training is training where a person learns skills and knowledge that are specific to a particular job, profession, trade, or vocation, which will enhance their ability to do that job.

job training, skills, education, employment, low income, vocation, career, mentor, mentoring, work experience, professional training, occupational training, adults, teacher training, internships, technical, professional development, lifelong learning

Worker Centers (54): Worker centers are community-based mediating organizations that provide support to low-wage workers through service delivery, advocacy, and organizing. Services and advocacy may include legal representation to recover unpaid wages, English classes, worker rights education, providing access to health clinics, researching legislation, working with government agencies to improve grievance processes, and loans. Organizing is directed at global/local issues such as sweatshops and linking student, community, and worker organizations internationally.

wages, worker dignity, labor laws, people of color, low-wage workers, unskilled workers, immigrant workers, worker rights, grievance processes, sweatshops, job advancement, worker education, training programs, labor law, labor unions

Worker Health and Safety (450): *Worker health and safety* refers to the working conditions of employees in all workplaces, where employees have the right to remain safe and healthy both at work and as a result of their work.

workplace, labor, protection, prevention, accident, insurance, employee, standards, certification, accountability, sweatshop, rights, management, precaution, EHS, ISO 14000, worker, occupational health, toxic exposure, worker protection, enforcement, employment law, OSHA

Worker Rights (1,052): Worker rights vary from nation to nation and industry to industry. In impoverished nations, any paying job will be taken and no government regulation of worker conditions (e.g. hours, age, safety, health) may exist. Job security is not an issue, let alone a right. The government and military may enforce rules to prevent collective bargaining and the right to organize collectively. In developed nations, many workers have achieved many benefits—minimum wage rates, earned income tax credit, child tax credit, medical assistance programs, food stamps, paid vacation time, pregnancy leave, sick leave, and government monitoring of workplace conditions. In developed nations, the lowest wage and unskilled workers have the most difficult time achieving a living wage, job security, and a place in the social safety net. In both developing and developed nations, worker rights are a contentious arena of contested power.

job, labor law, sweatshops, working hours, workers, worker safety, working conditions, living wages, human rights, occupational hazards, right to organize, unionize, solidarity, employment law, worker centers, unskilled workers, low-wage earners, employment age, child labor, gender discrimination, racial discrimination, pregnancy leave, sick leave, vacation pay, ILO, labor rights

ACKNOWLEDGMENTS

Blessed Unrest is about the kindness of strangers, the magnanimity of people everywhere. This book and the WiserEarth.org project are the result of the kindness of friends, associates, and volunteers. The brevity of these acknowledgments beggars the generosity bestowed upon the entire project and me.

Isabel Stirling gave unstintingly in research for the book; she is an amanuensis of the digital age who provided troves of literature, PDFs, and papers from the University of California library at Berkeley, California. Although I practically grew up in the library (my father worked there), nothing I learned then could match Isabel's skills in ferreting out information and inspiration from the library's 10 million volumes. I cannot express appreciation for her devotion to this project commensurate to what I have received. Kent Bicknell, a scholar and collector of books owned by the Transcendentalists, provided extraordinary insight in the form of original letters, annotated books from Thoreau's personal library, correspondence from Gandhi, and from his own research. His contribution amplified and explored the influence of Asian spiritual traditions on this global movement, a vein so rich that it is worthy of a book in itself. William Kovarik's work on the chronology of the social justice and environmental movements provided a critical orientation to a subject that I believe will increase in historical importance as twenty-first-century cultures reconsider what constitutes value.

I was equally blessed by feedback from a diverse group of readers who, taking me at my word, were unflinching in their comments and criticisms of early drafts. They brought an original perspective, surprising insights, and a breadth of knowledge from the field that no book or reading could touch. I cannot imagine writing a book without such a band of intellects and editorial advisers. They include John Maybury, Kenny Ausubel, Janine Benyus, Chhaya Bhanti, Amira Diamond, Mark DuBois, Michael Tweed, Diana Cohn, Daphne Beckett, Tony Clarke, Greg Colando, Jonathan Miller, Anuradha Mittal, Morgan Williams, Peter Teague, Lynne and Bill Twist, Belvie Rooks, Wolfgang Sachs, Melissa Nelson, Randy Hayes, Jodie Evans, Yomi Abiola, and Paul Saffo. I want to especially thank Peter Coyote, who applied astute editorial skills to the last draft during a respite in his schedule, an act of diligence that made a profound difference in the final manuscript. Mike Bryan was an editorial keepsake on two previous books and once again provided invaluable feedback with his wry and commonsense insights. Nina Simons and Kenny Ausubel graciously provided numerous opportunities to first explore these ideas at the annual Bioneers conference, which they organize every year. Kenny was dogged in his enthusiasm for the project and provided a stream of ideas and resources.

Rick Kot has been my editor for three successive volumes, and his brilliance and staying power is attested by the fact that I have followed him faithfully in and out of four publishing houses, affirming in our own way that what truly counts in this world is people, not corporations. To all the staff at Viking known and unknown who have been responsive to the needs of the project, I offer my deepest gratitude. To Peter and Mimi Buckley, I am indebted for the editorial sanctuary so kindly provided at the end of the process. It made all the difference.

I am privileged and fortunate to work with a talented and extraordinary group of people who covered and accommodated me as my schedule morphed to create time to write. At the Natural Capital Institute, I have been supported every step of the way by my longtime associate Kelly Costa, as well as Peggy Duvette, Betsy Power, Liz Roberts, Diana Redmond, Kelsang Aukatsang, and Melinda Kramer. WiserEarth and Wiser-Business are their creation. At the Pax Group I thank Erica Linson, Kim Shekar, Jay Harman, Francesca Bertone, and Trevor Daughney. They have been a joy to work with, and are astounding in their patience, support, and understanding.

Creating the WiserEarth.org project and database entailed a significant amount of research and data entry, none of it by rote. The staff of employees, volunteers, and interns worked tirelessly to design and populate a Web site, database, and platform that now belongs to the global community of civil society. Our staff come from all over the world (we have yet to meet face-to-face) and have been extraordinary in their dedication and selflessness. The Natural Capital Institute staff listed above led this effort. Biologist Peter Warshall made a spectacular contribution to WiserEarth by writing and creating definitions to the "Areas of Focus" in the appendix of this book and employed on the Web site, an effort that complemented Liz Roberts's earlier seminal work. I am indebted to the staff who worked together to create the Wiser projects. They include Michael Spalding, Daniel Lavelle, Molly Doyle, Adam Burkett, Stephen Lamm, Edward Fong, Jonathan Waldman, Brooking Gatewood, Hilary Mandel, Rina Horiuchi, and Camilla Burg. The team of researchers include A. Guna Jothi, Agnes Wierzbicki, Alexis Hyatt, Ali Kafaii, Luis Jimenez, Janet Smartt, Negin Tahvildary, Alison Loomis, Amy Williams, Anne-Claire Thieulon, Arnas Palaima, Elizabeth Hanson, Shezad Lakhani, Brenda Harriman, C. Senthil Kumar, Cheryl Anseeuw, D. Joniganesh, David Jay, Debbie Cheng, Deborah Fleischer, Diana Walsh, Emundo Vintel, Eunice Lee, G. Bino Sundar, Jacob Conley, Jennifer De Bann, Lydia Dixon, Heike Bridgwater, Kevin Davis, Arielle Segal, Paul Bisazza, Stephanie Auer, Jessica Bremmer, John Deans, Anton Prakash, Annie Prakash, Jonathan Hawken, Julia Brown, K. Deepak, Kirsten Rose Jardine, Kristen Steck, Kristen Weiss, Laura Pelcher, Lauren Ayers, Magdalena Szpala, Mara Gross, Matthew Borish, Meng-Li Wang, Michele Ross, Nabin Rimal, Nushin Sarkarati, P. K. Yamuna, P. Sharmeela, P. Vadivukarasi, S. Ramya, Tod Chubrich, Tricia Powell, S. Sundaresan, and Wendy Furry. Cheryl, Daniel, Danielle, Khadejhia, Kyle, Mya, Netiya, Noel, Sarah, and Jasmine delighted us with their work on WiserEarth as part of a class project for the Presidio School in San Francisco. From the Movimiento Mi Cometa in Quito, Ecuador, we were blessed by a wonderful group including Roberto, Isaac, Ruth Semenario, Carlos Valles, and Wilmer Rodriguez. We are utterly beholden to our wonderful friend Jon Ramer, who leads the Wiser Commons project with support from more than one hundred NGOs. Polymath Irwin Miller did masterful design work for the group. However, Wiser simply would not have happened were it not for the faith and support given to us all by one person, Hans Schoepflin. It is awkward to thank someone who does not want to be acknowledged, yet Hans believed in this project when no one else rightfully could have. To any and all who may benefit from what has been created, the path of gratitude leads directly to the faith and foresight of Hans. We also want to extend our profound appreciation to the Nathan Cummings Foundation, the Garfield Foundation, and the Marion Institute, all of whom support this work.

Thank you, Patsy Northcutt, Michael Baldwin, John Picard, Rick Fedrizzi, Craig Merrillees, China Galland, Leila Conners Petersen, Leda Dederich, Darian Rodriguez Heyman, Mark Valentine, and Susan Clark for gifts you have given all of us. To the Pink Drummer, Iona, Anastasia, Calliope, Tansy, Aidan, and Palo, there could be no better reason to sit indoors on warm sunny days and write. Last and foremost, I bow to Gayathri Roshan for believing in everything and everyone.

THE BEGINNING

1. Adrienne Rich, "Natural Resources," *The Dream of a Common Language: Poems 1974–1977* (New York: W. W. Norton, 1993), p. 60.
2. Mary Oliver, "Journey," *Dream Work* (New York: Atlantic Monthly Press, 1986).
3. Adam Hochschild, *Bury the Chains: Prophets and Rebels in the Fight to Free an Empire's Slaves* (Boston: Houghton Mifflin, 2005), p. 5.
4. Peter Coyote, as quoted in Louise Steinman, book review of *Sleeping Where I Fall, Los Angeles Times*, June 4, 1998.
5. Peter Morville, *Ambient Findability: What We Find Changes Who We Become* (Sebastopol, Calif.: O'Reilly, 2005), p. 64.
6. Private correspondence with Belvie Rooks.
7. I am indebted to Wolfgang Sachs for these observations about various historical perspectives.

BLESSED UNREST

1. Agnes de Mille, *Dance to the Piper* (New York: Da Capo Press, 1980), p. 335–36. De Mille wrote Graham's comments down after a dinner in which she discussed how the musical *Rodeo*, which she choreographed, had degenerated over time due to cast turnover. De Mille was downcast and it was to this that Graham responded. The full quote, as remembered by de Mille, is

> "There is a vitality, a life-force, an energy, a quickening that is translated through you into action and because there is only one of you in all of time, this expression is unique. And if you block it, it will never exist through any other medium and be lost. The world will not have it. It is not your business to determine how good it is nor how valuable nor how it compares with other expressions. It is your business to keep it yours clearly and directly, to keep the channel open. You do not have to believe in yourself or your work. You have to keep open and aware directly to the urges that motivate you. Keep the channel open. As for you, Agnes, you have a peculiar and unusual gift and you have so far used about one third of your talent."
>
> "But," I said, "when I see my work I take for granted what other people value in it. I see only its ineptitude, inorganic flaws, and crudities. I am not pleased or satisfied."
>
> "No artist is pleased."
>
> "But then there is no satisfaction?"
>
> "No satisfaction whatever at any time," she cried passionately. "There is only a queer divine dissatisfaction, a blessed unrest that keeps us marching

and makes us more alive than the others. And at times I think I could kick you until you can't stand."

2. From *Small Pieces, Loosely Joined*, David Weinberger's book about the Web (New York: Perseus, 2002).

3. Eduardo Galeano, *Upside Down: A Primer for the Looking-Glass World* (New York: Metropolitan Books, 2000), p. 40.

4. Micheline R. Ishay, *The History of Human Rights: From Ancient Times to the Globalization Era* (Berkeley: University of California Press, 2004), p. 2.

5. Ibid.

6. Robert Neuwirth, *Shadow Cities: A Billion Squatters, a New Urban World* (New York: Routledge, 2006) p. xiii.

7. Arundhati Roy, "Shall We Leave it to the Experts?," *The Nation*, Feb 18, 2002.

8. Galeano, *Upside Down*, p. 13.

9. Bob Herbert, "The Young and the Restless," *The New York Times*, May 12, 2005.

10. J. B. Deregowski, *Illusions, Patterns and Pictures: A Cross-Cultural Perspective* (London: Academic Press, 1980), p. 113.

11. Louis Menand, *The Metaphysical Club: A Story of Ideas in America* (New York: Farrar, Straus and Giroux, 2001), p. xii.

12. Ed Hunt, "Retool Your Mind," *Tidepool*, Ecotrust, http://www.tidepool.org/hp/hpbigidea1.cfm, Jan. 21, 2000.

13. This thesis was put forth brilliantly in Adam Curtis's three-part BBC documentary *The Power of Nightmares*.

14. *The State of the World's Children* (New York: UNICEF, 1996).

15. Galeano, *Upside Down*, pp. 116–18.

16. As quoted in Ori Brafman and Rod A. Beckstrom, *The Starfish and the Spider* (New York: Portfolio, 2006), p. 103.

17. "How to Design a Kludge," *Datamation*, Feb. 1962, pp. 30–31.

18. Virginia Postel, "Friedrich the Great," *The Boston Globe*, Jan. 11, 2004.

19. Bruce Mau with Jennifer Leonard, *Massive Change* (London: Phaidon Press, 1994), p. 97.

20. W. Warren Wagar, "The Colors of World History," *Binghamton Journal of History*, Fall 2001.

21. Arundhati Roy, *Power Politics* (Cambridge, Mass.: South End Press, 2001), p. 3.

22. Adam Hochschild, *Bury the Chains: Prophets and Rebels in the Fight to Free an Empire's Slaves* (Boston: Houghton Mifflin, 2005), p. 2.

23. Ibid., p. 7.

24. Michael Pollan, *The Botany of Desire: A Plant's-Eye View of the World* (New York: Random House, 2001), p. xxi.

25. William Kittredge, *The Nature of Generosity* (New York: Vintage Books, 2000), p. 9.

THE LONG GREEN

1. This is a paraphrase of Robert Oppenheimer's invocation of Shiva from the Bhagavad Gita, murmured after watching the detonation of the first atomic bomb: "I am become Death, the shatterer of worlds."

2. Shepard Krech III, John Robert McNeill, and Carolyn Merchant, eds., *Encyclopedia of World Environmental History*, vol. 1 (New York: Routledge, 2004), p. ix.

3. Carl von Linné, *A General System of Nature*, vol. 1. (London: Allen Lackington, 1806), p. 2.

4. E. Janet Browne, *Charles Darwin: Voyaging* (Princeton, N.J.: Princeton University Press, 1995), pp. 147–48.

5. Peter R. Grant, "Natural Selection and Darwin's Finches," *Scientific American*, Oct. 1991, p. 84.

6. "When we consider and reflect upon Nature at large or the history of mankind or our own intellectual activity, at first we see the picture of an endless entanglement of relations and reactions, permutations and combinations, in which nothing remains what, where and as it was, but everything moves, changes, comes into being and passes away. We see, therefore, at first the picture as a whole, with its individual parts still more or less kept in the background; we observe the movements, transitions, connections, rather than the things that move, combine and are connected. This primitive naïve but intrinsically correct conception of the world is that of ancient Greek philosophy, and was first clearly formulated by Heraclitus; everything is and is not, for everything is fluid, is constantly changing, constantly coming into being and passing away." Friedrich Engels, *Natural Dialectics*, as quoted in Derek Wall, ed., *Green History* (New York: Routledge, 1994), p. 96.

7. Jonathan Weiner, *The Beak of the Finch* (New York: Vintage Books, 1994). No one has taken the minutiae of scientific research and more elegantly explored it with a reader. A not-to-be missed exploration of evolutionary science.

8. T. B. Hayes, et al., "Hermaphroditic, demasculinized frogs after exposure to the herbicide, atrazine, at low ecologically relevant doses," *Proceedings of the National Academy of Sciences* (2002): 99:5476–5480.

9. Colin McEvedy and Richard Jones, *Atlas of World Population History* (New York: Facts on File, 1978), pp. 342–51.

10. Donald Worster, *Nature's Economy: The Roots of Ecology* (San Francisco: Sierra Club, 1977).

11. Robert D. Richardson, Jr., *Emerson: The Mind on Fire* (Berkeley: University of California Press, 1995), pp. 11–17.

12. Lyman Littlefield, "Sights from the Long Tree," *Nauvoo Times and Season*, Nov. 15, 1841.

13. http://www.dickinson.edu/~nicholsa/Romnat/anxiety.htm

14. Richardson, *Emerson: The Mind on Fire*, p. 230

15. *The Early Lectures of Ralph Waldo Emerson*, ed. Stephen E. Whicher, Robert E. Spiller, and Wallace E. Williams, 3 vols. (Cambridge, Mass.: Harvard University Press, 1959–72), as quoted in Richardson, *Emerson: The Mind on Fire*, p. 155.

16. Hans Huth, *Nature and the Americas* (Berkeley: University of California Press, 1957), pp. 85–88.

17. Mary Oliver, *Long Life: Essays and Other Writings* (Cambridge, Mass.: Da Capo Press, 2004), p. 51.

18. A. Bronson Alcott, *Concord Days* (Boston: Roberts Brothers, 1872), pp. 11–12. http://www.walden.org/institute/thoreau/contemporaries/A/Alcott_Amos%20Bronson/Thoreau.htm

19. J. M. Hutchings, *Scenes of Wonder and Curiosity in California*, excerpted in Peter Johnstone, ed., *Giants in the Earth* (Berkeley, Calif.: Heyday Books, 2001).

20. Josiah D. Whitney, *The Yosemite Book, A Description of the Yosemite Valley and the Adjacent Region of the Sierra Nevada, and of the Big Trees of California, illustrated by maps and photographs* (New York: Julius Bien, 1869).

21. Richard Hartesveldt, et al., *The Giant Sequoia of the Sierra Nevada* (Washington, D.C.: U.S. Department of the Interior, 1975), p. 3; see also George Dollar,

"Timber Titans" (1898), rept. in *Current Literature*, vol. XXIV, no. 2 (1898–1912), APS Online, p. 165.

22. "The Big Trees of California," *Harper's Weekly*, June 5, 1858.
23. *Description of the Mammoth Tree from California* (London: R. S. Francis, 1857).
24. Ibid.
25. *Gleason's Pictorial Drawing-Room Companion*, Oct. 10, 1853 (Boston).
26. Mark Neuzil and William Kovarik, *Mass Media and Environmental Conflict* (Thousand Oaks, Calif.: Sage Publications, 1996).
27. See Simon Schama, *Landscape and Memory* (New York: Alfred A. Knopf, 1995), pp. 191–93.
28. Robert D. Richardson, Jr., *Henry Thoreau: A Life of the Mind* (Berkeley: University of California Press, 1986), p. 243.
29. Loren Eiseley, *Darwin's Century: Evolution and the Men Who Discovered It* (Garden City, N.Y.: Anchor Books, 1961), p. 353.
30. Bradley P. Dean, "Thoreau and Horace Greeley Exchange Letters on the 'Spontaneous Generation of Plants,' " *New England Quarterly*, vol. 66, no. 4 (Dec. 1993), pp. 630–38.
31. The Library of Congress uses Marsh's 1847 lecture as the starting point of "The Evolution of the Conservation Movement." See http://lcweb2.loc.gov/ammem/amrvhtml/conshome.html
32. David Lowenthal, *George Perkins Marsh: Prophet of Conservation* (Seattle: University of Washington Press, 2000), p. 268.
33. Ibid., quoting Senator Jacob Collamer, p. 13.
34. Lewis Mumford, *The Brown Decades: A Study of the Arts in America, 1865–1895* (New York: Dover Publications, 1955), pp. 5–34.
35. Ibid.
36. Philip Shabecoff, *A Fierce Green Fire: The American Environmental Movement* (New York: Hill and Wang, 1993), pp. 55–59.

THE RIGHTS OF BUSINESS

1. Laura Orlando, "Industry Attacks Dissent: From Rachel Carson to Oprah," *Dollars and Sense*, no. 240, March–April 2002.
2. From her acceptance speech at the Goldman Awards, 2004.
3. Linda Lear, *Rachel Carson: Witness for Nature* (New York: Henry Holt, 1997), p. 395.
4. Ibid., pp. 312–19.
5. Rachel Carson, *Silent Spring* (Boston: Houghton Mifflin, 1962), p. 2.
6. H. Patricia Hynes, *The Recurring Silent Spring* (New York: Pergamon Press, 1989), p. 116.
7. As quoted in Jonathan Norton Leonard, "Rachel Carson Dies of Cancer, 'Silent Spring' Author Was 56," *The New York Times*, April 15, 1964.
8. *Monsanto Magazine*, vol. 42, no. 4 (1962), p. 4.
9. Lear, *Rachel Carson: Witness for Nature*, p. 429.
10. Ibid., p. 462.
11. Hynes, *The Recurring Silent Spring*, p. 127.
12. Lear, *Rachel Carson: Witness for Nature*, p. 432.
13. Hynes, *The Recurring Silent Spring*, p. 121.
14. Ibid., pp. 127–28.
15. Rachel Carson, "The Pollution of Our Environment," reprinted in *Lost Woods: The*

Discovered Writings of Rachel Carson, ed. Linda Lear (Boston: Beacon Press, 1998), p. 231.

16. Lear, *Rachel Carson: Witness for Nature*, pp. 416–19.

17. P. R. Ehrlich, A. H. Ehrlich, and J. P. Holdren, *Ecoscience, Population, Resources, Environment* (San Francisco: W. H. Freeman and Co., 1977).

18. Merril Eisenbud, *Environment, Technology and Health: Human Ecology in Historical Perspective* (New York: New York University Press, 1978), p. 60.

19. E. Royston Pike, "Hard Times," *Human Documents of the Industrial Revolution* (New York: Frederick A. Praeger, 1966), p. 209.

20. Kevin Binfield, ed., *Writings of the Luddites* (Baltimore and London: The Johns Hopkins University Press, 2004).

21. Timothy Holtz, "The 1984 Bhopal Gas Disaster," chapter 10 in *Dying for Growth*, ed. J. Millen, et al. (Monroe, Me.: Common Courage Press, 2000).

22. *Clouds of Injustice: Bhopal Disaster 20 Years On*, Amnesty International, 2004.

23. To see the full list of all 124 groups that ExxonMobil funds or that work with ExxonMobil-funded groups to confuse or deny climate change science, go to ExxonSecrets.org, http://www.exxonsecrets.org

24. Chris Mooney, "Some Like It Hot," *Mother Jones*, May/June 2005.

25. Steven F. Hayward, "Don't Worry, Be Happy," American Enterprise Institute, April 26, 2004, http://www.aei.org/publications/pubID.20355,filter.all/pub_detail.asp

26. Mooney, "Some Like It Hot."

27. Ibid.

28. Ibid.

29. Ibid.

EMERSON'S SAVANTS

1. *The Journals and Miscellaneous Notebooks of Ralph Waldo Emerson*, ed. William H. Gilman et al. (Cambridge, Mass.: Harvard University Press, 1960–1982), Cabot 41; 5:50, as quoted in Robert D. Richardson, Jr., *Emerson: The Mind on Fire* (Berkeley: University of California Press, 1995), p. 42.

2. Ross Gelbspan, "Katrina's Real Name," *The Boston Globe*, Aug. 30, 2005.

3. Richardson, *Emerson*, p. 110.

4. Ibid., pp. 140–42.

5. Ibid., p. 53.

6. *The Early Lectures of Ralph Waldo Emerson*, 3 vols., eds. Stephen E. Whicher, Robert E. Spiller, and Wallace E. Williams (Cambridge, Mass.: Harvard University Press, 1959–72), vol. 1; p. 26, as quoted in Richardson, *Emerson*, p. 155.

7. Stephanie Pace Marshall, "The Transformative Power of Learning," *Shift: At the Frontiers of Consciousness*, no. 8, Sept.–Nov. 2005, p. 11.

8. Susan Sutton Smith, ed., *The Topical Notebooks of Ralph Waldo Emerson*, vol. 1 (Columbia, Mo.: University of Missouri Press, 1990), p. 67.

9. John Albee, *Remembrances of Emerson* (New York: Robert G. Cooke, 1901), pp. 18–19, 22, as quoted in Richardson, *Henry Thoreau*, p. 282.

10. Evan Carton, "The Price of Privilege: Civil Disobedience at 150," *The American Scholar*, Autumn 1998, iss. 4, p. 107. Carton gives the dates of July 24 and 25; Richardson uses the earlier two dates.

11. Ethan Allen Hitchcock, *Fifty Years in Camp and Field*, ed. W. A. Crofutt (New York: G. P. Putnam's Sons, 1909), p. 192.
12. Richard Lebeaux, *Thoreau's Seasons* (Amherst: University of Massachusetts Press, 1984), p. 79.
13. Carton, "The Price of Privilege," p. 108.
14. Ibid., p. 112.
15. Robert D. Richardson, Jr., *Henry Thoreau: A Life of the Mind* (Berkeley: University of California Press, 1986), p. 172.
16. Fritz Oehlschlaeger, "Another Look at the Text and Title of Thoreau's 'Civil Disobedience,' " *ESQ: A Journal of the American Renaissance*, vol. 36 (1990): 240.
17. Mohandas K. Gandhi, *Satyagraha in South Africa* (Ahmadabad, India: Mavajivan, 1961), p. 102, as quoted in Judith M. Brown, *Gandhi: Prisoner of Hope* (New Haven, Conn.: Yale University Press, 1989), p. 55.
18. Brown, *Gandhi: Prisoner of Hope*, p. 55.
19. George Hendrick, "The Influence of Thoreau's 'Civil Disobedience' on Gandhi's Satyagraha," *The New England Quarterly*, vol. 29, no. 4 (Dec. 1956), p. 466.
20. Mohandas K. Gandhi, *Speeches and Writings of Mahatma Gandhi* (Madras: G. A. Natesan, 1933), 4th edition, p. 227.
21. Mohandas K. Gandhi, *Indian Opinion*, Oct. 26, 1907, p. 438, as quoted in Hendrick, "The Influence of Thoreau's 'Civil Disobedience,' "p. 464.
22. Stewart Burns, *To the Mountaintop: Martin Luther King Jr.'s Sacred Mission to Save America 1955–1968* (San Francisco: Harper, 2004), p. 19.
23. Rosa Parks, *Rosa Parks: My Story* (New York: Dial Books, 1992), pp. 77–79.
24. Ibid., p. 101.
25. Ibid., p. 105.
26. Burns, *To the Mountaintop*, pp. 9–11.
27. Steven Millner, "The Montgomery Bus Boycott: A Case Study in the Emergence and Career of a Social Movement," in *The Walking City: The Montgomery Bus Boycott, 1955–56*, ed. David Garrow (New York: Carlson, 1989), p. 461.
28. Burns, *To the Mountaintop*, pp. 14–16.
29. Ibid., pp. 31–32.
30. Ibid., p. 44.
31. Interview with Donald T. Ferron, Montgomery, Ala., Feb. 4, 1956, *King Papers*, 3:125, as quoted in Burns, *To the Mountaintop*, p. 82.
32. Burns, *To the Mountaintop*, p. 89.
33. Richard Gregg, "Mohandas Gandhi and the Strategy of Nonviolence," *The Journal of American History*, vol. 91, no. 4 (Mar. 2005).
34. Burns, *To the Mountaintop*, pp. 92–94.
35. Oliver, *Long Life*, p. 51.
36. Jo Ann Robinson, *The Montgomery Bus Boycott and the Women Who Started It: The Memoir of Jo Ann Gibson Robinson* (Knoxville: University of Tennessee Press, 1987), pp. 53–67.
37. Henry David Thoreau, *Civil Disobedience and Other Essays* (New York: Dover Books, 1993), p. 9.

INDIGENE

1. "Delta Protests Against ChevronTexaco Continue: Delta Women in Their Own Words," Field Report #105, Environmental Rights Action/Friends of the Earth,

Benin, Nigeria, 2002. The report was filed when more than three thousand Ijaw women from the Gbaramatu Clan occupied Chevron flow stations on the Niger River delta demanding compensation for spills, cessation of pollution, and cleanup of their environment. See http://www.seen.org/pages/urgent_niger_delta_women.shtml

2. Wade Davis, "The Ethnosphere and the Academy," speech given at the Indigenous Knowledges: Transforming the Academy Conference, May 27, 2004, Pennsylvania State University.

3. There is reason to believe that the sailors who had gone on an exploratory trip on the whaleboat and returned on a raft had scapegoated the natives as a means to cover up their own incompetence. Captain Fitzroy had only recently taken command due to the suicide of Captain Pringle-Stockes and was loath to question his men.

4. Adrian Desmond and James Moore, *Darwin: The Life of a Tormented Evolutionist* (New York: W. W. Norton and Company, 1991), p. 132–35.

5. E. Janet Browne, *Charles Darwin: Voyaging* (Princeton, N.J.: Princeton University Press, 1995), chapter 10.

6. Charles Darwin, *The Voyage of the Beagle: Journal of Researches into the Natural History and Geology of the Countries Visited During the Voyage of H.M.S. Beagle Round the World* (New York: Modern Library, 2001), p. 122.

7. Richard Keynes, *Fossils, Finches and Fuegians: Darwin's Adventures and Discoveries on the Beagle* (New York: Oxford University Press, 2003), p. 120.

8. L.-F. Martial, *Mission Scientifique du Cap Horn, 1882–1883, Tome Ier, Histoire du Voyage* (Paris: Gauthier-Villars, 1888).

9. P. Hyades and J. Deniker, *Mission Scientifique du Cap Horn, 1882–1883, Tome VII, Anthropologie, Ethnographie* (Paris: Gauthier-Villars, 1891).

10. From *Beagle Diary: Charles Darwin's Beagle Diary*, ed. Richard Keynes (Cambridge, UK: Cambridge University Press, 1988), pp. 122–25, as quoted in Keynes, *Fossils, Finches and Fuegians*, p. 121.

11. http://www.aboutdarwin.com/voyage/voyage04.html

12. The Yámana are often referred to as *Yaghan* in the literature. The name *Yaghan* was coined by Thomas Bridges when he studied their language. He had chosen the Murray Narrows area as his base for study, a region the Yámana called Yahgashaga, meaning Mountain Valley Channel. Bridges shortened the name of the residents, Yahgashagalumoala, to Yaghan. Like many indigenous cultures, they referred to themselves as "the people," Yámana.

13. See the Shakespeare Database Project, http://www.shkspr.uni-muenster.de/index.php

14. Bill Bryson, *Mother Tongue: English and How It Got That Way* (New York: Avon, 1991), p. 148.

15. Thomas Bridges, *Yámana-English Dictionary: A Dictionary of the Speech of Tierra del Fuego*, ed. Ferdinand Hestermann and Martin Gusinde (Mödling, Austria: Missiondruckerei St. Gabriel, 1933).

16. In *Language Death* (Cambridge, UK: Cambridge University Press, 2000), author David Crystal reported that in 1992, the International Linguistics Congress issued the following statement: "As the disappearance of any one language constitutes an irretrievable loss to mankind, it is for UNESCO a task of great urgency to respond to this situation by promoting and, if possible, sponsoring programs of linguistic organizations for the description in the form of grammars, dictionaries and texts, including the recording of oral literatures, of hitherto unstudied or inadequately documented endangered and dying languages."

17. Jack Hitt, "Say No More," *The New York Times*, Feb. 29, 2004; and Daniel Nettle and Suzanne Romaine, *Vanishing Voices: The Extinction of the World's Languages* (New York: Oxford University Press, 2000), p. ix.
18. William J. Sutherland, "Parallel extinction risk and global distribution of languages and species," *Nature* 423 (2003), pp. 276–79.
19. Ibid.
20. Lewis Thomas, *Lives of a Cell: Notes of a Biology Watcher* (New York: The Viking Press, 1974), pp. 134–35.
21. Davis, "The Ethnosphere and the Academy."
22. Marianne Mithun, quoted in Crystal, *Language Death*, p. 19.
23. Crystal, *Language Death*, p. 19.
24. Ibid.
25. As quoted in Charles C. Mann, *1491: New Revelations of the Americas Before Columbus* (New York: Knopf, 2005), p. 15.
26. Ibid.
27. Christopher Hitchens, "Minority Report," *The Nation* 225, Oct. 19, 1992, as quoted in Angela Miller, "The Soil of an Unknown America: New World Lost Empires and the Debate over Cultural Origins," *American Art*, vol. 8, no. 3–4, (1994), pp. 8–27.
28. Kenan Malik, "Let Them Die," *Prospect*, Nov. 2000.
29. Nettle and Romaine, *Vanishing Voices*, pp. 41–49.
30. Luisa Maffi and Tove Skutnabb-Kangas, "Language Maintenance and Revitalization," *Cultural and Spiritual Values of Biodiversity*, ed. Darrel Posey (London: Intermediate Technology Publications, 1999), p. 37.
31. Diana Parsell, "Explorer Wade Davis on Vanishing Cultures," *National Geographic News*, Washington, D.C., June 28, 2002.
32. Nettle and Romaine, *Vanishing Voices*, p. 69.
33. Mann, *1491*, pp. 101–2.
34. Crystal, *Language Death*, p. 72.
35. Barry Lopez, *The Rediscovery of America* (New York: Vintage Books, 1990), pp. 6–7.
36. Ibid.
37. Mann, *1491*, p. 71.
38. Ibid., pp. 360–61.
39. Ibid., p. 197.
40. Jack Weatherford, *Indian Givers: How the Indians of the Americas Transformed the World* (New York: Ballantine, 1989), p. 63.
41. John Mohawk, "Subsistence and Materialism," *Paradigm Wars, Indigenous People's Resistance to Globalization*, ed. Jerry Mander and Victoria Tauli-Corpuz (San Francisco: International Forum on Globalization, Committee on Indigenous Peoples, 2005), pp. 22–24.
42. George Steiner, *Language and Silence: Essays on Language, Literature, and the Inhuman* (New York: Atheneum, 1967), as quoted in Crystal, *Language Death*, p. 35.
43. Silas Tertius Rand, *Legends of the Micmacs* (New York: Longmans, Green, and Co., 1894), pp. 253–58.
44. R. E. Johannes, *Words of the Lagoon: Fishing and Marine Lore in the Palau District of Micronesia* (Berkeley: University of California Press, 1981), pp. 5–9.
45. Davis, "The Ethnosphere and the Academy."
46. Red Cloud. See Nettle and Romaine, *Vanishing Voices*, p. 99.
47. Darrell Posey, "Biological and Cultural Diversity: The Inextricable Linked by

Language and Politics," as quoted in *Iatiku*, Newsletter for the Foundation for Endangered Languages, #4, Jan. 31, 1997.

48. Mander, *Paradigm Wars 2005*, p. 4.
49. Tar or oil sands exist around the world, but the only place they can be surface-mined is Canada. The sands are similar to what is used to pave roads, only in this case, instead of mixing oil, sand, and rock, the sand and rock are subjected to steam treatment to release the oil. The sands lie under muskeg and overburden that can be as deep as 240 feet. The deposits in Athabasca represent 174 billion barrels of oil, and if fully extracted, would fuel the world economy for six years and two months. The importance of the reserves is attested to by the large inflows of capital coming from outside the country, including India and China, which has already invested in a pipeline.
50. Mander, *Paradigm Wars*.
51. "Berbers: The Proud Raiders," *BBC World Service*, Monday, April 23, 2001, http://www.bbc.co.uk/worldservice/people/highlights/010423_berbers.shtml
52. Miguel San Sebastián and Anna-Karin Hurtig, "Oil exploitation in the Amazon basin in Ecuador: a public health emergency," *Public Health* 15(3), 2004, pp. 205–11.
53. Ibid.
54. Brigid McMenamin, "Bring Me Your Tired, Your Poor, Your Litigious," *Forbes*, Nov. 15, 1999.
55. Anna-Karin Hurtig and Miguel San Sebastián, "Incidence of Childhood Leukemia and Oil Exploitation in the Amazon Basin of Ecuador," *International Journal of Occupational Environmental Health*, 2004, 10:245–250.
56. Judith Kimerling, "Corporate Responsibility in Ecuador: The Many Faces of Oxy," draft, California Global Corporate Responsibility Project, May 2001.
57. Moises Naim, "An Indigenous World," *Foreign Policy*, Nov.–Dec. 2003.

WE INTERRUPT THIS EMPIRE

1. Lester Thurow, *American Fiscal Policy: Experiment for Prosperity* (Englewood Cliffs, N.J.: Prentice-Hall, 1967), p. 125.
2. As quoted in Naomi Klein, *No Logo: Taking Aim at Brand Bullies* (New York: Picador, 2000), p. 325.
3. Joseph Stiglitz, *Globalization and Its Discontents* (New York: W. W. Norton, 2002), p. 35.
4. Thomas Friedman, *The Lexus and the Olive Tree* (New York: HarperCollins, 2000), p. 464.
5. "The Prophet of Prison," *The Economist*, Sept. 3, 2005, p. 58.
6. William Cobbett, "Such Slavery, Such Cruelty," *Political Register*, vol. LII, Nov. 20, 1824, as quoted in "Hard Times," *Human Documents of the Industrial Revolution*, ed. E. Royston Pike (New York: Frederick A. Praeger, 1966), p. 62.
7. Julia Gabrielle, "Coalition of Imokalee Workers," at http://www.ciw-online.org/slavery.html
8. Madeleine Bunting, "Stop, I Want to Get Off," *Guardian Weekly*, Nov. 29, 1999.
9. Thomas Frank, *One Market Under God* (New York: Doubleday, 2000), pp. xiv, xv.
10. The metaphor of the Green and Red Zones belongs entirely to the editors of *Global Civil Society* (London: Sage Publications, 2005), Mary Kaldo, Helmut Anheier, and Marlies Glasius, who wrote about it in the introduction to the 2004/5 edition.

11. William Langewiesche, "Welcome to the Green Zone," *Atlantic Monthly*, Nov. 2004.
12. Rajiv Chandrasekaran, *Imperial Life in the Emerald City: Inside Iraq's Green Zone* (New York: Alfred A. Knopf, 2006). This is the most current and disconcerting description of the cultural hubris of the Green Zone.
13. S. Johnson, "Lost in the Green Zone," *Newsweek*, Sept. 20, 2004.
14. Langewiesche, "Welcome to the Green Zone."
15. Lori Wallach and Michelle Sforza, *Whose Trade Organization? Corporate Globalization and the Erosion of Democracy* (Washington, D.C.: Public Citizen, 1999), pp. 142–43.
16. Robert Evans, "Green Push Could Damage Trade Body—WTO Chief," Reuters, May 15, 1998, as quoted in Wallach and Sforza, *Whose Trade Organization?*, p. 13.
17. This subject was brilliantly portrayed in Herbert Sauper's Academy Award–nominated documentary *Darwin's Nightmare*, in which he portrayed the daily lives of the pilots, the glue-sniffing children, the AIDs-infected men, and the self-important ministers and entrepreneurs who benefit from fish exports.
18. Michael Goldman, *Imperial Nature, the World Bank, and Struggles for Social Justice in the Age of Globalization* (New Haven: Yale University Press, 2005), p. 232.
19. Ibid.
20. Eduardo Galeano, "Where People Voted Against Fear," *The Progressive*, Jan. 2005.
21. Goldman, *Imperial Nature*, p. 45.
22. David Held, "Toward a new consensus: answering the dangers of globalization. (PERSPECTIVES)," *Harvard International Review* 27, 2 (Summer 2005):14(4).
23. Stewart Brand, *The Clock of the Long Now: Time and Responsibility* (New York: Basic Books, 1999), pp. 34–37.
24. Mike Davis, *Planet of Slums* (London: Verso, 2006), p. 151.
25. Mike Davis, "Planet of Slums," *New Left Review* 26, March–April 2004.
26. Robert Neuwirth, *Shadow Cities, a Billion Squatters, a New Urbanworld* (New York: Routledge, 2006), p. xiii.
27. Davis, *Planet of Slums*, pp. 204–5.
28. Ibid., p. 19.
29. Pankaj Mishra, *The End of Suffering* (New York: Farrar, Straus and Giroux, 2004), pp. 392–93.
30. Clarissa Pinkola Estes, *Women Who Run with the Wolves* (New York: Ballantine, 1996).

IMMUNITY

1. Gerald N. Callahan, *Faith, Madness, and Spontaneous Human Combustion: What Immunology Can Teach Us About Self-Perception* (New York: St. Martin's, 2002), p. 227.
2. Kenny Ausubel, ed., *Nature's Operating Instructions* (San Francisco: Sierra Club Books, 2004), p. xiv.
3. Lewis Thomas, *The Lives of a Cell: Notes of a Biology Watcher* (New York: Viking Press, 1974), pp. 113–14.
4. From the *New Oxford Dictionary*, quoted in *The Historical Atlas of Immunology*, ed. Julius M. Cruse and Robert E. Lewis (London: Taylor and Francis, 2005), p. 1.
5. Fritjof Capra, *The Web of Life* (New York: HarperCollins, 1996), p. 278.
6. Callahan, *Faith, Madness, and Spontaneous Human Combustion*, pp. 4–9.
7. Ibid., p. 9.

8. Capra, *Web of Life*, p. 279.

9. Francesco Varela and Antonio Coutinho, "Second Generation Immune Networks," *Immunology Today*, 1991, 12:159–66.

10. Paul Farmer, *Pathologies of Power: Health, Human Rights and the New War on the Poor* (Berkeley: University of California Press, 2005), p. xxviii.

11. From opening statement to secretariat of the Convention on Biological Diversity, by Ahmed Djog-hlaf, Curitiba, Brazil, March 27, 2006.

12. Bill McKibben, "Born Again, Again! Will Evangelicals Help Save the Earth?" *GRIST*, Oct. 5, 2006, www.GRIST.org/news/maindish/2006/10/05/McKibben/index.html.

13. David Graeber, "The New Anarchists," *The New Left Review* 13 (Jan.-Feb. 2002), p. 62.

14. http://www.nydailynews.com/news/politics/story/225853p-193988c.html

15. Rupert Steiner, "Bill Gates is just a figurehead. I am actively engaged," *Spectator*, July 16, 2006.

16. George Soros, "America's global role: Why the fight for a worldwide open society begins at home," *The American Prospect*, June 2003, p. 29.

17. Irwin M. Stelzer, "The Gates-Buffett Merger," *The Weekly Standard*, July 17, 2006, pp. 14–16.

18. Suzanne Wooley and Jessi Hempel, "Top Givers," *BusinessWeek*, Nov. 28, 2005, pp. 58–60.

19. Richard Dowden, "Bribing African leaders to leave power on time," *The New Vision*, Uganda, http://www.newvision.co.ug/

20. Michael S. Malone, "The India Movie Mogul," *Wired*, Feb. 2006.

21. Daniel Ben-Horin, "More power for nonprofit organizations: CompuMentor," *Electronic Democracy*, *Whole Earth Review*, Summer 1991, p. 14.

22. Mitch Nauffts, *Philanthropy News Digest*, interview posted May 2, 2006, http://foundationcenter.org/pnd/newsmakers

23. Gil Friend, "The 2030 Climate Challenge and West Coast Green," *Worldchanging*, Oct. 2, 2006, http://www.worldchanging.com/archives/005005.html

24. Michael Chabon, "The Omega Glory," *Details*, Jan. 2006.

25. John O'Connell, "Slow down, you eat too fast: A European culinary revolt against the grazing culture is simmering," *New Statesman*, Sept. 24, 2001.

26. Ibid.

27. Carol Ness, "Slow Food movement has global outreach," *San Francisco Chronicle*, Oct. 30, 2006, p. 1.

28. "Interview: Carlo Petrini," *The Ecologist*, vol. 34, issue 3 (April 2004), pp. 50–53.

29. Stacy Schiff, "Know It All: Can Wikipedia Conquer Expertise?" *The New Yorker*, July 31, 2006, p. 38.

30. *Nature* 438 (2005): 900–901.

31. See http://corporate.britannica.com/britannica_nature_response.pdf

32. Jeff Howe and Mark Robinson, "The Rise of Crowdsourcing," *Wired*, June 2006. The term coined by the authors is used in contrast to *outsourcing*, and in the article applies itself only to the for-profit world. But it is just as applicable in the nonprofit world where people work together to solve problems that are not commercial but social and environmental.

33. Steve Stecklow, "Virtual Battle: How a Global Web of Activists Gives Coke Problems in India," *The Wall Street Journal*, June 7, 2005, A1.

34. C. Surendranath, "Coca-Cola: Continuing Battle in Kerala," India Resource Center, July 10, 2003, http://www.indiaresource.org/campaigns/coke/

35. Stecklow, "Virtual Battle," A1.
36. "The Street Fight," *Down to Earth: Science and Environment Online*, Aug. 15, 2006, Center for Science and the Environment, New Delhi.
37. Stecklow, "Virtual Battle," A1.
38. Philip Ball, *Critical Mass, How One Thing Leads to Another* (New York: Farrar, Straus and Giroux, 2004), pp. 459–60.
39. Callahan, *Faith, Madness, and Spontaneous Human Combustion*, p. 63.

RESTORATION

1. Julia Whitty, "The Fate of the Oceans," *Mother Jones*, March/April 2006.
2. Richard Fortey, *Life: A Natural History of the First Four Billion Years of Life on Earth* (New York: Vintage, 1997), pp. 36–37.
3. As quoted in Franklin M. Harold, *The Way of the Cell* (New York: Oxford University Press, 2001), p. 101.
4. John Steinbeck, *The Log from the Sea of Cortez* (New York: Viking Press, 1951), p. 217.
5. Stuart Kauffman, *At Home in the Universe, The Search for the Laws of Self-Organization and Complexity* (Oxford: Oxford University Press, 1995), p. 45.
6. Ibid.
7. Harold, *Way of the Cell*, pp. 66–68.
8. Robinson Jeffers, "De Rerum Virtute," *Selected Poems* (New York: Vintage, 1965), pp. 176–78.
9. As quoted in Lynn Margulis and Dorion Sagan, *Microcosmos, Four Billion Years of Evolution from Our Microbial Ancestors* (New York: Summit Books, 1986), p. 18.
10. See The Resilience Alliance (http://www.resalliance.org), a project created by C. S. Hollings, the brilliant Canadian ecologist who pioneered the concepts of resilience, adaptive systems, and panarchy.
11. Walter Reid, Harold Mooney, Angela Cropper, et al., *Millennium Ecosystem Assessment Synthesis Report* (Pre-publication final draft), Millennium Assessment Board, March 23, 2005.
12. "The state of the world? It is on the brink of disaster," *The Independent* (UK), March 30, 2006.
13. "A blast from the past: Climate change," *The Economist*, Feb. 25, 2006, p. 90.
14. "The Heat Is On," *The Economist*, Sept. 7, 2006.
15. This insight goes back to Hillaire Belloc's verse as quoted in Harold, *Way of the Cell*, p. 9: "The man behind the microscope, Has this advice for you: Never ask what something Is, Just Ask, what does it Do?"
16. Mahlon Hoagland and Bert Dodson, *The Way Life Works* (New York: Times Books, 1995), p. 2.
17. Walter Truett Anderson, *All Connected Now: Life in the First Global Civilization* (Westview Press, 2001). Anderson makes a very compelling case for the globalization of everything, not as an advocate, but as fact. He distinguishes between economic globalization and evolutionary globalization, the coming together of the world through increased mobility, communications, trade, biology (including disease), information technology, and organizations (NGOs), with the latter being inevitable and largely helpful.
18. Kenneth E. Boulding, "Earth as a Spaceship," speech given at Washington State University, May 10, 1965.

19. Hoagland and Dodson, *Way Life Works*, p. 26.
20. Albert Camus, "Helen's Exile," *Lyrical and Critical Essays* (New York: Alfred A. Knopf, 1969), pp. 149–50.
21. Karen Armstrong, *The Great Transformation: The Beginning of Our Religious Traditions* (New York: Alfred A. Knopf, 2006), pp. xi–xiv.
22. Ibid., p. xiv.
23. Martin Luther King, Jr., "Beyond Vietnam," address delivered to the Clergy and Laymen Concerned about Vietnam, at Riverside Church, New York City, April 4, 1967.
24. James Carse, *Finite and Infinite Games: A Vision of Life as Play and Possibility* (New York: Ballantine, 1986).
25. After I had written this passage, I saw a better list from David James Duncan's book *God Laughs and Plays* (Great Barrington, Mass: Triad, 2006): "I hold the evangelical truth of the matter to be that contemporary fundamentalists, including first and foremost those aimed at Empire and Armageddon, need us nonfundamentalists, mystics, ecosystem activists, unprogrammable artists, agnostic humanitarians, incorrigible writers, truth-telling musicians, incorruptible scientists, organic gardeners, slow food farmers, gay restaurateurs, wilderness visionaries, pagan preachers of sustainability, compassion-driven entrepreneurs, heart-broken Muslims, grief-stricken children, loving believers, loving disbelievers, peace-marching millions, and the Ones who love us all in such a huge way that it is not going too far to say *they need us for their salvation*."
26. Duncan, *God Laughs and Plays*, p. 118.
27. From an e-mail from Wolfgang Sachs.
28. David Orr, *The Last Refuge: Patriotism, Politics, and the Environment in an Age of Terror* (Washington, D.C.: Island Press, 2004), pp. 74–77.
29. Mary Oliver, *Dream Work* (Boston: Atlantic Monthly Press, 1986), p. 38. The last lines read: "But little by little, as you left their voices behind, the stars began to burn / through the sheets of clouds, and there was a new voice / which you slowly / recognized as your own, that kept you company / as you strode deeper and deeper / into the world, / determined to do / the only thing you could do—determined to save / the only life you could save."
30. Harold, *Way of the Cell*, p. 101.

BIBLIOGRAPHY

Anderson, Sarah, ed. *Views from the South: The Effects of Globalization and the WTO on Third World Countries.* Oakland, CA: Food First, 2000.

Anderson, Walter Truett. *All Connected Now: Life in the First Global Civilization.* Boulder, CO: Westview Press, 2004.

Anheier, H., M. Glasius, and M. Kaldor. eds. *Global Civil Society 2001.* Oxford: Oxford University Press, 2001.

———. *Global Civil Society 2002.* Oxford: Oxford University Press, 2002.

———. *Global Civil Society 2003.* Oxford: Oxford University Press, 2003.

———. *Global Civil Society 2004/5.* Oxford: Oxford University Press, 2005.

Armstrong, Karen. *The Great Transformation: The Beginning of our Religious Traditions.* New York: Alfred A. Knopf, 2006.

Ausubel, Kenny, and J. P. Harpignies, eds. *Nature's Operating Instructions: The True Biotechnologies.* San Francisco: Sierra Club Books, 2004.

Bakshi, S. R. *Gandhi and Salt Satyagraha.* Malayattoor, Kerala: Vishwavidya, 1981.

Ball, Philip. *Critical Mass, How One Thing Leads to Another.* New York: Farrar, Straus and Giroux, 2004.

Barlow, Maude, and Tony Clarke. *Global Showdown: How the New Activists Are Fighting Global Corporate Rule.* Toronto: Stoddart Publishing, 2002.

Bergreen, Laurence. *Over the Edge of the World: Magellan's Terrifying Circumnavigation of the World.* New York: HarperCollins 2003.

Bhagwati, Jagdish. *In Defense of Globalization.* Oxford: Oxford University Press, 2004.

Bird, Junius. "Antiquity and Migrations of the Early Inhabitants of Patagonia." *Geographical Review,* 28. no. 2 (April 1938).

Bornstein, David. *How to Change the World: Social Entrepreneurs and the Power of New Ideas.* Oxford: Oxford University Press, 2004.

Bosco, Ronald A., and Joel Myerson, eds. *Emerson in His Own Time.* Iowa City: University of Iowa Press, 2003.

Bourdieu, Pierre. *Acts of Resistance: Against the Tyranny of the Market.* New York: The New Press, 1998.

Bowler, Peter J. *Evolution: The History of an Idea.* Berkeley: University of California Press, 2003.

Brand, Stewart. *The Clock of the Long Now: Time and Responsibility.* New York: Basic Books, 1999.

Brecher, Jeremy, and Tim Costello. *Global Village or Global Pillage: Economic Reconstruction from the Bottom Up.* Cambridge, MA: South End Press, 1998.

Brecher, Jeremy, Tim Costello, and Brendan Smith *Globalization from Below: The Power of Solidarity* Cambridge, MA: South End Press, 2000.

Breton, Mary Joy. *Women Pioneers for the Environment.* Boston: Northeastern University Press, 1998.

Bridges, E. Lucas. *Uttermost Part of the Earth.* New York: E. P. Dutton, 1949.

Bridges, Thomas. *Yámana-English Dictionary: A Dictionary of the Speech of Tierra del Fuego.* Buenos Aires: Zagier y Urruty Publicaciones, 1987.

Brinkley, Douglas. *Mine Eyes Have Seen the Glory: The Life of Rosa Parks.* London: Weidenfeld & Nicolson, 2000.

Brodsky, Anne E. *With All Our Strength: The Revolutionary Association of the Women of Afghanistan.* New York: Routledge, 2004.

Browne, E. Janet. *Charles Darwin, Voyaging.* Princeton: Princeton University Press, 1998.

Bryson, Bill. *The Mother Tongue: English & How it Got That Way.* New York: Avon Books, 1990.

Buell, Lawrence. *Emerson.* Cambridge, MA: Harvard University Press, 2003.

Burns, Stewart. *To the Mountaintop: Martin Luther King Jr.'s Sacred Mission to Save America: 1955–1968.* New York: HarperCollins, 2004.

Burt, Ronald S. *Brokerage and Closure: An Introduction to Social Capital.* Oxford: Oxford University Press, 2005.

Callahan, Gerald N. *Faith, Madness, and Spontaneous Human Combustion.* New York: St. Martin's Press, 2002.

Camus, Albert. *Lyrical and Critical Essays.* New York: Alfred A. Knopf, 1969.

Capra, Fritjof. *The Web of Life: A New Scientific Understanding of Living Systems.* New York: Anchor Books, 1996.

———. *The Hidden Connections—Integrating the Biological, Cognitive, and Social Dimensions of Life into a Science of Sustainability.* New York: Doubleday, 2002.

Carse, James. *Finite and Infinite Games: A Vision of Life as Play and Possibility.* New York: Ballantine Books, 1986.

Chaisson, Eric. *The Life Era: Cosmic Selection and Conscious Evolution.* New York: W. W. Norton, 1987.

Chandhoke, Neera. *The Conceits of Civil Society.* New Delhi: Oxford University Press, 2003.

Chasteen, John Charles. *Born in Blood and Fire.* New York: W. W. Norton, 2001.

Chernus, Ira. *American Nonviolence: The History of an Idea.* Maryknoll, NY: Orbis Books, 2004.

Chua, Amy. *World on Fire, How Exporting Free Market Democracy Breeds Ethnic Hatred and Global Instability.* New York: Doubleday, 2003.

Coates, Ken S. *A Global History of Indigenous Peoples.* Basingstoke, UK: Palgrave Macmillan, 2004.

Coclanis, Peter A., and Stuart Bruchey. *Ideas, Ideologies, and Social Movements: The United States Experience Since 1800.* Columbia: University of South Carolina Press, 1999.

Crofutt, W. A. ed. *Fifty Years in Camp and Field: Diary of Major-General Ethan Allen Hitchcock, U.S.A.* New York: G. P. Putnam's Sons, 1909.

Crosby, Jr., Alfred W. *The Columbian Exchange.* Westport, CT: Greenwood Press, 1972.

———. *Ecological Imperialism: The Biological Expansion of Europe, 900–1900.* Cambridge, UK: Cambridge University Press, 1986.

Crossley, Nick. *Making Sense of Social Movements.* Buckingham, UK: Open University Press, 2002.

Cruse, Julius M., and Robert E. Lewis. *Historical Atlas of Immunology.* London: Taylor & Francis, 2005.

Crystal, David. *Language Death.* Cambridge: Cambridge University Press, 2000.

Dalton, Russell J. *The Green Rainbow: Environmental Groups in Western Europe.* New Haven: Yale University Press, 1994.

Darwin, Charles. *The Voyage of the Beagle: Journal of Researches into the Natural History and Geology of the Countries Visited During the Voyage of H.M.S. Beagle Round the World.* New York: Modern Library, 2001.

Davis, Mike. *Planet of Slums.* London: Verso, 2006.

Davis, Wade. *Light at the Edge of the World.* Washington, DC: National Geographic, 2002.

de Duve, Christian. *Vital Dust: The Origin and Evolution of Life on Earth.* New York: Basic Books, 1995.

———. *Life Evolving: Molecules, Mind, and Meaning.* Oxford: Oxford University Press, 2002.

Denevan, William M., ed. *The Native Population of the Americas.* Madison, WI: The University of Wisconsin Press, 1992.

Dennett, Daniel C. *Darwin's Dangerous Idea: Evolution and the Meanings of Life.* New York: Simon & Schuster, 1995.

Desai, Ashwin. *We Are the Poors: Community Struggles in Post-Apartheid South Africa.* New York: Monthly Review Press, 2002.

Desfor, Gene, Deborah Barndt, and Barbara Rahder, eds. *Just Doing It: Popular Collective Action in the Americas.* Montreal: Black Rose Books, 2002.

Desmond, Adrian, and James Moore. *Darwin: The Life of a Tormented Evolutionist.* New York: W. W. Norton, 1991.

Diamond, Jared. *Collapse: How Societies Choose to Fail or Succeed.* New York: Viking, 2005.

Drake, Thomas E. *Quakers and Slavery in America.* New Haven: Yale University Press, 1950.

Drinnon, Richard. *Facing West: The Metaphysics of Indian Hating and Empire Building.* New York: Schocken Books, 1980.

Dryzek, John S. et al. *Green States and Social Movements, Environmentalism in the United States, United Kingdom, Germany, and Norway.* Oxford: Oxford University Press, 2003.

Duncan, David James. *God Laughs & Plays: Churchless Sermons in Response to the Preachments of the Fundamentalist Right.* Great Barrington, MA: The Triad Institute, 2006.

Economy, Elizabeth C. *The River Runs Black: The Environmental Challenge to China's Future.* Ithaca, NY: Cornell University Press, 2004.

Eisenbud, Merril. *Environment, Technology, and Health: Human Ecology in the Historical Perspective.* New York: New York University Press, 1978.

Ellwood, Wayne, and John McMurtry. *The No-Nonsense Guide to Globalization.* London: New Internationalist, 2001.

Farmer, Paul. *Pathologies of Power—Health, Human Rights, and the New War on the Poor.* Berkeley: University of California Press, 2005.

Felton, Keith Spencer. *Warriors' Words: A Consideration of Language and Leadership.* Westport, CT: Praeger, 1995.

Fischer, Louis. *Mahatma Gandhi: His Life and Message for the World.* New York: New American Library, 1954.

Flannery, Tim. *The Weather Makers.* New York: Atlantic Monthly Press, 2005.

Fort, Meredith, Mary Anne Mercer, and Oscar Gish. *Sickness and Wealth: The Corporate Assault on Global Health.* Cambridge, MA: South End Press, 2004.

Fortey, Richard. *Life: A Natural History of the First Four Billion Years of Life on Earth.* New York: Vintage Books, 1999.

Frank, Dana. *Bananeras.* Cambridge, MA: South End Press, 2005.

Frank, Thomas. *One Market Under God: Extreme Capitalism, Market Populism, and the End of Economic Democracy.* New York: Doubleday, 2000.

Freire, Paulo. *Pedagogy of the Oppressed.* New York: Continuum, 1998

Galeano, Eduardo. *Upside Down: A Primer for the Looking Glass World.* New York: Henry Holt, 2000.

———. *Memory of Fire: Genesis.* New York: Pantheon Books, 1985.

Garrow, David J. *The Walking City: The Montgomery Bus Boycott, 1955–1956.* Brooklyn, NY: Carlson Publishing, 1989.

Gentry, Carole M., and Donald A. Grinde, Jr. eds. *The Unheard Voices: American Indian Responses to the Columbian Quincentenary 1492–1992.* American Indian Studies Center, Los Angeles: University of California Press, 1994.

Glacken, Clarence J. *Traces on the Rhodian Shore.* Berkeley: University of California Press, 1967.

Glave, Dianne D., and Mark Stoll. *To Love the Wind and the Rain: African Americans and Environmental History.* Pittsburgh: University of Pittsburgh Press, 2006.

Goldman, Michael. *Imperial Nature: The World Bank and Struggles for Social Justice in the Age of Globalization.* New Haven: Yale University Press, 2005.

Goldschmidt, Tijs. *Darwin's Dreampond: Drama in Lake Victoria.* Cambridge, MA: MIT Press, 1998.

Goodsell, David S. *The Machinery of Life.* New York: Springer-Verlag, 1998.

Gottlieb, Robert. *Forcing the Spring: The Transformation of the American Environmental Movement.* Washington, DC: Island Press, 2005.

Guattari, Félix. *The Three Ecologies.* London: Athlone Press, 2000.

Guidry, John A., Michael D. Kennedy, Mayer N. Zald, eds. *Globalization and Social Movements: Culture, Power, and the Transnational Public Sphere.* Ann Arbor: University of Michigan Press, 2000.

Gunderson, Lance H., and C. S. Holling. *Panarchy: Understanding Transformations in Human and Natural Systems.* Washington, DC: Island Press, 2002.

Harding, Walter. *The Days of Henry Thoreau.* New York: Alfred A. Knopf, 1965.

Hardt, Michael, and Antonio Negri. *Empire.* Cambridge: Harvard University Press, 2000.

———. *Multitude: War and Democracy in the Age of Empire.* New York: Penguin Press, 2004.

Harold, Franklin M. *The Way of the Cell.* Oxford: Oxford University Press, 2001.

Harris, Sam. *The End of Faith: Religion, Terror, and the Future of Reason.* New York: W. W. Norton and Company, 2005.

Hawkins, David R. *Power vs. Force: An Anatomy of Consciousness.* Sedona, AZ: Veritas Publishing, 1995.

Hayden, Tom, ed. *The Zapatista Reader.* New York: Thunder's Mouth Press, 2001.

Heat-Moon, Least. *PrairyErth.* Boston: Houghton Mifflin, 1991.

Henige, David. *Numbers from Nowhere.* Norman: University of Oklahoma Press, 1998.

Hillstrom, Kevin, and Laurie Collier Hillstrom. *Asia: A Continental Overview of Environmental Issues.* Santa Barbara: ABC-CLIO, 2003.

———. *Latin America and the Caribbean: A Continental Overview of Environmental Issues.* Santa Barbara: ABC-CLIO, 2004.

Hines, Colin. *Localization: A Global Manifesto.* London: Earthscan, 2000.

Hoagland, Mahlon, and Bert Dodson. *The Way Life Works.* New York: Times Books, 1995.

Hochschild, Adam. *Bury the Chains: Prophets and Rebels in the Fight to Free an Empire's Slaves.* Boston: Houghton Mifflin, 2005.

Hofstadter, Richard. *The Age of Reform: From Bryan to F.D.R.* New York: Alfred A. Knopf, 1969.

Holland, John H. *Emergence: From Chaos to Order.* Reading, PA: Perseus Books Group, 1999.

Horton, Myles. *The Long Haul.* New York: Doubleday, 1990.

Howarth, William Lewis. "Turning the Tide." *American Scholar,* 74 no. 3 (2005).

Hunt, Helen LaKelly. *Faith and Feminism: A Holy Alliance.* New York: Atria Books, 2004.

Huth, Hans. *Nature and the American: Three Centuries of Changing Attitudes.* Berkeley: University of California Press, 1957.

Hutton, Drew, and Libby Connors. *A History of the Australian Environment Movement.* Cambridge, UK: Cambridge University Press, 1999.

Hyades, P., and J. Deniker. *Mission Scientifique du Cap Horn, 1882–1883, Tome VII, Anthropologie, Ethnographie.* Paris: Gauthier-Villars, 1891.

Hynes, Patricia H. *The Recurring Silent Spring.* New York: Pergamon Press, 1989.

Ishay, Micheline R. *The History of Human Rights: From Ancient Times to the Globalization Era.* Berkeley: University of California Press, 2004.

IUCN Intercommission Task Force on Indigenous Peoples, prepared by Darrell Posey and Graham Dutfield. *Indigenous People and Sustainability: Cases and Actions.* Utrecht, Netherlands: International Books, 1997.

Jacoby, Susan. *Freethinkers: A History of American Secularism.* New York: Henry Holt, 2004.

Johnson, Steven. *Emergence: The Connected Lives of Ants, Brains, Cities and Software.* New York: Scribner, 2001.

Kaplan, Robert D. *The Coming Anarchy: Shattering the Dreams of the Post Cold War.* New York: Random House, 2000.

———. *The Ends of the Earth: A Journey to the Frontiers of Anarchy.* New York: Vintage Books, 1996.

Kauffman, Stuart. *At Home in the Universe: The Search for the Laws of Self-Organization and Complexity.* Oxford: Oxford University Press, 1995.

Keck, Margaret E., and Kathryn Sikkink. *Activists Beyond Borders.* Ithaca, NY: Cornell University Press, 1998.

Kennedy, Jr., Robert F. *Crime Against Nature.* New York: HarperCollins, 2004.

Kidder, Tracy. *Mountains Beyond Mountains: The Quest of Dr. Paul Farmer: A Man Who Would Cure the World.* New York: Random House, 2003.

Kim, Jim Yong, Joyce V. Millen, Alec Irwin, and John Gershman. *Dying for Growth: Global Inequality and the Health of the Poor.* Monroe, ME: Common Courage Press, 2000.

Kimmerling, Judith et al. *Amazon Crude.* New York: Natural Resources Defense Council, 1991.

King, Mary. *Mahatma Gandhi and Martin Luther King Jr.: The Power of Nonviolent Action.* Paris: UNESCO Publishing, 1999.

Kingsworth, Paul. *One No, Many Yeses: A Journey to the Heart of the Global Resistance Movement.* London: Free Press, 2003.

Kittredge, William. *The Nature of Generosity.* New York: Vintage Books, 2000.

Klein, Naomi. *No Logo: Taking Aim at Brand Bullies.* New York: Picador, 2000.

Krech III, Shephard, J. R. McNeill, and Carolyn Merchant, eds. *World Environmental History, vols. 1–3.* New York: Routledge, 2004.

LaDuke, Winona. *All Our Relations.* Cambridge, MA: South End Press, 1999.

————. *Recovering the Sacred: The Power of Naming and Claiming.* Cambridge, MA: South End Press, 2005.

Langer, Erick D., and Elena Muñoz, eds. *Contemporary Indigenous Movements in Latin America.* Wilmington, DE: Scholarly Resources, 2003.

Lansing, J. Stephen. *Perfect Order: Recognizing Complexity in Bali.* Princeton: Princeton University Press, 2006.

Lapham, Lewis, and Peter T. Struck. *The End of the World.* New York: St. Martin's Press, 1997.

Lear, Linda. *Rachel Carson: Witness for Nature.* New York: Henry Holt, 1997.

Lebeaux, Richard. *Thoreau's Seasons.* Amherst: University of Massachusetts Press, 1984.

Lewis, Thomas, Fari Amini, and Richard Lannon. *A General Theory of Love.* New York: Random House, 2000.

Ling, Gilbert N. *Life at the Cell and Below-Cell Level.* New York: Pacific Press, 2001.

Lopez, Barry. *The Rediscovery of North America.* New York: Vintage Books, 1992.

Lovelock, James. *The Revenge of Gaia.* London: Allen Lane, 2006.

————. *Gaia: The Practical Science of Planetary Medicine.* London: Gaia Books, 1991.

Lowenthal, David. *George Perkins Marsh: Prophet of Conservation.* Seattle: University of Washington Press, 2000.

Mabee, Carleton. *Sojourner Truth: Slave, Prophet, Legend.* New York: New York University Press, 1993.

Macy, Joanna. *World as Lover, World as Self.* Berkeley: Parallax Press, 1991.

Madeley, John. *Big Business, Poor Peoples—The Impact of Transnational Corporations on the World's Poor.* London: Zed Books, 1999.

Mander, Jerry, and Victoria Tauli-Corpuz. *Paradigm Wars: Indigenous Peoples' Resistance to Economic Globalization.* International Forum on Globalization. San Francisco: Sierra Club Books, 2006.

Marcos, Subcomandante, and Juana Ponce de León, ed. *Our Word Is Our Weapon.* New York: Seven Stories Press, 2001.

Margulis, Lynn, and Dorion Sagan. *Microcosmos: Four Billion Years of Evolution from Our Microbial Ancestors.* New York: Summit Books, 1986.

Margulis, Lynn, and Karlene V. Schwartz. *Five Kingdoms: An Illustrated Guide to the Phyla of Life on Earth.* New York: W. H. Freeman, 1998.

Martial, L.-F. *Mission Scientifique du Cap Horn, 1882–1883, Tome Ier, Histoire du Voyage.* Paris: Gauthier-Villars, 1888.

McCann, James C. *Green Land, Brown Land, Black Land: An Environmental History of Africa, 1800–1990.* Portsmouth, NH: Heinemann, 1999.

McCarthy, Ronald M., and Gene Sharp. *Nonviolent Action: A Research Guide.* New York: Garland Publishing, 1997.

McEwean, Colin, Luis A. Borrero, and Alfredo Prieto, eds. *Patagonia: Natural History, Prehistory, and Ethnography at the Uttermost Ends of the Earth.* Princeton: Princeton University Press, 1997.

McPhee, John. *Encounters with the Archdruid.* New York: Farrar, Straus and Giroux, 1971.

Menand, Louis. *The Metaphysical Club: A Story of Ideas in America.* New York: Farrar, Straus and Giroux, 2001.

Meyer, Aubrey. *Contraction & Convergence: The Global Solution to Climate Change.* Devon: Green Books, 2000.

Miller, Angela. "The Soil of an Unknown America: New World, Lost Empires and the Debate over Cultural Origins." *American Art,* 8, no. 3/4 (1994).

Miller, Randall M., and Paul A. Cimbala, eds. *American Reform and Reformers.* Westport, CT: Greenwood Press, 1996.

Mohawk, John C. *Utopian Legacies: A History of Conquest and Oppression in the Western World.* Santa Fe, NM: Clear Light Publishers, 2000.

Monbiot, George. *Manifesto for a New World Order.* New York: The New Press, 2003.

Morville, Peter. *Ambient Findability: What We Find Changes Who We Become.* Sebastapol, CA: O'Reilly, 2005.

Murphy, Daniel. *Rounding the Horn.* New York: Basic Books, 2004.

Nace, Ted. *Gangs of America: The Rise of Corporate Power and the Disabling of Democracy.* San Francisco: Berrett-Koehler Publishers, 2003.

Nettle, Daniel, and Suzanne Romaine. *Vanishing Voices: The Extinction of the World's Languages.* Oxford: Oxford University Press, 2000.

Neuwirth, Robert. *Shadow Cities: A Billion Squatters, A New Urban World.* New York: Routledge, 2006.

Neuzil, Mark, and William Kovarik. *Mass Media & Environmental Conflict: America's Green Crusades.* Thousand Oaks, CA: Sage Publications, 1996.

Oates, Stephen B. *Let the Trumpet Sound: The Life of Martin Luther King, Jr.* New York: New American Library, 1982.

Oliver, Mary. *Dream Work.* New York: Atlantic Monthly Press, 1986.

———. *Long Life: Essays and Other Writings.* Cambridge, MA: Da Capo Press, 2004.

Orr, David. *The Last Refuge: Patriotism, Politics, and the Environment in an Age of Terror.* Washington, DC: Island Press, 2004.

Orth, Ralph H. ed. *The Topical Notebooks of Ralph Waldo Emerson.* Columbia: University of Missouri Press, 1990.

Osterhammel, Jürgen, and Niels P. Petersson. *Globalization: A Short History.* Princeton: Princeton University Press, 2005.

Parks, Rosa, and Jim Haskins. *Rosa Parks: My Story.* New York: Dial Books, 1992.

Peattie, Donald Culross. *Green Laurels: The Lives and Achievements of the Great Naturalists.* New York: Simon & Schuster, 1936.

Pigafetta, Antonio. *Magellan's Voyage Around the World, vol. I.* Cleveland: Arthur H. Clark, 1906.

Pike, E. Royston, ed. *"Hard Times": Human Documents of the Industrial Revolution.* New York: Frederick A. Praeger, 1966.

Pilger, John. *The New Rulers of the World.* London: Verso, 2002.

Playfair, John. *Living with Germs: In Sickness and in Health.* Oxford: Oxford University Press, 2004.

Pollan, Michael. *The Botany of Desire: A Plant's-Eye View of the World.* New York: Random House, 2001.

Posey, Darrell Addison, ed. *Cultural and Spiritual Values of Biodiversity.* London: Intermediate Technology Publications, 1999.

Rand, Silas Tertius. *Legends of the Micmacs.* New York: Longmans, Green, 1894.

Redman, Charles L. *Human Impact on Ancient Environments.* Tucson: University of Arizona Press, 1999.

Rensberger, Boyce. *Life Itself: Exploring the Realm of the Living Cell.* Oxford: Oxford University Press, 1996.

Rheingold, Howard. *Smart Mobs: The Next Social Revolution.* New York: Perseus, 2002.

Rich, Bruce. *Mortgaging the Earth: The World Bank: Environmental Impoverishment, and the Crisis of Development.* Boston: Beacon Press, 1994.

Richardson Jr., Robert D. *Henry Thoreau: Life of the Mind.* Berkeley: University of California Press, 1986.

———. *Emerson: The Mind on Fire.* Berkeley: University of California Press, 1995.

Robinson, Jo Ann. *The Montgomery Bus Boycott and the Women Who Started It: The Memoir of Jo Ann Gibson Robinson.* Knoxville: University of Tennessee Press, 1987.

Rose, Michael R. *Darwin's Spectre: Evolutionary Biology in the Modern World.* Princeton: Princeton University Press, 1998.

Roy, Arundhati. *Power Politics.* Cambridge, MA: South End Press, 2001.

Salamon, Lester M., Helmut K. Anheier, Regina List, Stefan Toepler, S. Wojciech Sokolowski. *Global Civil Society: Dimensions of the Non-Profit Sector.* Baltimore: Johns Hopkins Center for Civil Society Studies, 1999.

Salamon, Lester M., S. Wojciech Sokolowski, et al. *Global Civil Society: Dimensions of the Non-Profit Sector, vol. 2.* Bloomfield, CT: Kumarian Press, 2004.

Sattlemeyer, Robert. *Thoreau's Reading: A Study in Intellectual History.* Princeton: Princeton University Press, 1988.

Schama, Simon. *Landscape and Memory.* New York: Alfred A. Knopf, 1995.

Schell, Jonathan. *The Unconquerable World: Power, Nonviolence, and the Will of the People.* New York: Henry Holt, 2003.

Schelling, Thomas C. *Choice and Consequence: Perspectives of an Errant Economist.* Cambridge, MA: Harvard University Press, 1984.

———. *The Strategy of Conflict.* Cambridge, MA: Harvard University Press, 1980.

Schirato, Tony, and Jen Webb. *Understanding Globalization.* London: Sage Publications, 2003.

Schmidt, Leigh Eric. *Restless Souls: The Making of American Spirituality.* New York: HarperCollins, 2005.

Sen, Amartya. *Development as Freedom.* New York: Anchor Books, 1999.

Shabecoff, Phillip. *A Fierce Green Fire: The American Environmental Movement.* New York: Hill and Wang, 1993.

Shah, Ghanshyam. *Social Movements in India.* New Delhi: Sage Publications, 2004.

Sharp, Gene. *The Politics of Nonviolent Action: Part Two: The Methods of Nonviolent Action.* Boston: Porter Sargeant, 1973.

———. *The Politics of Nonviolent Action: Part One: Power and Struggle.* Boston: Porter Sargeant, 1973.

———. *The Politics of Nonviolent Action: Part Three: The Dynamics of Nonviolent Action.* Boston: Porter Sargeant, 1985.

Shiva, Vandana. *Earth Democracy: Justice, Sustainability, and Peace.* Cambridge, MA: South End Press, 2005.

Sicherman, Barbara. *Alice Hamilton: A Life in Letters.* Cambridge, MA: Harvard University Press, 1984.

Slater, José Perich. *Indigenous Extinction in Patagonia.* Puntas Arena, Chile: Impresos Vanic, 1997.

Smith, Andrea. *Conquest: Sexual Violence and American Indian Genocide.* Cambridge, MA: South End Press, 2005.

Smith, John Maynard, and Eörs Szathmáry. *The Origins of Life: From the Birth of Life to the Origin of Language.* Oxford: Oxford University Press, 1999.

Sober, Elliott, and David Sloan Wilson. *Unto Others: The Evolution and Psychology of Unselfish Behavior.* Cambridge, MA: Harvard University Press, 1998.

Stearns, Peter N. *The Impact of the Industrial Revolution: Protest and Alienation:* Englewood Cliffs, NJ: Prentice-Hall, 1972.

————. *The Industrial Revolution in World History.* Boulder, CO: Westview Press, 1993.

Stewart, James Brewer. *Holy Warriors: The Abolitionists and American Slavery.* New York: Hill and Wang, 1997.

Stiglitz, Joseph. *Globalization and Its Discontents.* New York: W. W. Norton, 2002.

Tarnas, Richard. *Cosmos and Psyche: Intimations of a New World View.* New York: Viking, 2006.

Terkel, Studs. *Hope Dies Last.* New York: The New Press, 2003.

Thomas, Lewis. *The Lives of a Cell: Notes of a Biology Watcher.* New York: Viking Press, 1974.

Thomas, Jr., William L. ed., with Carl O. Sauer, Marston Bates, and Lewis Mumford. *Man's Role in Changing the Face of the Earth, vol. 2.* Chicago: University of Chicago Press, 1956.

Thomis, Malcolm I. *The Luddites: Machine-Breaking in Regency England.* Hampden, UK: Archon Books, 1970.

Thomis, Malcolm I., and Jennifer Grimmett. *Women in Protest 1800–1850.* London: Croom Helm, 1982.

Thoreau, Henry D. *Walden and Resistance to Civil Government.* 2nd ed. William Rossi ed. New York: W. W. Norton, 1992.

True, Michael. *An Energy Field More Intense Than War: The Nonviolent Tradition and American Literature.* Syracuse, NY: Syracuse University Press, 1995.

Uchitelle, Louis. *The Disposable American: Layoffs and Their Consequences.* New York: Alfred A. Knopf, 2006.

Van Cott, Donna Lee, ed. *Indigenous People and Democracy in Latin America.* New York: St. Martin's Press, 1994.

Vermot-Mangold, Ruth-Gaby. *1000 Peace Women Across the Globe.* Zurich: Scalo, 2005.

Volk, Tyler. *Gaia's Body: Toward a Physiology of Earth.* Cambridge, MA: MIT Press, 2003.

Wade, Nicholas. *Before the Dawn: Recovering the Lost History of Our Ancestors.* New York: Penguin Press, 2006.

Wagar, W. Warren. "The Colors of World History." *Binghamton Journal of History* (Fall 2001).

Wall, Derek. *Green History: A Reader in Environmental Literature, Philosophy and Politics.* London: Routledge, 1994.

Weatherford, Jack. *Native Roots: How the Indians Enriched America.* New York: Ballantine Books, 1991.

————. *Indian Givers: How the Indians of the Americas Transformed the World.* New York: Crown Publishers, 1988.

Weinberger, David. *Small Pieces, Loosely Joined: A Unified Theory of the Web.* New York: Perseus, 2002.

Weiner, Jonathan. *The Beak of the Finch.* New York: Vintage Books, 1994.

Wheeler, Michael, ed. *Ruskin and Environment: The Storm-Cloud of the Nineteenth Century.* Manchester, UK: Manchester University Press, 1995.

Wilson, Edward O. *Consilience: The Unity of Knowledge.* New York: Vintage Books, 1999.

Wilson, James. *The Earth Shall Weep: A History of Native America.* New York: Atlantic Monthly Press, 1999.

Wolf, Martin. *Why Globalization Works.* New Haven: Yale University Press, 2004.

Woodin, Michael, and Caroline Lucas. *Green Alternatives to Globalisation: A Manifesto.* London: Pluto Press, 2004.

Worster, Donald. *Nature's Economy: The Roots of Ecology.* San Francisco: Sierra Club Books, 1977.

Worster, Donald, ed. *The Ends of the Earth: Perspectives on Modern Environmental History.* Cambridge, UK: Cambridge University Press, 1988.

Wright, Donald. *Stolen Continents: 500 Years of Conquest and Resistance in the Americas.* Boston: Houghton Mifflin, 1992.

Wurm, Stephen A., ed. *Atlas of the World's Languages in Danger of Disappearing.* Paris: UNESCO Publishing, 2001.

Zinn, Howard. *A People's History of the United States.* New York: HarperCollins, 1980.

INDEX

Evangelical Climate Initiative, 148
evolution: as constantly occurring, 33; and
    creationism, 42, 141; cyclical nature of,
    179; Darwin's theory of, 31–32, 33, 72;
    definition of, 25, 171–72; and
    differentiation, 32–33; as dynamic
    process, 100; and emergence of biology
    as science, 31–32; Grant study of,
    32–33; of immune system, 164, 165;
    and indigenous people, 91, 99;
    measurement and observation of,
    32–33; as optimism in action, 25; result
    of, 25; of social systems, 172
extractive industries, 102
Exxon/ExxonMobil, 23, 64, 65–66, 105,
    148, 177

faith, 174, 177, 184, 188
Farmer, Paul, 145, 176
farming, 165, 174, 178, 179, 195–200. See
    also agriculture
fast food, 155
Faster Food Bill, 161
Father of the Forest, 38
Fedrizzi, Rick, 153
Fellowship of Reconciliation, 84
fish/fisheries, 57, 130, 163, 237–38
Fitzroy, Robert, 90
food, 55, 129, 155–61, 178, 238–40
Forbes, Steve, 149
Foreman, Dave, 163–64
Forests and European Union Resource
    Network (FERN), 109
forests/forestry, 1, 37–40, 45, 98, 104,
    106, 165, 240–43. See also logging;
    timber
Fortey, Richard, 167
fossil fuels, 53, 67
Frank, Thomas, 124
Free Trade Agreement of America (Quebec
    City, 2001), 148–49
Friedman, Thomas, 118, 123, 124
Friends of the Earth, 47, 158, 159
friends organizations, 147
Fuller, Buckminster, 179, 182
fundamentalism, 124, 169, 187
future, pessimism or optimism about, 4, 8,
    190

G-6 nations, 120
Gaia hypothesis, 141
Galbraith, John Kenneth, 115
Gale, George, 38
Galleano, Eduardo, 131
games, infinite and finite, 186–87

Gandhi, Mohandas, 7, 78–79, 80, 83, 84,
    122, 152
gardens, 180
Garfield Foundation, 162–63
Garrison, William Lloyd, 77
gas. See greenhouse gases; oil and gas; specific
    gas
Gates, Bill, 23, 138, 149, 150
Gates Foundation, 150
GE, 146, 152
Geneva Conventions, 187
genocide, 93, 97
George, Susan, 126
Ghana, 131
giantness, 23
global climate change. See climate change
Global Climate Coalition, 65
global warming, 4, 6, 66–67
globalization: beginning of, 125; and climate
    change, 118; and communities, 118;
    and corporations, 7, 117–38; and
    corporatization of the commons, 121;
    definition and keywords associated with,
    244–46; and desires to imitate U.S.,
    123–24; and diversity, 113, 118, 122;
    and economic security, 118; and
    education, 118; and food, 155, 156; and
    human rights, 113, 126; and immune
    system metaphor, 144; and indigenous
    people, 7, 102–14; and
    internationalization, 127; and land,
    102–14; liabilities of, 118; and
    localization, 157; and NGOs, 64, 126;
    opposition to, 103, 135–36; and
    pollution, 29, 118; positive effects of,
    118; and poverty, 118; protests against,
    121; and roots of the movement, 12, 13;
    and social justice, 126; of trade, 118,
    120–38; and work, 118, 119–20
gold, 104–5, 107
Golden Rule, 184, 186
Goldtooth, Tom, 47
Goodall, Jane, 47, 147, 186
Google Foundation, 151, 152
Gore, Al, 151
Gottfried, David, 153
governance, 134–35, 150–51, 246–48
government: and citizen-based organizations,
    164–65; as disease, 145; and food safety
    standards, 161; and human rights, 13
Graham, Martha, 9
Grameen Bank, 152, 176
Grand Canyon, 45, 46, 47
Grant, Peter and Rosemary, 32–33
Greeley, Horace, 39–40, 41, 42

global history, 22–23; World Bank loans
to, 131
Indigenous Peoples Coalition Against
Biopiracy, 126
individuality, 167
individuals, importance of, 85, 164, 174–75,
183
industrial capitalism, 14
Industrial Revolution, 15
industrialism, history of, 59–61
industry. *See* business; corporations; greening
of industry; *specific industry*
information: life as organized with, 177–78;
and localization, 157–58; technology,
12; and totalitarianism, 22
Institute for Policy Studies, 126
Interface, 67
Intergovernmental Panel on Climate Change,
147–48
International Campaign for Justice, 62
International Forum on Globalization, 126
International Monetary Fund (IMF), 21, 104,
111, 112, 117, 137
International Rivers Network, 47
International Year of the World's Indigenous
People, 87
internationalization, 127
Inuit people, 101–2, 137–38
Iraq War, 4, 24–25, 99
Islam, radical, 136–37
Itsero, Mrs. Felicia, 87

Jacob, François, 167
James, William, 73
Japan, 133
Jeffers, Robinson, 170
Jefferson, Thomas, 29
Johannes, R. E., 101
Johnson, Robert Underwood, 46
Jussieu family, 7, 72–73
justice: and reimaging the world, 189–90. *See
also* social justice

Kaiser Foundation, 57
Kant, Immanuel, 141
Kaplan, Robert, 172
Keating, Paul, 87
keeper groups, 146
Keller, Helen, 189
Kelly, Kevin, 143, 144
Kennedy, Robert Jr., 59, 146
Keynes, John Maynard, 100
Khor, Martin, 126
King, Coretta Scott, 83–84
King, Martin Luther Jr., 69, 81–84, 186

King, Patricia, 126–27
King, Thomas Starr, 40
Kitlope rain forest, 41
Kittredge, William, 25
Klein, Naomi, 12
kludges, 20
Kpanan'Ayoung Siakor, Silas, 11
Kyoto Protocols, 53, 65, 66, 174

Lamarck, Jean-Baptiste de, 31
land: conservation of, 37–41; and
globalization, 102–14; history of U.S.,
37–41; of indigenous people, 6,
102–14; protection of, 102–14; reform,
7; rights to, 6, 103
languages, 91, 92–95, 96, 99–100, 101, 135
Lasn, Kalle, 149
law, 263–67
leadership, 126–27
Leadership in Energy and Environmental
Design (LEED), 153
Leopold, Aldo, 20, 47, 58
Lewis, C. S., 23
life: and aim of movement with no name, 68;
as assembling into chains, 175–76;
beginning of, 169; as building from
bottom up, 175; first form of, 169–70;
forks in road of, 84; as generating many
variations, 176; inevitability of, 169; as
optimizing rather than maximizing, 183;
as organized with information, 177–78;
qualities common to all, 175–83;
sacredness of, 186; as time frame,
134–35. *See also* interconnectedness
Lincoln, Abraham, 40, 41, 42, 76
Linnaeus, 30–31
lobbyists, 19
localization, 21, 156–58
Locke, John, 29
logging, 38–39, 45. *See also* forests/forestry
Lonestar Energy, 105
Lonestar Oil Company, 105
Long Now Foundation, 154
Lopez, Barry, 9
love, 188
Lovelock, James, 47, 141
Lowell, James Russell, 39–40
Lowenthal, David, 43
Luddites, 60–61
Lyell, Charles, 31
Lyons, Oren, 47

Maathai, Wangari, 68, 174, 186
McDonald's, 65, 155
Machado, Antonio, 85